# Geophysical Monograph Series

Including
**IUGG Volumes**
**Maurice Ewing Volumes**
**Mineral Physics Volumes**

# Geophysical Monograph Series

Geophysical Monograph 180

# Arctic Sea Ice Decline: Observations, Projections, Mechanisms, and Implications

Eric T. DeWeaver
Cecilia M. Bitz
L.-Bruno Tremblay
*Editors*

American Geophysical Union
Washington, DC

**Library of Congress Cataloging-in-Publication Data**

Arctic sea ice decline : observations, projections, mechanisms, and implications / Eric T. DeWeaver, Cecilia M. Bitz, L.-Bruno Tremblay, editors.

     p. cm. — (Geophysical monograph, ISSN 0065-8448 ; 180)

    ISBN 978-0-87590-445-0

  1. Sea ice—Arctic regions. 2. Climatic changes—Environmental aspects—Arctic Regions. 3. Environmental impact analysis—Arctic regions. 4. Arctic regions—Climate. I. DeWeaver, Eric T., 1964- II. Bitz, Cecilia M. III. Tremblay, L.-Bruno.

    GB2595.A735 2008

    551.34′3091632—dc22

                                         2008045367

ISBN: 978-0-87590-445-0
ISSN: 0065-8448

**Cover Photo:** Courtesy of Don Perovich (CRELL) during the 2005 Arctic Basin Transect HORTRAX field campaign.

# CONTENTS

## Section IV: The Threat to Polar Bears From Sea Ice Decline

# PREFACE

Prospects for Arctic sea ice are grim and apparently worsening. Following decades of decline, the September 2007 sea ice extent shattered all previous record lows with a 39% loss in ice cover relative to the 1979–2000 September mean. September 2007 also saw, for the first time on record, the opening of the northern branch of the Northwest Passage, the route through the Canadian Archipelago pioneered by Sir William Parry in 1819. These observations are accompanied by equally alarming climate model projections of sea ice decline due to greenhouse gas increases. Based on this evidence, expert predictions of the first appearance of an ice-free Arctic have advanced steadily.

The dramatic decline has prompted speculation in both the popular and scientific press that the Arctic sea ice may have passed a "tipping point," beyond which the complete destruction of perennial sea ice cover is inevitable. While this claim may seem reasonable in light of the strong positive ice-albedo feedback, its validity has not been scientifically demonstrated. It does, however, suggest a set of questions regarding the climate sensitivity of Arctic sea ice: Will Arctic summer sea ice cover disappear completely? What factors control the rate of decline? Is there true threshold behavior? What can we learn about the climate sensitivity of Arctic sea ice from observations, both from the modern instrumented and paleo–proxy records? Are there meaningful idealized models that can be used to identify and isolate the dominant feedback mechanisms and assess climate sensitivity? Are there diagnostic analysis tools that can help quantify the declining trajectory of the sea ice? Can we quantify the climate impacts of sea ice decline?

This book is a collection of papers addressing these questions. It seeks to describe the climate sensitivity of Arctic sea ice in simplest terms, identifying the most prominent mechanisms and assessing their ability to produce a true, irreversible tipping point. The book is a combination of new results from original research and review material identifying key results from the literature that are insightful for understanding Arctic sea ice decline. The review material also serves to broaden the appeal of the book, which is intended for an audience including both specialists and interested nonspecialists with some background in physical climatology.

The book is divided into four sections. The first section, "Arctic sea ice in the instrumented and paleo–proxy records" sets the stage. Chapter 1 gives an assessment of sea ice cover change over recent decades, with emphasis on the contribution of wind forcing to the downward trend in sea ice cover. This discussion is followed in Chapter 2 by a reconstruction of sea ice conditions from the early Holocene, a period of reduced Arctic sea ice cover, using dinocyst assemblages and isotope ratios.

The second section, "Factors in sea ice sensitivity" considers mechanisms responsible for sea ice sensitivity in observations and climate model simulations. Chapter 3 considers the impact of clouds on the longwave and shortwave surface radiative fluxes over sea ice, and illustrates the contribution of cloudiness changes to sea ice loss in one climate change simulation. Chapter 4 considers the importance of the sea ice–albedo feedback as a source of spread in climate model simulations of sea ice decline, which is found to be small compared to the role of ice thickness differences already present at the start of these simulations. Chapter 5 continues the discussion of the large intermodel spread of Arctic sea ice thickness in climate model simulations, giving a simple argument for the intrinsic sensitivity of ice thickness to surface energy fluxes, and showing how differences in surface energy flux lead to the large spread in simulated sea ice thickness. Chapter 6 looks at a different kind of sensitivity: the sensitivity of an atmospheric circulation model subject to realistic reductions in summer Arctic sea ice.

The third section, "Rapid loss versus abrupt transition," examines the extent to which true threshold behavior is implicated in climate model simulations that produce a seasonally ice-free Arctic in the 21st century. The notion of an abrupt, irreversible transition to ice-free Arctic summers, brought about by the destabilizing sea ice–albedo feedback, has become commonplace. But rapid loss does not necessarily imply a tipping point: strong sea ice sensitivity can combine with natural variability to produce rapid loss in the absence of a critical threshold between ice-free and ice-covered states. Chapter 7 considers the ability of the

Arctic Sea Ice Decline: Observations, Projections, Mechanisms, and Implications
Geophysical Monograph Series 180
Copyright 2008 by the American Geophysical Union.
10.1029/180GM01

sea ice–albedo feedback to produce abrupt Arctic climate change, looking in particular at the role of the "small ice cap instability" in simulations in which Arctic sea ice vanishes in winter as well as summer. Chapter 8 analyzes simulations with a global climate model in which September Arctic sea ice vanishes as early as 2040, a rapid sea ice loss which apparently occurs through a combination of natural variability and anthropogenic forcing. Alternatively, chapter 9 presents a simple model in which abrupt loss can occur as a nonlinear transition between stable states after "tuning" to match one of the integrations described in the previous chapter. Chapter 10 quantifies the popular notion of the "trajectory" of Arctic sea ice, and shows trajectories of sea ice decline in a simplified phase space with dimensions representing ice cover of different thicknesses. A trajectory model in this phase space produces abrupt transitions when strong sea ice–albedo feedback is added. Chapter 11 gives an account of extreme loss events that occur as part of the natural variability of unforced climate model simulations. The characteristics of these unforced extreme events can shed light on the role of natural variability in forced sea ice decline.

Sea ice cover is a defining element of the Arctic ecosystem, and its loss will clearly have severe consequences for all life in the region. The final section of the book, "The threat to polar bears from sea ice decline," describes a new modeling approach for assessing the impact of sea ice decline on the Arctic's most iconic species. The model, presented in chapter 12, uses a Bayesian framework to determine the likely fate of polar bears in four distinct eco-regions of the Arctic, based in part on climate model projections of sea ice decline. The model was used to inform the recent decision by the Department of the Interior to list the polar bear as a threatened species under the Endangered Species Act.

This book grew out of a special session at the 2006 annual meeting of the American Geophysical Union titled "Rapid transition from perennial to seasonal Arctic sea ice," in which many of the results in the chapters were presented. All chapters underwent a formal, anonymous review process. We gratefully acknowledge the efforts of the many individuals who served as reviewers for this monograph. Finally, we thank the chapter authors for their generous contributions, which made this book possible.

*Eric T. DeWeaver*
*University of Wisconsin-Madison*

*Cecilia M. Bitz*
*University of Washington*

*L.-Bruno Tremblay*
*McGill University*

# Arctic Sea Ice Decline: Introduction

Eric T. DeWeaver

*Center for Climate Research, University of Wisconsin-Madison, Madison, Wisconsin, USA*

## 1. THE GREAT DECLINE OF 2007

By any measure, the loss of Arctic sea ice cover in September 2007 was spectacular. The National Snow and Ice Data Center (NSIDC) called it a loss "the size of Alaska and Texas combined," in comparison to the 1979–2000 September mean. Record-breaking minima in sea ice extent are not unexpected, given the declining trend of the past 30 years and its recent acceleration [e.g., *Meier et al.*, 2007; *Deser and Teng*, this volume]. But the 2007 minimum was remarkable even compared to the decline, a full four standard deviations below the trend line (H. Stern, quoted by *Schweiger et al.* [2008]). *Kerr* [2007] reported an Alaska-sized loss compared to the previous record low in 2005, which was itself an Alaska-sized retreat from the value at the beginning of the satellite era in 1979. Deser and Teng point out that the loss between September 2006 and September 2007 is as large as the entire September extent loss from 1979 to 2006.

Following the 2007 melt season there was some cause for optimism that 2008 could see a partial recovery. Writing at the end of the melt season, *Comiso et al.* [2008, paragraph 5] noted that the ice was "rebounding with a rapid early autumn growth." Following a cold winter, the April 2008 maximum ice extent reported by NSIDC was relatively high by recent standards, although still below the long-term mean. But while the temperatures were cooperating, the winds were not. In early February I. Rigor noted that buoys embedded in multiyear ice flows were "streaming out" of the Arctic, flushed through Fram Strait along with their ice floes by circumpolar wind anomalies [*Kizzia*, 2008]. Also, as discussed by *Maslanik et al.* [2007], ice cover following the 2007

Arctic Sea Ice Decline: Observations, Projections, Mechanisms, and Implications
Geophysical Monograph Series 180

minimum was unusually thin and thus vulnerable to melting away. The July and August extent were somewhat higher than in 2007, but the daily loss rate accelerated in early August after storms broke apart thin ice in the Beaufort and Chukchi seas. Southerly winds following the storms further promoted opening by pushing the ice away from the eastern Siberian coast (information from the "Arctic sea ice news and analysis" Web pages for 11 August through 4 September 2008 at http://www.nsidc.org).

With approxmiately 2 weeks left in the 2008 melt season, Arctic sea ice extent is now very close to the 2007 minimum. While the lack of recovery is discouraging, the 2008 loss could have been worse. In May the NSIDC suggested, based on the prevalence of thin first year ice cover in the Arctic, that the North Pole could become ice free in 2008, a prediction more commonly made for the middle of the century. Three researchers contributing to the May Sea Ice Outlook (produced by the interagency Study of Environmental Arctic Change (SEARCH)), anticipated a return toward the long-term trend of summer sea ice loss, six argued that 2008 September extent should be close to 2007, and five expected losses exceeding those in 2007.

## 2. RESEARCH ON THE CAUSE OF THE LOSS

Research on the causes of the 2007 loss is already well underway. Surface wind anomalies are generally identified as the proximal cause [*Nghiem et al.*, 2007; *Stroeve et al.*, 2008; *Deser and Teng*, this volume; *Overland et al.*, 2008; *Zhang et al.*, 2008], as the transpolar winds dubbed the "Polar Express" by Nghiem et al. pushed ice away from the Alaskan and eastern Siberian coastlines and out of the Arctic. *Kay et al.* [2008] claim an additional role for enhanced summer melting as high pressure and sunny skies persisted over the western Arctic Ocean. Their claim is disputed by *Schweiger et al.* [2008], who note that the sunny skies are not well collocated with the largest sea ice loss in the Chukchi and East Siberian seas. On the other hand, a strong role for insolation as a positive feedback is not in dispute. *Perovich et al.* [2008] find a 500% increase in January to September solar heat input to the Beaufort Sea compared to the 1979—2005 climatology. They further determine that the large increase is due to the large area of low-albedo ocean surface exposed by the dramatic sea ice retreat. Accompanying the increased heat uptake, they report a sixfold increase in bottom melt measured by an ice mass buoy in the Beaufort Sea. Their results thus document the classical sea ice–albedo feedback, presumably initiated by wind-driven opening of the ice pack. The modeling study of *Zhang et al.* [2008] concludes that 70% of the 2007 loss anomaly was due to amplified melting while 30% resulted from ice motion.

In these studies, the meteorology of 2007 is generally given less prominence than the vulnerability of the 2007 sea ice cover. Maslanik et al. and Nghiem et al. document the long-term change from multiyear sea ice to younger floes which are thinner and more prone to breakup and melting. Overland et al. and Kay et al. also question the novelty of the 2007 meteorological conditions. They relate the offshore winds and sunny skies of 2007 to a surface high over the western Arctic Ocean, a rare but not unprecedented occurrence. Four years with comparable high pressure can be seen in the 50-year record shown by Overland et al. (their Figure 11), while Kay et al. find four additional years with sunnier skies than 2007. The older, thicker ice in these earlier years was not dramatically affected by the adverse meteorology.

The contribution of greenhouse warming in producing sharp, single-year declines is not easily quantified, since warming favors these events indirectly as it helps precondition the ice to a thinner state (e.g., Overland et al.). However, *Stroeve et al.* [2007] point out the consistency of the 2007 event with the periods of rapid loss found by *Holland et al.* [2006] in global warming simulations. Stroeve et al. note in particular the similarity between the March 2007 thickness estimates of Maslanik et al. and the mean Arctic thickness in simulations analyzed by Holland et al. The analysis of Holland et al. is expressed in terms of a three-part conceptual framework in which ice is first preconditioned for rapid loss by decades of thinning, after which loss is "triggered" by natural variability and then amplified by the sea ice–albedo feedback. The "preconditioning, trigger, feedback" framework was developed by *Lindsay and Zhang* [2005] to account for the observed sea ice decline from 1988 to 2003, and the same framework was invoked in the Zhang et al. study of the 2007 event. Thus, while the 2007 loss was unprecedented, descriptions of it are quite consistent with descriptions of the longer-term Arctic losses of the recent past and the rapid declines found in simulations of future Arctic change.

## 3. PAST THE TIPPING POINT?

Has the Arctic sea ice passed a tipping point? This is perhaps the most consistently asked question in news accounts about the 2007 and 2008 losses. Understandably, respondents to the question have not voiced much hope for reversal: "It's hard to see how the system may come back (I. Rigor, quoted by *Kizzia* [2008])"; "I'm much more open to the idea that we might have passed a point where it's becoming essentially irreversible" (J. M. Wallace, quoted by *Revkin* [2007]); "It's tipping now. We're seeing it happen now" (M. Serreze quoted by *Borenstein and Joling* [2008]). No doubt, the tipping point terminology aptly captures the precipitous loss of 2007 and lack of recovery in 2008. But questions remain as to how literally the tipping language should be taken. In a formal sense, tipping refers to a sudden and irreversible transition between two stable states of a system (e.g., right side up versus overturned), occurring as the system crosses some threshold value of a key parameter (e.g., angle to the local vertical). The scientific challenge would then be to find and characterize the stable states and threshold values of the Arctic sea ice system.

The idea of an unstable transition between ice-covered and ice-free Arctic states has a long history (see references of *Winton and Merryfield et al.* [this volume]), and such behavior does occur in simple energy balance models with diffusive heat transfer (the "small ice cap instability" of *North* [1984]). However, unstable transitions are somewhat elusive in global climate models, as *Winton* [this volume] shows. The rapid loss events in simulations of the Community Climate System Model (CCSM) shown by *Holland et al.* [2006] are commonly compared to the recent Arctic losses, yet threshold values for sea ice cover and thickness were not found for the CCSM events. Instead, the authors argue that rapid loss can occur through the superposition of natural variability and a steady downward trend (further analysis of these simulations is given by *Holland et al.* [this volume], *Merrifield et al.* [this volume], *Stern et al.* [this volume], and *Gorodetskaya and Tremblay* [this volume]. The lack of identifiable thresholds in CCSM is significant, since the yearly sea ice losses during CCSM rapid declines are larger than the 2007 loss observed by Holland et al., despite the absence of easily identifiable tipping points.

The primary motivation for claims of a tipping point comes from the destabilizing effect of the sea ice–albedo feedback. No doubt this is a strong feedback, but there is some subtlety in assessing its strength. Gorodetskaya and Tremblay point out that the effect of sea ice removal is mitigated by the cloudiness of the Arctic in summer, and note that the presence of sea ice reduces the top-of-atmosphere albedo by only 10 to 20%, despite the large albedo contrast between ice and open water. This finding is consistent with *Winton*'s [2006] conclusion that the sea ice albedo feedback is not dominant as a cause of polar amplification in climate models.

Moreover, the stability of the Arctic sea ice cover depends on the sign of the net feedback, with instability occurring when the positive sea ice–albedo feedback overwhelms the negative feedbacks which stabilize sea ice cover under colder conditions. *Bitz* [this volume] performed CCSM experiments in which Arctic Ocean surface albedo is held fixed even when sea ice cover is reduced by greenhouse gas increases, so that sea ice–albedo feedback is effectively disabled. The sea ice decline which occurs in the absence of sea ice–albedo feedback is not dramatically different from the sea ice decline in the control run. An explanation for this result is given by *Winton* [this volume], who performed model experiments in which sea ice cover was artificially removed. In these experiments increases in solar absorption due to increased open water area are offset by increases in turbulent heat flux from the ocean because of the removal of the insulating ice cover. The implication of these results is that the net feedback due to opening can still be negative, despite the strong positive sea ice–albedo feedback. Further support for this conclusion (at least in climate models) comes from *Cullather and Tremblay*'s [this volume] analysis of naturally occurring sea ice loss anomalies in a long CCSM control run with 1990 levels of greenhouse gases. Despite the sea ice–albedo feedback, sea ice cover rebounded within 1 to 3 years of each anomaly.

## 4. CLIMATE IMPACTS: POLAR BEAR LISTING DECISION

Of course, the implications of rapid sea ice loss go well beyond academic interest in climate stability. Policy makers are particularly challenged by Arctic sea decline, since they must plan for future sea ice conditions which are without precedent in the instrumented record. Faced with the lack of observed analogs, policy makers can seek guidance from global climate model (GCM) simulations of anthropogenic greenhouse warming. Such guidance can be quite valuable provided that two essential issues are addressed: first, the policy-relevant climate impacts of the simulated sea ice decline must be determined and, second, the uncertainty inherent in GCM projections of sea ice loss must be adequately assessed and incorporated. An important case in point is the research conducted by the U.S. Geological Survey (USGS) to advise the U.S. Fish and Wildlife Service (USFWS) on the impact of sea ice decline on polar bears. The research, which was presented in nine USGS administrative reports (online at www.usgs.gov/newsroom/special/polar_bears),

was comissioned to help the USFWS decide whether to list the polar bear as a threatened species under the Endangered Species Act (ESA). Coincidentally, the results of this were presented to the USFWS in September 2007, as the Arctic sea ice cover approached its record low.

It is clear even upon superficial consideration that sea ice decline is bad for polar bears, given their dependence on sea ice as a platform for hunting and other activities (see references of *Amstrup et al.* [this volume]). However, the threat to polar bears from sea ice decline cannot be rigorously assessed without an understanding, based on observational field biology, of the sea ice needs of polar bears. *Durner et al.* [2008] quantified the habitat value of sea ice using observations of radio-collared polar bears over 2 decades. The characteristics that make sea ice desirable as polar bear habitat could be identified and quantified based on this data. In particular, polar bears were found to prefer sea ice over the shallow, productive waters of the continental shelf. The decline of pan-Arctic sea ice extent matters less than the retreat of sea ice from the shelf areas, as the habitat value of ice remaining over the deep Arctic basin is low.

Durner et al.'s resource selection functions (RSFs) quantify the value of sea ice as polar bear habitat, expressed as the frequency of occupation by polar bears, in terms of simple parameters including distance to shore, ocean depth, and sea ice concentration. The RSF methodology can be applied with equal ease to sea ice decline in observations and climate model projections. Thus, they enable researchers to provide guidance to policy makers in terms of the policy-relevant impact, in this case the loss of polar bear habitat, rather than generic statements regarding the overall sea ice decline. Further use of field data combined with model projections in the USGS reports comes from *Hunter et al.* [2007] who used data from a capture-release study to estimate declines in polar bear population as a function of reductions in sea ice availability.

Projections of future sea ice loss and its impacts will inevitably be accompanied by substantial uncertainty, given the evident sensitivity of the Arctic climate system. As discussed by *Amstrup et al.* [this volume], the USGS research accounted for model uncertainty by using a subset of 10 climate models which satisfy a selection criterion based on present day sea ice simulation quality. Projections from this subset show a range of September sea ice loss from 30 % to complete loss by mid century (sources of uncertainty in sea ice projections are discussed by *Bitz and DeWeaver et al.* [this volume]). The uncertainty represented by the range of model simulations was propagated through the USGS analysis by applying techniques like the RSF calculation to the whole subset, so that ensemble spread in sea ice simulations leads to ensemble spread in polar bear outcomes. However, Amstrup et al. note that these projections may be overly optimistic, given *Stroeve et al.*'s [2007] finding that real-world Arctic sea ice has declined at almost twice the rate found in model simulations of the recent past. The USGS efforts culminated in Amstrup et al.'s synthesis report, which uses a Bayesian framework to assess the probability of decline in polar bear population based on consideration of sea ice decline and other factors. Despite the uncertainties of the research, none of the outcomes were favorable for polar bears; in effect, they run the gamut from bad to extremely bad.

In May 2008 the polar bear was listed as a threatened species under ESA, after considerable delay. It is clear from the final rule [*U.S. Fish and Wildlife Service*, 2008] that the policy makers understood and considered the scientific guidance. Consideration of the science is also evident from the announcement of the decision (www.doi.gov/secretary/speeches/081405_speech.html), which included a prominent display of *Stroeve et al.*'s [200X] work on observed and simulated sea ice trends, and maps of Arctic sea ice showing the change in coverage by old (at least 5 years) and new (less than 5 years old) ice, apparently from the drift model of *Rigor and Wallace* [2004]. But while the effort to provide scientific input for the listing decision was successful in some sense, it remains to be seen if the listing will have any direct effect on the status of the polar bear (see analysis of *Revkin* [2008]).

## 5. CONCLUSION

The events of 2007 and 2008 highlight the need for improved understanding of sea ice sensitivity and the impacts of sea ice decline. Perhaps, if we are fortunate, our understanding of the Arctic sea ice and climate system can evolve fast enough to keep pace with the changes occurring there.

*Acknowledgments.* The author's research is supported by the Office of Science (BER), U.S. Department of Energy, grant DE-FG02-03ER63604. I thank Cecilia Bitz, Steven Amstrup and members of the NCAR Polar Climate Working Group for helpful conversations. I am indebted to the chapter authors for contributing their best work to this monograph.

## REFERENCES

Amstrup, S. C., B. G. Marcot, and D. C. Douglas (2008), A Bayesian network model approach to forecasting the 21st century worldwide status of polar bears, this volume.

Bitz, C. M. (2008), Some aspects of uncertainty in predicting sea ice thinning, this volume.

Borenstein, S., and D. Joling (2008), Arctic sea ice drops to 2nd lowest level on record, *San Francisco Chron.*, 27 Aug.

Comiso, J. C., C. L. Parkinson, R. Gersten, and L. Stock (2008), Accelerated decline in the Arctic sea ice cover, *Geophys. Res. Lett.*, *35*, L01703, doi:10.1029/2007GL031972.

Deser, C., and H. Teng (2008), Recent trends in Arctic sea ice and the evolving role of atmospheric circulation forcing, 1979–2007, this volume.

de Vernal, A., C. Hillaire-Marcel, S. Solignac, T. Radi, and A. Rochon (2008), Reconstructing sea-ice conditions in the Arctic and subarctic prior to human observations, this volume.

DeWeaver, E. T., E. C. Hunke, and M. M. Holland (2008), Sensitivity of Arctic sea ice thickness to intermodel variations in the surface energy budget, this volume.

Durner, G. M., et al. (2008), Predicting 21st century polar bear habitat distribution from global climate models, *Ecol. Monogr.*, in press.

Gorodetskaya, I. V., and L.-B. Tremblay (2008), Arctic cloud properties and radiative forcing from observations and their role in sea ice decline predicted by the NCAR CCSM3 model during the 21st century, this volume.

Holland, M. M., C. M. Bitz, and B. Tremblay (2006), Future abrupt reductions in the summer Arctic sea ice, *Geophys. Res. Lett.*, *33*, L23503, doi:10.1029/2006GL028024.

Holland, M. M., C. M. Bitz, B. Tremblay, and D. A. Bailey (2008), The role of natural versus forced change in future rapid summer Arctic ice loss, this volume.

Kay, J. E., T. L'Ecuyer, A. Gettelman, G. Stephens, and C. O'Dell (2008), The contribution of cloud and radiation anomalies to the 2007 Arctic sea ice extent minimum, *Geophys. Res. Lett.*, *35*, L08503, doi:10.1029/2008GL033451.

Kerr, R. A. (2007), Is battered sea ice down for the count?, *Science*, *318*, 33–34.

Kizzia, T. (2008), Polar ice pack loss may break 2007 record, *Anchorage Daily News*, 12 Feb.

Lindsay, R. W., and J. Zhang (2005), The thinning of Arctic sea ice, 1988–2003: Have we passed a tipping point?, *J. Clim.*, *18*, 4879–4894.

Maslanik J. A., C. Fowler, J. Stroeve, S. Drobot, J. Zwally, D. Yi, and W. Emery (2007), A younger, thinner Arctic ice cover: Increased potential for rapid, extensive sea-ice loss, *Geophys. Res. Lett.*, *34*, L24501, doi:10.1029/2007GL032043.

Meier, W. N., J. Stroeve, and F. Fetterer (2007), Wither Arctic sea ice? A clear signal of decline, regionally, seasonally, and extending beyond the satellite record, *Ann. Glaciol.*, *46*, 428–434.

Merryfield, W. J., M. M. Holland, and A. H. Monahan (2008), Multiple equilibria and abrupt transitions in Arctic summer sea ice extent, this volume.

Nghiem, S. V., I. G. Rigor, D. K. Perovich, P. Clemente-Colón, J. W. Weatherly, and G. Neumann (2007), Rapid reduction of Arctic perennial sea ice, *Geophys. Res. Lett.*, *34*, L19504, doi:10.1029/2007GL031138.

North, G. R. (1984), The small ice cap instability in diffusive climate models, *J. Atmos. Sci.*, *41*, 3390–3395.

Overland, J. E., M. Wang, and S. Salo (2008), The recent Arctic warm period, *Tellus, Ser. A*, *60*, 589–597.

Perovich, D. K., J. A. Richter-Menge, K. F. Jones, and B. Light (2008), Sunlight, water, and ice: Extreme Arctic sea ice melt during the summer of 2007, *Geophys. Res. Lett., 35*, L11501, doi:10.1029/2008GL034007.

Revkin, A. C. (2007), Arctic melt unnerves the experts, *N. Y. Times*, 2 Oct.

Revkin, A. C. (2008), Polar bear is made a protected species, *N. Y. Times*, 15 May.

Rigor, I. G., and J. M. Wallace (2004), Variations in the age of Arctic sea-ice and summer sea-ice extent, *Geophys. Res. Lett., 31*, L09401, doi:10.1029/2004GL019492.

Schweiger, A. J., J. Zhang, R. W. Lindsay, and M. Steele (2008), Did unusually sunny skies help drive the record sea ice minimum of 2007?, *Geophys. Res. Lett., 35*, L10503, doi:10.1029/2008GL033463.

Stern, H. L., R. W. Lindsay, C. M. Bitz, and P. Hezel (2008), What is the trajectory of Arctic sea ice?, this volume.

Stroeve J., M. M. Holland, W. Meier, T. Scambos, and M. Serreze (2007), Arctic sea ice decline: Faster than forecast, *Geophys. Res. Lett., 34*, L09501, doi:10.1029/2007GL029703.

Stroeve, J., M. Serreze, S. Drobot, S. Gearheard, M. Holland, J. Maslanik, W. Meier, and T. Scambos (2008), Arctic sea ice extent plummets in 2007, *Eos Trans. AGU, 89*(2), 13.

U.S. Fish and Wildlife Service (2008), Endangered and threatened wildlife and plants; determination of threatened status for the polar bear (*Ursus maritimus*) throughout its range, *Fed. Regist., 73*, 28,211–28,303.

Winton, M. (2006), Amplified Arctic climate change: What does surface albedo feedback have to do with it?, *Geophys. Res. Lett., 33*, L03701, doi:10.1029/2005GL025244.

Winton, M. (2008), Sea ice–albedo feedback and nonlinear Arctic climate change, this volume.

Zhang, J., R. Lindsay, M. Steele, and A. Schweiger (2008), What drove the dramatic retreat of Arctic sea ice during the summer 2007?, *Geophys. Res. Lett., 35*, L11505, doi:10.1029/2008GL034005.

E. T. DeWeaver, Center for Climate Research, University of Wisconsin-Madison, 1225 West Dayton Street, Madison, WI 53706, USA. (deweaver@aos.wisc.edu)

# Recent Trends in Arctic Sea Ice and the Evolving Role of Atmospheric Circulation Forcing, 1979–2007

Clara Deser and Haiyan Teng

*National Center for Atmospheric Research, Boulder, Colorado, USA*

This study documents the evolving trends in Arctic sea ice extent and concentration during 1979–2007 and places them within the context of overlying changes in the atmospheric circulation. Results are based on 5-day running mean sea ice concentrations (SIC) from passive microwave measurements during January 1979 to October 2007. Arctic sea ice extent has retreated at all times of the year, with the largest declines ($0.65 \times 10^6$ km$^2$ per decade, equivalent to 10% per decade in relative terms) from mid July to mid October. The pace of retreat has accelerated nearly threefold from the first half of the record to the second half, and the number of days with SIC less than 50% has increased by 19 since 1979. The spatial patterns of the SIC trends in the two halves of the record are distinctive, with regionally opposing trends in the first half and uniformly negative trends in the second half. In each season, these distinctive patterns correspond to the first two leading empirical orthogonal functions of SIC anomalies during 1979–2007. Atmospheric circulation trends and accompanying changes in wind-driven atmospheric thermal advection have contributed to thermodynamic forcing of the SIC trends in all seasons during the first half of the record and to those in fall and winter during the second half. Atmospheric circulation trends are weak over the record as a whole, suggesting that the long-term retreat of Arctic sea ice since 1979 in all seasons is due to factors other than wind-driven atmospheric thermal advection.

## 1. INTRODUCTION

The accelerating retreat of Arctic sea ice in recent decades, evident in all months of the year, is one of the most dramatic signals of climate change worldwide (see *Serreze et al.* [2007], *Meier et al.* [2007], and *Stroeve et al.* [2007] for recent overviews; ongoing updates on Arctic sea ice may be obtained from the National Snow and Ice Data Center (available at http://nsidc.org)). Although climate models

Arctic Sea Ice Decline: Observations, Projections, Mechanisms, and Implications
Geophysical Monograph Series 180

predict that Arctic sea ice will decline in response to atmospheric greenhouse gas increases [*Holland et al.*, 2006], the current pace of retreat at the end of the melt season is exceeding the models' forecasts by approximately a factor of 3 [*Stroeve et al.*, 2007]. Long-term records of summer sea ice extent within the central Arctic Ocean dating back to 1900 exhibit large multidecadal variations [*Polyakov et al.*, 2003], a factor which must be taken into account when interpreting the recent sea ice retreat.

The physical mechanisms underlying the Arctic sea ice decline are not fully understood but include dynamical processes related to changes in winds and ocean currents and thermodynamic processes involving changes in air temperature, radiative and turbulent energy fluxes, ocean heat storage, and ice-albedo feedback [*Serreze et al.*, 2007; *Stroeve and*

*Maslowski*, 2008; *Francis and Hunter*, 2006, 2007; *Shimada et al.*, 2006; *Perovich et al.*, 2007]. A better understanding of these mechanisms and their relationship to increasing greenhouse gas concentrations is an important step for assessing future predictions of Arctic climate change.

Numerous studies indicate that the atmospheric circulation played an important role in driving Arctic sea ice declines from the 1960s to the early 1990s [e.g., *Deser et al.*, 2000; *Rigor et al.*, 2002; *Hu et al.*, 2002; *Rigor and Wallace*, 2004; *Rothrock and Zhang*, 2005; *Stroeve et al.*, 2007; *Serreze and Francis*, 2006; *Ukita et al.*, 2007]. In particular, the declines during this period were due in part to a trend in the dominant pattern of wintertime atmospheric circulation variability over the high-latitude Northern Hemisphere known variously as the "North Atlantic Oscillation," "Arctic Oscillation," or "Northern Annular Mode" [*Hurrell*, 1995; *Deser*, 2000; *Thompson and Wallace*, 2000], collectively referred to hereinafter as the "NAM." In particular, the anomalous cyclonic wind circulation associated with the upward trend in the winter NAM flushed old, thick ice out of the Arctic via Fram Strait, causing the winter ice pack to thin, which, in turn, preconditioned the summer ice pack for enhanced melt.

Since the early 1990s, however, the trend in the NAM has reversed sign, yet Arctic sea ice has continued to decline [*Overland and Wang*, 2005; *Comiso*, 2006; *Serreze and Francis*, 2006; *Maslanik et al.*, 2007; *Serreze et al.*, 2007; *Stroeve and Maslowski*, 2008]. This has led to speculation that the Arctic climate system has reached a "tipping point" whereby strong positive feedback mechanisms such as those associated with ice albedo and open water formation efficiency are accelerating the thinning and retreat of Arctic sea ice [e.g., *Lindsay and Zhang*, 2005; *Holland et al.*, 2006]. These positive feedback mechanisms leave the ice pack more vulnerable to forcing from other processes, natural and anthropogenic. For example, enhanced downward longwave radiation associated with increases in air temperature, water vapor and cloudiness over the Arctic Ocean [*Francis and Hunter*, 2006] along with enhanced ocean heat transport into the Arctic [*Polyakov et al.*, 2005; *Shimada et al.*, 2006; *Stroeve and Maslowski*, 2008] and positive ice-albedo feedback [*Perovich et al.*, 2007] have become dominant factors driving summer sea ice extent declines since the mid-to-late 1990s. There is also evidence that the winter atmospheric circulation has continued to affect the winter sea ice distribution since the mid-1990s [*Comiso*, 2006; *Maslanik et al.*, 2007; *Francis and Hunter*, 2007].

The purpose of this study is to revisit the issue of Arctic sea ice trends from 1979 to present in the context of evolving atmospheric circulation conditions. In addition to examining trends over the entire period of record, we investigate trends

over the two halves separately as a simple way of characterizing their evolution. We note that the first half coincides with an upward trend in the NAM, while the second half coincides with a downward trend. We are particularly interested in assessing the evolving role of thermodynamic atmospheric circulation forcing of sea ice concentration trends, taking into account any seasonal dependencies. We use 5-day running mean sea ice concentration data on a 25 km × 25 km grid derived from passive microwave measurements from 1 January 1979 through 31 October 2007. Early results were presented by *Deser and Teng* [2008] for the winter and summer seasons only.

Our study is organized as follows. Section 2 describes the data sets and methodology. Section 3.1 provides results on trends in Arctic sea ice extent throughout the annual cycle, as well as derived quantities such as the timing of the seasonal cycle. Section 3.2 presents the spatial patterns of sea ice concentration, sea level pressure, and wind-induced atmospheric thermal advection trends for the two halves of the study period and for the record as a whole, stratified by season. Air temperature, sea surface temperature (SST), and net surface downward longwave radiation trends from 1979 to present are also shown. Section 4 provides a summary and discussion of the results.

## 2. DATA AND METHODS

Daily sea ice concentrations (SIC) on a 25 km × 25 km grid for the period 1 January 1979 to 31 October 2007 were obtained from the National Snow and Ice Data Center. These data are derived from the Nimbus 7 Scanning Multichannel Microwave Radiometer and Defense Meteorological Satellite Program (DMSP) F8, F11, and F13 Special Sensor Microwave/Imager radiances using the NASA team algorithm [*Cavalieri et al.*, 1999].

In addition to daily SIC, we use monthly mean sea level pressure (SLP), 1000 hPa zonal and meridional wind components, 2-m and 1000-hPa air temperatures on a 2.5° × 2.5° latitude grid from the National Centers for Environmental Prediction/National Center for Atmospheric Research (NCEP/NCAR) reanalysis project [*Kalnay et al.*, 1996] for the period January 1979 through October 2007. We also use monthly mean sea surface temperature data from the HadISST1 data set [*Rayner et al.*, 2003] on a 1° × 1 ° latitude grid, updated through December 2006. The NCEP/NCAR reanalysis and HadISST1 data sets were obtained from the Data Support Section at NCAR. Finally, we make use of daily net surface downward longwave radiative fluxes derived from the NASA-NOAA Television and Infrared Observation Satellite (TIROS) Operational Vertical Sounder (TOVS) polar pathfinder data set [*Francis and Hunter*,

2007]. These data are available from July 1979 through December 2005 on a 100-km² grid north of 55°N.

## 3. RESULTS

### 3.1. Arctic Sea Ice Extent

A map of the locations of the Arctic and sub-Arctic seas referred to in this study are shown in Figure 1. These place names are superimposed upon the long-term mean distributions of maximum and minimum sea ice extent (defined as marine areas within which sea ice concentrations equal or exceed 15%), based on the period 1979–2006. The maxi-

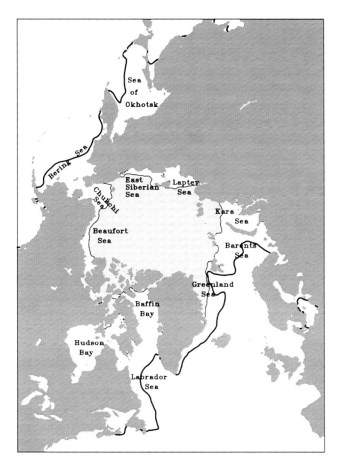

**Figure 1.** Locations of the Arctic and sub-Arctic seas referred to in this study, superimposed upon the long-term (1979–2007) mean sea ice extent at month of maximum (thick black contour) and month of minimum (thin black contour and light shaded areas). The maximum (minimum) sea ice extent is defined as the 30-day average centered on the mean date of maximum (minimum) extent, 7 March (17 September).

mum sea ice extent is defined as the 30-day average centered on the mean date of maximum extent, 7 March, and the minimum extent is defined as the 30-day average centered on the mean date of minimum extent, 17 September. At maximum extent, all of the Arctic and sub-Arctic seas are ice covered (sea ice concentrations >15%), while at minimum extent, only the Greenland and Beaufort seas and the Arctic basin (central Arctic Ocean) are ice covered.

Arctic sea ice extent serves as a useful starting point for describing the temporal character of sea ice over the Arctic as a whole. Following convention, we have defined Arctic sea ice extent as the area of the ocean covered by at least 15% sea ice concentration based on 5-day running mean data; note that Hudson Bay and the Baltic Sea have been excluded from this calculation. Plate 1 shows the 5-day running mean Arctic sea ice extent during the period 1 January 1979 to 31 October 2007, with each year overlaid in a different color (see color scale to the right of Plate 1). Plate 1 conveys the regularity of the seasonal cycle throughout the period of record, with maximum values (14–16 × 10⁶ km²) occurring in early March and minimum values (5–7.5 × 10⁶ km² excluding 2007) in the middle of September. In addition to the regularity of the seasonal cycle, Plate 1 conveys the systematic reduction of Arctic sea ice extent over time, with the 1980s exhibiting the highest values (red hues) and the 2000s showing the lowest values (blue hues). This systematic retreat of Arctic sea ice extent has occurred in all months of the year. Arctic sea ice extent reached unprecedented minimum values in August–October of 2007 (purple curve). The change in September sea ice extent between 2006 and 2007 alone (~1.5 × 10⁶ km²) is approximately equivalent to the entire change that occurred between September 1979 and September 2006 (~1.3 × 10⁶ km²).

The 5-day running mean Arctic sea ice extent data shown in Plate 1 are replotted in Figure 2 (top) as a single continuous time series starting on 1 January 1979 and ending on 31 October 2007. The same record after removing the long-term 5-day running mean seasonal cycle is shown in Figure 2 (bottom). The former depiction serves to emphasize that the seasonal cycle of Arctic sea ice extent is still the most prominent feature of the record, while the latter underscores the accelerating downward trend of Arctic sea ice extent over time. Arctic sea ice extent has declined at a rate of −0.52 × 10⁶ km² per decade, or −1.76 × 10⁶ km² over the period 1 January 1979 to 1 January 2007; this trend is significant at the 99% level. The magnitude of the downward linear trend has increased from −0.35 × 10⁶ km² per decade to −0.9 × 10⁶ km² per decade from the first half (January 1979 to December 1993) to the second half (January 1993 to October 2007) of the record (similar results were found by *Comiso et al.* [2008]).

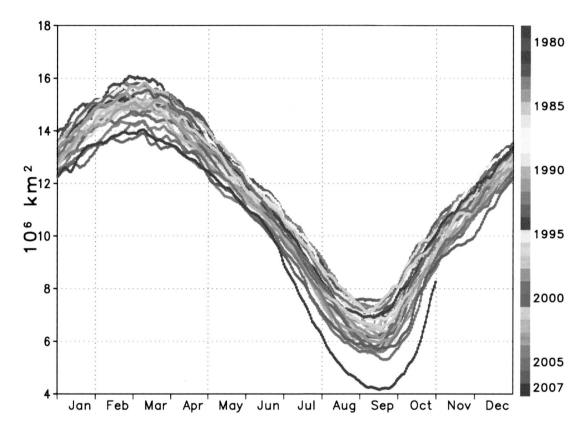

**Plate 1.** Five-day running mean Arctic sea ice extent ($10^6$ km²) during the period 1 January 1979 to 31 October 2007, with each year overlaid in a different color (see color scale at right). Arctic sea ice extent is defined as the area of the ocean covered by at least 15% sea ice concentration.

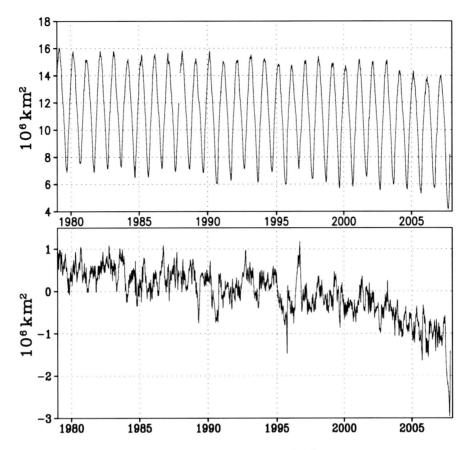

**Figure 2.** Time series of 5-day running mean Arctic sea ice extent ($10^6$ km$^2$) from 1 January 1979 to 31 October 2007 (top) with and (bottom) without the long-term mean seasonal cycle.

Figure 3 isolates the behavior of the maximum and minimum values of sea ice extent shown in Plate 1 and their dates of occurrence during 1979–2007. Minimum and maximum values were determined by comparing adjacent 5-day running means. Linear trend lines in the dates and values of maximum and minimum Arctic sea ice extent, determined by linear least squares "best fit" regression lines to the 5-day running mean data during January 1979 to June 2007 (note that data after June 2007 were purposefully omitted from this calculation), are superimposed on the original time series in Figure 3. Figure 3 (left) shows that there has been a downward trend in both the maximum and minimum sea ice extent values ($-0.5 \times 10^6$ km$^2$ per decade and $-0.7 \times 10^6$ km$^2$ per decade, respectively: significant at the 99% level), with a corresponding increase in the amplitude of the annual cycle ($0.15 \times 10^6$ km$^2$ per decade, although this does not pass the 90% significance threshold). In terms of percentage of the mean maximum ($15.2 \times 10^6$ km$^2$) and mean minimum ($6.4 \times 10^6$ km$^2$) sea ice extent, the downward trends in maximum and minimum extent are $-9.0\%$ per decade and $-3.4\%$ per decade, respectively.

The time series of the dates of maximum and minimum sea ice extent (Figure 3, right) show that there has been a slight upward trend (indicative of a progressively later date) in the date of maximum sea ice extent of 4 days per decade (or 12 days over the period 1979–2007; significant at the 90% level), while there has been little change in the date of minimum ice extent (1 day per decade). The length of time between maximum and minimum extent has decreased slightly at a rate of 3 days per decade or 9 days over the period January 1979 to June 2007, but this does not pass the 90% significance threshold. The mean dates of maximum and minimum sea ice extent occur on 7 March and 17 September, respectively. It should be noted that the record low minimum sea ice extent in 2007 occurred on 16 September, close to the mean date of minimum sea ice extent.

The magnitude and sign of the linear trends in Arctic sea ice extent as a function of time of year are shown in Figure 4 based on 5-day running mean data for the period 1 January 1979 to 30 June 2007 (the record low values since June 2007 are purposefully omitted from the trend calculation). The trends are expressed in terms of actual magnitude (square

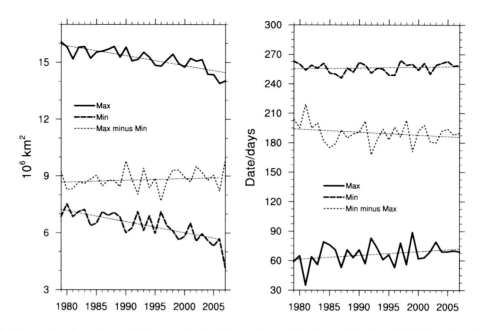

**Figure 3.** Time series of maximum (solid curves) and minimum (long dashed curves) values of (left) Arctic sea ice extent ($10^6$ km$^2$) and (right) their dates of occurrence, determined from 5-day running mean data. The dotted curves show the time series of the maximum-minus-minimum sea ice extent and the date of minimum extent minus the date of maximum extent. Linear trend lines are superimposed using data before 1 June 2007.

kilometers per decade) and relative magnitude (percent per decade, taken with respect to the long-term mean extent for each 5-day running mean period). All trend values are statistically significant at the 95% level. The linear trend is negative at all times of year, with the largest magnitudes (0.55 to $0.65 \times 10^6$ km$^2$ per decade or 7 to 10% per decade) from the end of July to the middle of October when the mean sea ice extent is smallest (recall Plate 1). The relative magnitude of the trend has a larger seasonal dependence than the actual magnitude, ranging from a maximum value of nearly $-10\%$ per decade in mid September to a minimum value of $-3$ to $-4\%$ per decade from November through June. The actual magnitude of the trend ranges from a maximum value of $-0.65 \times 10^6$ km$^2$ per decade in early October to a minimum value of $-0.4$ to $-0.5 \times 10^6$ km$^2$ per decade from November through June.

Trends in the dates of maximum and minimum sea ice extent were presented in Figure 3 (right). Another approach to characterizing the timing of the seasonal cycle is to examine the date when sea ice concentration first falls below 50% and when it first exceeds 50% at each grid point for each year. Plate 2 shows the geographical distributions of the trends in these dates along with the corresponding long-term mean

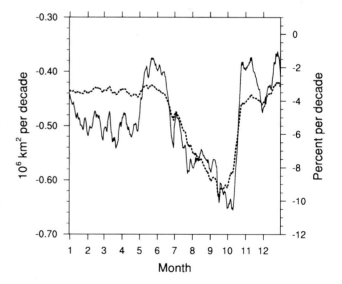

**Figure 4.** Linear trends in Arctic sea ice extent as a function of time of year based on 5-day running mean data for the period 1 January 1979 to 30 June 2007, expressed in terms of actual magnitude (km$^2$ per decade, solid curve) and relative magnitude (percent per decade, taken with respect to the long-term mean extent for each 5-day running mean period, dotted curve).

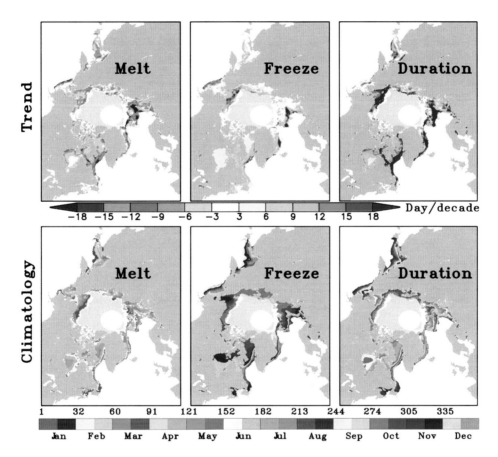

**Plate 2.** Geographical distributions of the (top) linear trends and (bottom) climatologies of the dates when sea ice concentration first falls below 50% ("melt"), first exceeds 50% ("freeze"), and their difference ("duration") based on 5-day running mean data during 1979–2006. The trend values are expressed in days per decade, and the climatological values are expressed in terms of calendar date (number of days for "duration"). The white ellipse around the North Pole indicates missing data.

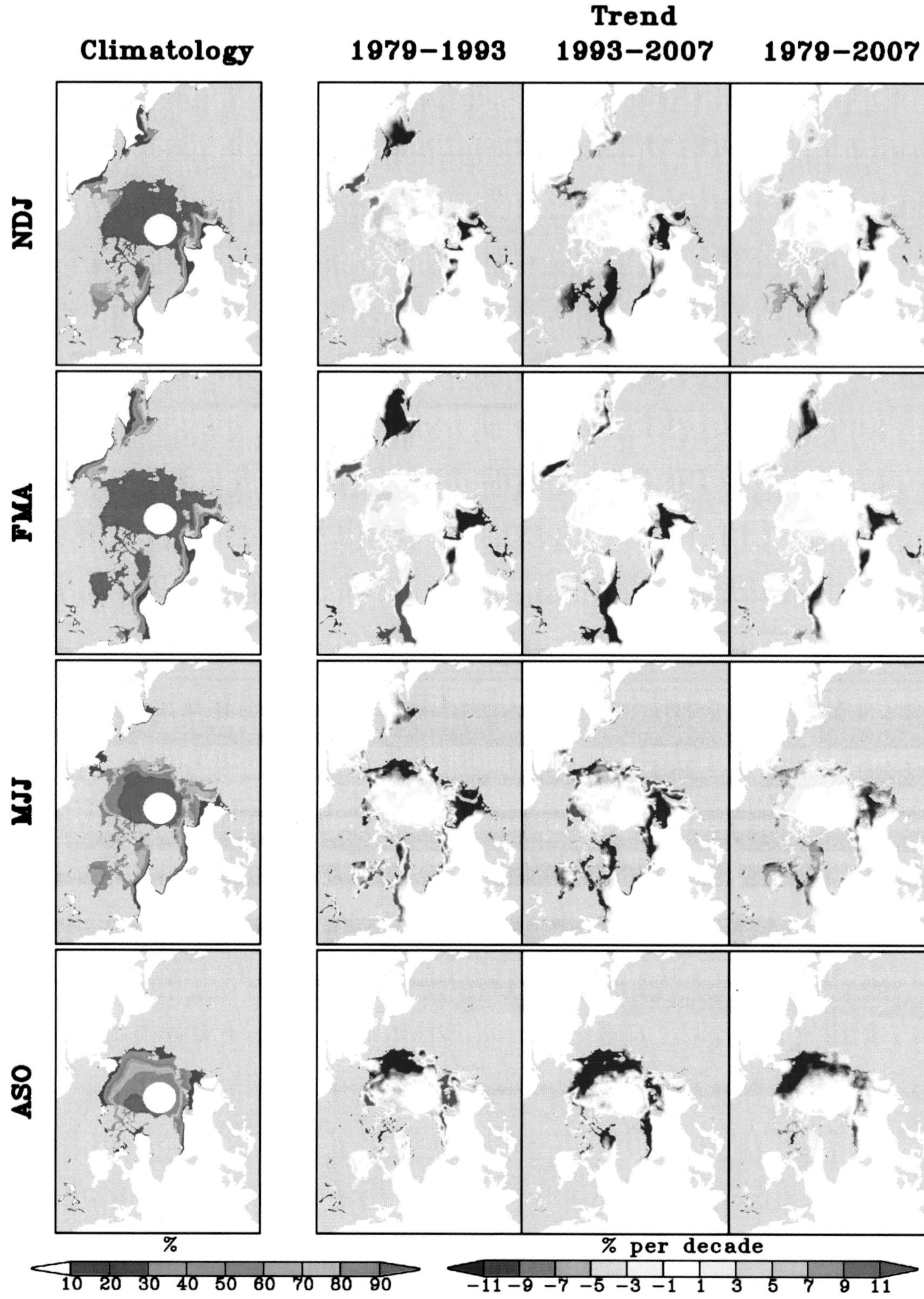

**Plate 3.** Seasonal mean sea ice concentration (far left panels) climatologies based on the period 1979–2007 (percent) and linear trends (percent per decade) during (left) 1979–1993, (middle) 1993–2007, and (right) 1979–2007. Data after 30 April 2006 are excluded. Seasons are defined as (first row) November–January (NDJ), (second row) February–April (FMA), (third row) May–July (MJJ), and (fourth row) August–October (ASO). The white ellipse around the North Pole indicates missing data.

values, based on 5-day running means during January 1979 to June 2007. The linear trends in the dates when SIC first falls below 50% are consistently negative throughout the Arctic, with amplitudes ~6–9 days per decade in many regions and even greater values in the Barents Sea (Plate 2, top left). Similarly, the linear trends in the dates when SIC first exceeds 50% are positive throughout the marginal ice zone except in the Bering Sea, with amplitudes ~6–12 days per decade (12–18 days per decade in the Barents and Chuckchi seas; Plate 2, top middle). These patterns result in large positive trends in the duration of SIC <50%, with values in excess of 18 days per decade in the Labrador, Greenland, Barents and Chuckchi seas (Plate 2, top right). The spatial uniformity of the trend values in all three quantities (Plate 2, top panels) contrasts with the large meridional gradients present in their background climatologies (Plate 2, bottom). Qualitatively similar results are obtained with thresholds of 30% and 70% (not shown).

Figure 5 shows the time series of the area-averaged dates when SIC first falls below or exceeds 50%. There is a trend toward an earlier (later) date of occurrence of sea ice concentrations first falling below (exceeding) 50% of −3.0 days per decade (3.7 days per decade). This results in an increasing trend in the duration of SIC <50% (defined as the difference between the dates when SIC first falls below 50% and first exceeds 50%) of 6.9 days per decade or 19 days over the period

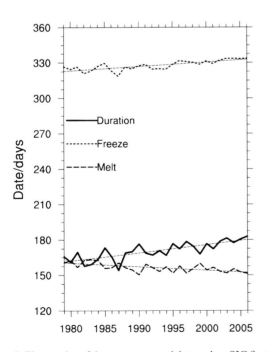

**Figure 5.** Time series of the area-averaged dates when SIC first falls below 50% ("melt," dashed curve), first exceeds 50% ("freeze," dotted curve), and their difference ("duration," solid curve).

1979–2006, statistically significant at the 99% level. The duration of the sea ice melt season, determined from emissivity changes associated with liquid and frozen water, has shown an even larger increase (approximately 2 weeks per decade [*Stroeve et al.*, 2006]) than the duration of SIC <50%.

### 3.2. Arctic Sea Ice Concentration

*3.2.1. SIC trends.* Up to now, we have focused on trends in sea ice extent for the Arctic as a whole. In this section, we consider the spatial patterns of recent trends in Arctic sea ice concentration (SIC), taking into account the seasonal dependence of the trends. To reduce the amount of information, we focus on 3-month seasons defined as follows: November–January (NDJ); February–April (FMA); May–July (MJJ); and August–October (ASO); we refer to these seasons as autumn, winter, spring, and summer, respectively. This choice of seasonal averaging retains the basic characteristics of the annual cycle and seasonal dependence of recent trends. The spatial patterns of the monthly SIC trends are highly coherent within each season (not shown).

The climatological SIC distributions for each season based on the period January 1979 to April 2007 are shown in the far left-hand column of Plate 3. In winter (FMA), the season of maximum sea ice extent, long-term mean SIC values between 10% and 90%, indicative of the location of the marginal ice zone, are found in the Labrador Sea, the Greenland and Barents seas, the Bering Sea, and the Sea of Okhotsk. In summer (ASO), the season of minimum sea ice extent, the marginal ice zone retreats northward to coastal regions of the Arctic Ocean and the Canadian Archipelago. The long-term mean SIC distribution in autumn (NDJ) resembles that in winter, albeit with reduced values in the peripheral seas, particularly the Bering Sea and the Sea of Okhotsk. The climatological SIC distribution in spring (MJJ) is not identical to that in autumn: although similar values prevail over the Atlantic sector, lower amounts occur within coastal regions of the central Arctic Ocean, and the Pacific marginal seas are nearly ice free.

These climatological SIC distributions provide a context for the spatial patterns of recent SIC trends shown in the left, middle, and right columns of Plate 3. In addition to showing the trends over the full period of record (January 1979 to June 2007), Plate 3 also shows the evolution of the trends from the first half of the record (January 1979 to December 1993) to the second (January 1993 to June 2007). Note that the magnitudes of the trends in each period may be directly compared as they are expressed in percent SIC per decade. The regions of largest-amplitude SIC trends in each season correspond to the marginal ice zone as depicted in the left-hand columns. As a general rule of thumb, SIC trends exceeding 3% per decade in absolute value (corresponding to the second level of red or blue shading in Plate 3) are statistically significant at the 95% level.

The pattern of winter SIC trends in the first half of the record (1979–1993) exhibits positive values in the Labrador and Bering seas and negative values in the Greenland and Barents seas and the Sea of Okhotsk. In contrast, winter SIC trends in the second half of the record (1993–2007) are negative throughout the marginal seas, with the largest declines in the Atlantic sector. The winter SIC trends over the full period of record (1979–2007) are also negative throughout the marginal seas, except in the Bering Sea where the trends are near zero. The change in pattern of winter SIC trends between the first and second halves of the record is notable and will be discussed further below in the context of evolving atmospheric circulation trends. The patterns of SIC trends in the autumn season are very similar to those in winter, with somewhat reduced magnitudes commensurate with the lower long-term mean SIC amounts.

In summer, SIC trends in the first half of the record are negative in the East Siberian Sea and positive in the Barents, Kara, and eastern Beaufort seas, with the area of reduced SIC outweighing that of increased SIC. In the second half of the record (1993–2006), the area of negative SIC trends has expanded to cover almost all longitudes. The summer SIC trend over the full period of record (1979–2006) is similar to that for the second half of the record, with the largest declines extending from the Laptev Sea eastward to the Beaufort Sea. The SIC trends in spring are a mixture of those in winter and summer. In particular, the trends in the Atlantic sector follow those in winter, while the trends in the central Arctic Ocean resemble those in summer but with weaker magnitudes. The hybrid nature of the spring SIC trends is consistent with that of the long-term mean SIC distribution discussed earlier.

It is instructive to relate the results shown in Plate 3 back to the behavior of sea ice extent for the Arctic as a whole. Figure 6 shows the linear trends in Arctic sea ice extent as a function of time of year based on 5-day running means for the two halves of the record separately (excluding data since June 2007). Trend values significant at the 95% level are shown with a bold line segment. Note that the magnitudes of the trends in each period may be directly compared as they are expressed in square kilometers per decade and percent of the period mean per decade. The negative trends in Arctic sea ice extent increased dramatically in magnitude between the first and second halves of the record during October–March, from values of 0 to $-0.3 \times 10^6$ km$^2$ (0 to $-2\%$) per decade to values of $-0.8$ to $-1.1 \times 10^6$ km$^2$ ($-6$ to $-8\%$) per decade. The weak and statistically insignificant trends in the first half of the record are due to the large degree of cancellation between negative and positive SIC trends in different regions not to a lack of SIC trends (recall Plate 3). In terms of actual magnitude (Figure 6, left), the seasonal dependence of the trend amplitudes is nearly opposite between the two

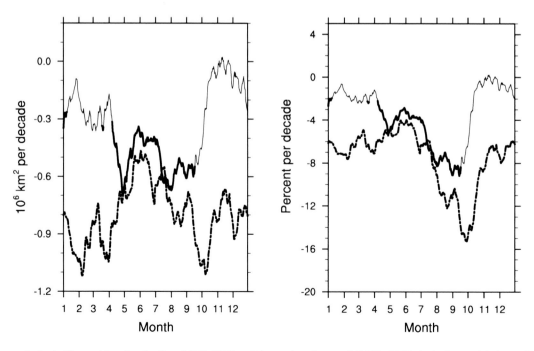

**Figure 6.** As in Figure 4 but for the first (1979–1993, solid curve) and second (1993–2007, dashed curve) halves of the record. Trend values significant at the 95% level are shown with a bold line segment.

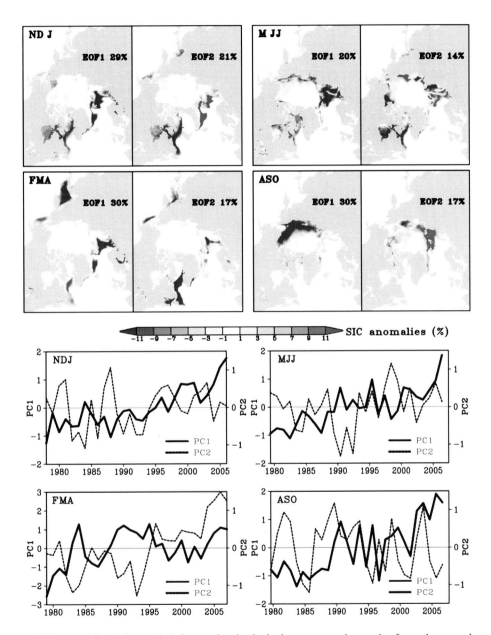

**Plate 4.** (top) EOFs 1 and 2 and (bottom) their associated principal component time series for each season based on the period January 1979 to April 2007. The percent variances explained are given in the EOF plots.

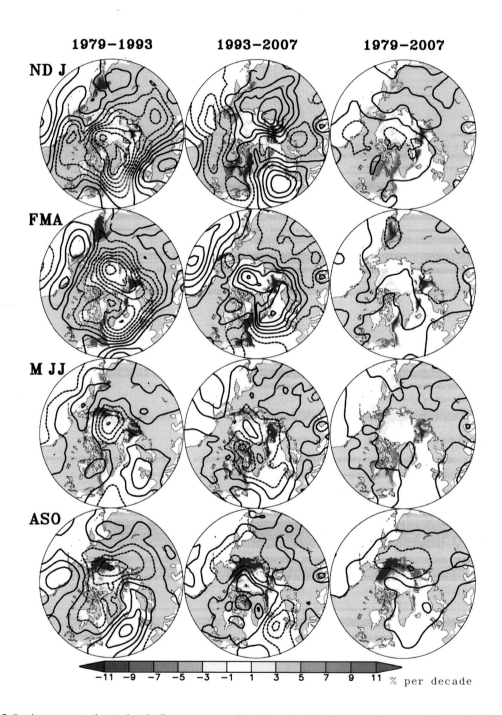

**Plate 5.** Sea ice concentration (color shading, percent per decade) and sea level pressure (contours, hPa per decade) trends during (left) 1979–1993, (middle) 1993–2007, and (right) 1979–2007 for each season as indicated. The contour interval for sea level pressure is 1 hPa per decade; negative values are dashed, and the zero and positive values are solid. Note that data after April 2007 are excluded.

halves of the record, with the smallest (largest) trends occurring during the cold season (October–March) during the first (second) half of the record. In terms of relative magnitude (Figure 6, right), the seasonality of the trends in Arctic sea ice extent remains similar between the two periods, with the largest negative trends occurring during summer (mid July to early October). The maximum relative trend amplitudes have shifted from mid September in the first half of the record to late September in the second half, and they have also amplified (9% per decade in the first half to 15% per decade in the second half).

*3.2.2. Empirical orthogonal function analysis of SIC anomalies.* Are the SIC trend patterns shown in Plate 3 preferred structures of variability, or are they simply a result of dividing the record into halves? To address this question, we have applied empirical orthogonal function (EOF) analysis to seasonal SIC anomaly fields over the full period of record, using a separate EOF analysis for each season. Note that the SIC anomalies have not been normalized by their standard deviation for this calculation. The two leading EOFs in each season and their associated principal component (PC) time series are shown in Plate 4. In both winter and summer, the first and second EOFs account for 30% and 17% of the variance, respectively. In autumn (spring), EOF1 accounts for 29% (20%) of the variance and EOF2 for 21% (14%) of the variance. In all seasons, EOFs 1 and 2 are well separated according to the criterion of *North et al.* [1982].

In winter, the leading EOF exhibits out-of-phase variations between the eastern and western Atlantic and between the eastern and western Pacific, strongly reminiscent of the trend pattern during the first half of the record (recall Plate 3). This EOF is nearly identical to that given by *Ukita et al.* [2007] based on February–March averages over the period 1979–2003 and consistent with results obtained using data sets beginning in the early 1950s [*Walsh and Johnson*, 1979; *Fang and Wallace*, 1994; *Deser et al.*, 2000]. The associated PC time series exhibits an upward trend from 1979 to 1995, near zero values from 1996 through 2004, and positive values from 2005 through 2007. EOF2 of winter SIC is characterized by uniform polarity throughout the Arctic marginal ice zones, with largest amplitudes in the Labrador Sea. This EOF resembles the trend pattern during the second half of the record (recall Plate 3). Its PC time series exhibits generally negative values before 1995 and positive values thereafter, indicative of a decreasing trend of winter SIC in the peripheral seas. It is notable that the first and second EOFs of winter SIC anomalies during 1979–2007 correspond to the winter SIC trend patterns in the first and second halves of the record, respectively, indicating that these two trend patterns dominate the variability over the period of study. To

our knowledge, the only other study documenting the spatial pattern associated with EOF2 of winter SIC variability is that of *Deser and Teng* [2008].

The leading EOF of summer SIC anomalies during 1979–2006 exhibits uniform polarity throughout most of the Arctic marginal ice zone, with largest amplitudes from the Laptev Sea eastward to the Beaufort Sea (Plate 4). This EOF resembles closely the patterns of summer SIC trends during 1979–2006 and 1993–2006 and projects substantially onto the trend pattern for 1979–1993 (recall Plate 3). This similarity is consistent with the fact that the leading PC time series exhibits an upward trend over the period of record. The second EOF of summer SIC anomalies consists of out-of-phase variations between the Barents/Kara seas and the East Siberian/Beaufort seas, with no discernible trend in its PC time series. This EOF does not correspond closely to any of the summer SIC trend patterns shown in Plate 3, although it captures some of the out-of-phase behavior evident in the early period.

Although the SIC trend patterns in autumn resemble those in winter (especially over the Atlantic sector), the ordering of the EOFs is reversed, with the leading (second) EOF in winter corresponding to the second (leading) EOF in autumn (Plate 4). The leading EOF in spring consists of negative values throughout the marginal ice zone, similar to the trend patterns during 1993–2006 and 1979–2006, while the second EOF resembles the trend pattern during 1979–1993. Thus, the leading EOF of SIC anomalies in spring, summer, and autumn (and the second EOF in winter) exhibit negative values throughout the marginal ice zone, similar to the trend patterns since 1993 (and 1979). We expect that if the current winter SIC declines continue, the leading EOF for that season will also eventually exhibit negative values throughout the peripheral seas.

*3.2.3. SIC trends in the context of atmospheric circulation trends.* As discussed in section 1, our motivation for examining the two halves of the record separately is not only to assess in a simple fashion the evolution of the SIC trends over time but also to examine the SIC trends in the context of a rising and falling NAM index. Recall that the NAM is the leading pattern of atmospheric circulation variability over the extratropical Northern Hemisphere in all seasons [*Portis et al.*, 2001], and in winter it exhibits a positive trend in the first half of the record and a negative trend in the second half. To aid our interpretation of the role of SLP forcing of SIC trends, we also consider trends in wind-induced atmospheric thermal advection due to trends in the 1000 hPa zonal and meridional wind components advecting the time-mean zonal and meridional 1000 hPa air temperature gradients, respectively, for each time period considered.

Plate 5 shows the SIC trends in the context of overlying trends in SLP for the first half (1979–1993), second half (1993–2007), and full period of record (1979–2007); note the larger domain compared to Plate 3. We have omitted data after 30 April 2007 from the trend calculations to avoid biasing the results because of the unusually large sea ice extent reductions in summer 2007 (recall Plate 1). The SLP trends are based on the same seasonal definitions used for SIC and are very similar to those leading by 1 month (e.g., October–December; January–March; April–June; July and September are not shown). The simultaneous (or 1-month lead) relationships between seasonal SLP and SIC trends provide an indication of the role of atmospheric circulation forcing of sea ice anomalies [see, e.g., Fang and Wallace, 1994]. However, there may also be a longer response time (e.g., seasonal and multiyear) associated with atmospheric forcing of sea ice thickness changes, which, in turn, feedback upon SIC [Rigor et al., 2002; Rigor and Wallace, 2004; Nghiem et al., 2007]. This component of the SIC response to atmospheric circulation forcing will not be addressed with our approach. The accompanying seasonal trends in 1000-hPa atmospheric thermal advection as defined above are shown in Plate 6.

We consider first the trends in autumn and winter. In these two seasons, SLP trends in the first half of the record resemble the positive phase of the NAM, with negative values over the Arctic and northern North Atlantic (maximum amplitudes ~4–6 hPa per decade) and positive values farther south (Plate 5). This pattern results in anomalous northwesterly winds over the enhanced sea ice cover in the Labrador and Bering seas and anomalous southerly winds over the regions of reduced ice cover in the Greenland and Barents seas and the Sea of Okhotsk. These wind anomalies advect cold air over the Labrador and Bering seas and warm air over the Greenland and Barents seas and the Sea of Okhotsk and thus contribute thermodynamically to forcing the pattern of SIC trends (Plate 6).

The pattern of winter SLP trends in the second half of the record is largely opposite to that in the first half over the Arctic and north Atlantic sectors, consistent with the behavior of the winter NAM index which reached a relative maximum in the early 1990s (the winter SLP trends over the north Pacific do not reverse sign between the two halves of the record). The inferred geostrophic wind trends are indicative of enhanced southeasterly (southwesterly) flow, which, in turn, results in anomalous warm advection, over the reduced SIC in the Labrador Sea (Bering Sea, Plate 6). Thus, the change in sign of the winter SIC trends in the Labrador and Bering seas between the two halves of the record may be attributed at least in part to the change in sign of the overlying wind and associated thermal advection trends. The persistence of negative winter SIC trends in the Barents seas in the second half of the

record is also consistent with the continued trend of enhanced warm air advection because of low-level wind changes.

Spring and summer SLP trends in the first half of the record exhibit a low-pressure center over the Arctic Ocean, a pattern which results in southerly geostrophic wind anomalies over the negative spring and summer SIC trends in the East Siberian, Chukchi and Beaufort seas (Plate 5). These wind anomalies, in turn, drive increased warm air advection, consistent with the notion that atmospheric circulation changes contribute to the SIC anomalies in these regions (Plate 6). The northerly wind trends that occur over the positive SIC trends in the Barents and Kara seas in summer in the first half of the record are also indicative of the role of atmospheric circulation forcing via enhanced cold air advection. However, the negative SIC trends in spring in these regions do not appear to be forced by wind-induced atmospheric thermal advection.

In the second half of the record, spring and summer SLP trends over the Arctic are relatively weak, as are the accompanying trends in low-level thermal advection (Plates 5 and 6). Thus, the large sea ice losses in the second half of the record in spring and summer do not appear to be due to trends in atmospheric thermal advection.

Over the record as a whole (1979–2007), SLP trends in all seasons are weak, with magnitudes generally less than 1 hPa per decade (Plate 5). The accompanying trends in wind-induced atmospheric thermal advection are also generally weak (the Barents Sea in fall is an exception) and even negative in many areas of the marginal ice zone (Plate 6).

As discussed in section 1, other factors besides atmospheric circulation forcing are contributing to the long-term Arctic sea ice loss, including higher air temperatures [Comiso, 2003; Serreze et al., 2007; Comiso et al., 2008], increased net downward longwave radiation [Francis and Hunter, 2006, 2007], warmer SSTs [Polyakov et al., 2005; Steele et al., 2008], positive ice-albedo feedback [Perovich et al., 2007], and thinning of the ice pack [Lindsay and Zhang, 2005; Nghiem et al., 2007]. Plate 7 confirms that the trends in near-surface (2 m) air temperatures, SSTs, and net downward longwave radiation during 1979–2007 are positive (and statistically significant at the 95% confidence level) over much of the Arctic and adjacent seas in each season. Air temperatures have risen by more than 1°C per decade in summer, fall, and winter and by up to 0.7°C per decade in spring. SSTs have warmed significantly in the North Atlantic, Pacific, and sub-Arctic seas, with the largest increases (up to 0.9°C per decade) in summer in the Labrador, East Siberian, and Chukchi seas. Net surface downward longwave radiation has also increased at high latitudes, with significant trends in all seasons except winter and the largest trends in summer (maximum values of 12–20 W m$^{-2}$ per decade). Collectively,

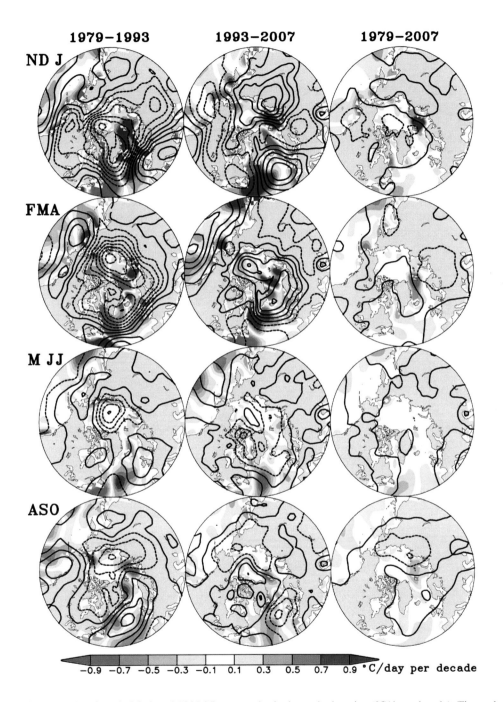

**Plate 6.** As in Plate 5 but for wind-induced 1000-hPa atmospheric thermal advection (°C/d per decade). Thermal advection trends over land are not shown.

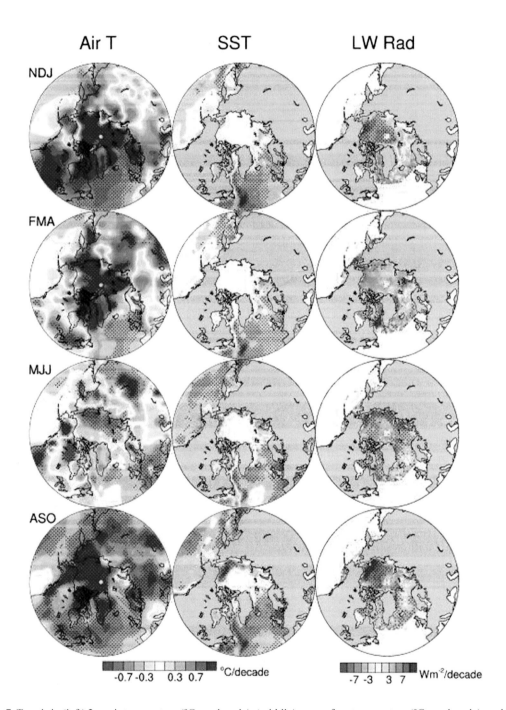

**Plate 7.** Trends in (left) 2-m air temperature (°C per decade), (middle) sea surface temperature (°C per decade), and (right) net surface downward longwave radiation (W m⁻² per decade) during 1979–2007 for each season as indicated. Stippling indicates trend values statistically significant at the 95% level. Note that data after April 2007 are excluded from the trend calculations (the last month of data for longwave radiation is December 2005). Note that the coverage of the longwave radiation data is limited to north of 55°N.

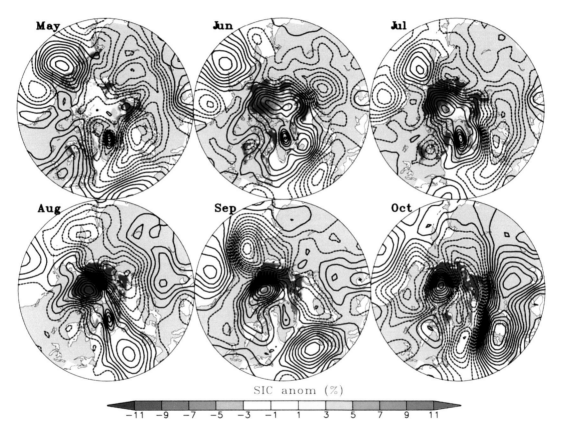

**Plate 8.** Monthly SLP (contours, hPa) and SIC (color shading, percent) anomaly maps for May through October 2007. The contour interval is 1 hPa; with negative values are dashed, and the zero and positive values are solid. Anomalies are defined relative to the 1979–2007 long-term monthly means.

these warming trends provide favorable environmental conditions for Arctic-wide sea ice losses since 1979, coupled with positive ice-albedo feedback and sea ice thinning.

*3.2.4. SIC and SLP anomaly maps for spring/summer 2007.* The drastic reduction of Arctic sea ice extent during the summer of 2007 deserves additional mention. Plate 8 shows the monthly SLP and SIC anomaly maps from May 2007 through October 2007, where anomalies are defined relative to the 1979–2006 long-term monthly means. Similar maps were presented by *Comiso et al.* [2008]. Large SIC losses within the central Arctic developed in June and reached peak amplitudes from late August to early September, leaving much of the eastern Arctic Ocean ice free [see also *Stroeve et al.*, 2008]. The SLP field was highly anomalous in all months of reduced SIC, featuring persistent high-pressure anomalies over the central Arctic Ocean and low-pressure anomalies over Eurasia and adjacent seas: this pattern resembles the distribution summer SLP trends during 1993–2007 (recall Plate 5). This configuration of SLP anomalies resulted in large geostrophic easterly wind anomalies over the marginal ice zone in June and July and strong southerly wind anomalies over the Beaufort, Chuckchi, and East Siberian seas in August through October where the largest SIC losses were observed. It is likely that the low-level wind anomalies contributed to the massive sea ice reductions during summer 2007, a point also made by *Slingo and Sutton* [2007]. Reduced cloudiness and associated enhancement of downwelling shortwave radiation also played an important role [*Kay et al.*, 2008].

We note that the role of atmospheric circulation forcing of the 2007 summer SIC anomalies does not negate our conclusion that the overall retreat of Arctic sea ice since 1979 is not directly controlled by long-term atmospheric circulation changes. Indeed, we expect that atmospheric circulation anomalies will continue to play an important role in individual years, especially as the ice pack continues to thin, but that over the long term they become less important compared to other factors such as the positive ice-albedo feedback mechanism and greenhouse gas–induced warming of the atmosphere and ocean.

## 4. SUMMARY AND DISCUSSION

The purpose of this study was to document aspects of the evolving trends in Arctic sea ice extent and concentration during 1979–2007 and to place them within the context of overlying changes in the atmospheric circulation. In addition to examining trends over the period as a whole, we investigated trends over the two halves of the record separately as a simple way of characterizing their evolution. It was noted that the first half coincides with an upward trend in the NAM, the leading pattern of winter atmospheric circulation variability over the extratropical Northern Hemisphere that is known to exert a strong influence on Arctic sea ice cover, while the second half coincides with a downward trend in the NAM. We used 5-day running mean sea ice concentration data on a 25 km × 25 km grid derived from passive microwave measurements from 1 January 1979 through 31 October 2007.

Our main findings are as follows. Arctic sea ice extent has been retreating throughout the year, with the largest declines occurring from mid July to mid October. Overall, the pace of retreat as estimated from linear least squares regression analysis is $-0.52 \times 10^6$ km$^2$ per decade ($\sim -5\%$ of the mean per decade) or $-1.76 \times 10^6$ km$^2$ in total during 1979–2007. The rate of retreat has accelerated from $-0.35 \times 10^6$ km$^2$ per decade in the first half of the record (1979–1993) to $-0.9 \times 10^6$ km$^2$ per decade in the second half of the record (1993–2007). The date of maximum (minimum) sea ice extent has increased by approximately 4 (1) days per decade, equivalent to a delay of approximately 10 (3) days in 2007 compared to 1979. The number of days with sea ice concentrations less than 50% over the Arctic as a whole has increased by 19 days from 1979 to 2007.

In each season, the spatial patterns of the SIC trends in the two halves of the record are distinctive. The first half is characterized by regional trends of opposing sign, and the second half is characterized by uniformly negative trends that resemble those over the full period. These distinctive trend patterns correspond in each season to the first two leading EOFs of SIC anomalies during 1979–2007. In spring, summer, and autumn, the leading (second) EOF corresponds to the trend pattern over the full record (first half), while in winter the order of the EOFs is reversed.

SIC trends in the first half of the record are characterized by positive values in the sub-Arctic seas of the western Atlantic and eastern Pacific and negative values in the peripheral seas of the eastern Atlantic and western Pacific in autumn, winter, and spring. In summer, the first half of the record exhibits positive SIC trends in the eastern Atlantic (Greenland and Barents seas) and negative trends in the Arctic (East Siberian, Chukchi, and eastern Beaufort seas). Atmospheric circulation trends, in particular a positive trend in the NAM, contributed to forcing the SIC trends in the first half of the record in all seasons via wind-induced low-level atmospheric thermal advection. In the second half of the record, the SIC declines in the Labrador, Barents, Kara, and Bering seas in fall and winter are associated with increased warm air advection in part because of a negative trend in the NAM. However, the pronounced SIC declines within the Arctic Ocean in spring and summer in the second half of the study period are not accompanied by commensurately large positive trends in low-level thermal advection.

During the period 1979–2007 as a whole, atmospheric circulation trends and associated changes in wind-induced low-level thermal advection are weak in all seasons; as a result, they are unlikely to have played a dominant role in the overall retreat of Arctic sea ice since 1979. However, Arctic air temperatures, SSTs, and net surface downward longwave radiation have all increased since 1979 and thus collectively provide a favorable environment for sea ice loss.

Our findings are in qualitative agreement with those of *Francis and Hunter* [2006, 2007], who examined some of the forcing factors for regional SIC variations in winter and summer. In particular, *Francis and Hunter* [2007] analyzed satellite-derived estimates of downwelling longwave radiation, low-level winds, and SSTs in relation to winter SIC variations in the Bering and Barents seas during 1979–2005. They found that wind anomalies were the dominant forcing mechanism in the Bering Sea, while a combination of wind and SST anomalies were the main factors in the Barents Sea. In a related paper, *Francis and Hunter* [2006] showed that downward longwave radiation was the primary cause of summer sea ice extent variations during 1979–2004 throughout the Arctic marginal ice zone. They also reported that wind anomalies played a role in forcing summer sea ice extent anomalies in the Barents and Chukchi seas before, but not after, 1991. The results shown here based on SLP and wind-induced atmospheric thermal advection trends are in general agreement with the findings of *Francis and Hunter* [2006, 2007], although we note that a direct comparison is not possible because of the different timescales of variability examined in the two studies (interannual and longer in the case of Francis and Hunter and trends in our case).

In summary, our results lend additional support to the findings of numerous studies that factors other than long-term atmospheric circulation trends are playing a dominant role in the overall retreat of Arctic sea ice since 1979. These include warming of the upper ocean [*Polyakov et al.*, 2005; *Shimada et al.*, 2006; *Stroeve and Maslowski*, 2008; *Steele et al.*, 2008] and lower atmosphere [*Comiso*, 2003; *Serreze et al.*, 2007; *Comiso et al.*, 2008], sea ice thinning and associated reduction in multiyear ice fraction [*Kwok*, 2007; *Rothrock et al.*, 2007; *Nghiem et al.*, 2007], increased oceanic absorption of solar radiation in summer associated with the mechanism of positive ice-albedo feedback [*Perovich et al.*, 2007] and enhanced downwelling longwave radiative flux due to increased water vapor, cloudiness and carbon dioxide concentration [*Francis and Hunter*, 2006, 2007]. A better understanding of these and other processes affecting Arctic sea ice remains an important task with relevance to future predictions of Arctic climate change.

*Acknowledgments.* We thank the two anonymous reviewers and Editor Eric DeWeaver for constructive suggestions. We are grateful to the National Snow and Ice Data Center for providing the sea ice concentration data. Clara Deser acknowledges support from the NSF Office of Polar Programs (ARCSS), and Haiyan Teng acknowledges support from the U.S. Department of Energy under Cooperative Agreement DE-FC02-97ER62402. The National Center for Atmospheric Research is sponsored by the National Science Foundation.

## REFERENCES

Cavalieri, D. J., C. L. Parkinson, P. Gloersen, J. C. Comiso, and H. J. Zwally (1999), Deriving long-term time series of sea ice cover from satellite passive-microwave multisensor data sets, *J. Geophys. Res.*, *104*, 15,803–15,814.

Comiso, J. C. (2003), Arctic warming signals from clear-sky surface temperature satellite observations, *J. Clim.*, *16*, 3498–3510.

Comiso, J. C. (2006), Abrupt decline in the Arctic winter sea ice cover, *Geophys. Res. Lett.*, *33*, L18504, doi:10.1029/2006GL027341.

Comiso, J. C., C. L. Parkinson, R. Gersten, and L. Stock (2008), Accelerated decline in the Arctic sea ice cover, *Geophys. Res. Lett.*, *35*, L01703, doi:10.1029/2007GL031972.

Deser, C. (2000), On the teleconnectivity of the "Arctic Oscillation," *Geophys. Res. Lett.*, *27*, 779–782.

Deser, C., and H. Teng (2008), Evolution of Arctic sea ice concentration trends and the role of atmospheric circulation forcing, 1979–2007, *Geophys. Res. Lett.*, *35*, L02504, doi:10.1029/2007GL032023.

Deser, C., J. E. Walsh, and M. S. Timlin (2000), Arctic sea ice variability in the context of recent atmospheric circulation trends, *J. Clim.*, *13*, 617–633.

Fang, Z., and J. M. Wallace (1994), Arctic sea ice variability on a timescale of weeks: Its relation to atmospheric forcing, *J. Clim.*, *7*, 1897–1913.

Francis, J. A., and E. Hunter (2006), New insight into the disappearing Arctic sea ice, *Eos Trans. AGU*, *87*(46), 509, doi:10.1029/2006EO460001.

Francis, J. A., and E. Hunter (2007), Drivers of declining sea ice in the Arctic winter: A tale of two seas, *Geophys. Res. Lett.*, *34*, L17503, doi:10.1029/2007GL030995.

Holland, M. M., C. M. Bitz, and B. Tremblay (2006), Future abrupt reductions in the summer Arctic sea ice, *Geophys. Res. Lett.*, *33*, L23503, doi:10.1029/2006GL028024.

Hu, A., C. Rooth, R. Bleck, and C. Deser (2002), NAO influence on sea ice extent in the Eurasian coastal region, *Geophys. Res. Lett.*, *29*(22), 2053, doi:10.1029/2001GL014293.

Hurrell, J. W. (1995), Decadal trends in the North Atlantic Oscillation: Regional temperatures and precipitation, *Science*, *269*, 676–679.

Kalnay, E., et al. (1996), The NCEP/NCAR 40-year reanalysis project, *Bull. Am. Meteorol. Soc.*, *77*, 437–471.

Kay, J. E., T. L'Ecuyer, A. Gettelman, G. Stephens, and C. O'Dell (2008), The contribution of cloud and radiation anomalies to the

2007 Arctic sea ice extent minimum, *Geophys. Res. Lett.*, *35*, L08503, doi:10.1029/2008GL033451.

Kwok, R. (2007), Near zero replenishment of the Arctic multiyear sea ice cover at the end of 2005 summer, *Geophys. Res. Lett.*, *34*, L05501, doi:10.1029/2006GL028737.

Lindsay, R. W., and J. Zhang (2005), The thinning of Arctic sea ice, 1988–2003: Have we reached the tipping point?, *J. Clim.*, *18*, 4879–4895.

Maslanik, J., S. Drobot, C. Fowler, W. Emery, and R. Barry (2007), On the Arctic climate paradox and the continuing role of atmospheric circulation in affecting sea ice conditions, *Geophys. Res. Lett.*, *34*, L03711, doi:10.1029/2006GL028269.

Meier, W. N., J. Stroeve, and F. Fetterer (2007), Whither Arctic sea ice?: A clear signal of decline regionally, seasonally and extending beyond the satellite record, *Ann. Glaciol.*, *46*, 428–434.

Nghiem, S. V., I. G. Rigor, D. K. Perovich, P. Clemente-Colón, J. W. Weatherly, and G. Neumann (2007), Rapid reduction of Arctic perennial sea ice, *Geophys. Res. Lett.*, *34*, L19504, doi:10.1029/2007GL031138.

North, G. R., T. L. Bell, R. F. Calahan, and F. J. Moeng (1982), Sampling errors in the estimation of empirical orthogonal functions, *Mon. Weather Rev.*, *110*, 699–706.

Overland, J. E., and M. Wang (2005), The Arctic climate paradox: The recent decrease of the Arctic Oscillation, *Geophys. Res. Lett.*, *32*, L06701, doi:10.1029/2004GL021752.

Perovich, D. K., B. Light, H. Eicken, K. F. Jones, K. Runciman, and S. V. Nghiem (2007), Increasing solar heating of the Arctic Ocean and adjacent seas, 1979–2005: Attribution and role in the ice-albedo feedback, *Geophys. Res. Lett.*, *34*, L19505, doi:10.1029/2007GL031480.

Polyakov, I. V., et al. (2003), Long-term ice variability in Arctic marginal seas, *J. Clim.*, *16*, 2078–2085.

Polyakov, I. V., U. S. Bhatt, R. Colony, D. Walsh, G. V. Alekseev, R. V. Bekryaev, V. P. Karklin, A. V. Yulin, and M. A. Johnson (2005), One more step toward a warmer Arctic, *Geophys. Res. Lett.*, *32*, L17605, doi:10.1029/2005GL023740.

Portis, D. H., J. E. Walsh, M. El Hamly, and P. J. Lamb (2001), Seasonality of the North Atlantic Oscillation, *J. Clim.*, *14*, 2069–2078.

Rayner, N. A., D. E. Parker, E. B. Horton, C. K. Folland, L. V. Alexander, D. P. Rowell, E. C. Kent, and A. Kaplan (2003), Global analyses of sea surface temperature, sea ice, and night marine air temperature since the late nineteenth century, *J. Geophys. Res.*, *108*(D14), 4407, doi:10.1029/2002JD002670.

Rigor, I. G., and J. M. Wallace (2004), Variations in the age of Arctic sea-ice and summer sea-ice extent, *Geophys. Res. Lett.*, *31*, L09401, doi:10.1029/2004GL019492.

Rigor, I. G., J. M. Wallace, and R. L. Colony (2002), Response of sea ice to the Arctic Oscillation, *J. Clim.*, *15*, 2648–2663.

Rothrock, D. A., and J. Zhang (2005), Arctic Ocean sea ice volume: What explains its recent depletion?, *J. Geophys. Res.*, *110*, C01002, doi:10.1029/2004JC002282.

Rothrock, D. A., D. B. Percival, and M. Wensnahan (2008), The decline in Arctic sea-ice thickness: Separating the spatial, annual, and interannual variability in a quarter century of submarine data, *J. Geophys. Res.*, *113*, C05003, doi:10.1029/2007JC004252.

Serreze, M. C., and J. A. Francis (2006), The Arctic amplification debate, *Clim. Change*, *76*, 241–264, doi:10.1007/s10584-005-9017-y.

Serreze, M. C., M. M. Holland, and J. Stroeve (2007), Perspectives on the Arctic's shrinking sea-ice cover, *Science*, *315*, 1533–1536.

Shimada, K., T. Kamoshida, M. Itoh, S. Nishino, E. Carmack, F. A. McLaughlin, S. Zimmermann, and A. Proshutinsky (2006), Pacific Ocean inflow: Influence on catastrophic reduction of sea ice cover in the Arctic Ocean, *Geophys. Res. Lett.*, *33*, L08605, doi:10.1029/2005GL025624.

Slingo, J., and R. Sutton (2007), Sea-ice decline due to more than warming alone, *Nature*, *450*, 27.

Steele, M., W. Ermold, and J. Zhang (2008), Arctic Ocean surface warming trends over the past 100 years, *Geophys. Res. Lett.*, *35*, L02614, doi:10.1029/2007GL031651.

Stroeve, J., and W. Maslowski (2008), Arctic sea ice variability during the last half century, in *Climate Variability and Extremes During the Past 100 Years*, *Adv. Global Change Res.*, vol. 33, pp. 143–154, Springer, New York.

Stroeve, J., T. Markus, W. Meier, and J. Miller (2006), Recent changes in the Arctic melt season, *Ann. Glaciol.*, *44*, 367–374.

Stroeve, J., M. M. Holland, W. Meier, T. Scambos, and M. Serreze (2007), Arctic sea ice decline: Faster than forecast, *Geophys. Res. Lett.*, *34*, L09501, doi:10.1029/2007GL029703.

Stroeve, J., M. Serreze, S. Drobot, S. Gearheard, M. Holland, J. Maslanik, W. Meier, and T. Scambos (2008), Arctic sea ice extent plummets in 2007, *Eos Trans. AGU*, *89*(2), 13, doi:10.1029/2008EO020001.

Thompson, D. W. J., and J. M. Wallace (2000), Annular modes in the extratropical circulation. Part I: Month-to month variability, *J. Clim.*, *13*, 1000–1026.

Ukita, J., M. Honda, H. Nakamura, Y. Tachibana, D. J. Cavalieri, C. L. Parkinson, H. Koide, and K. Yamamoto (2007), Northern Hemisphere sea ice variability: Lag structure and its implications, *Tellus, Ser. A*, *59*, 261–272, doi:10.111/j.1600-0870.2006.00223.x.

Walsh, J. E., and C. M. Johnson (1979), An analysis of Arctic sea ice fluctuations, 1953–1977, *J. Phys. Oceanogr.*, *9*, 580–591.

C. Deser and H. Teng, Climate and Global Dynamics Division, NCAR, P.O. Box 3000, Boulder, CO 80307, USA. (cdeser@ucar.edu)

# Reconstructing Sea Ice Conditions in the Arctic and Sub-Arctic Prior to Human Observations

Anne de Vernal, Claude Hillaire-Marcel, Sandrine Solignac, and Taoufik Radi

*GEOTOP, Université du Québec à Montréal, Montreal, Quebec, Canada*

André Rochon

*GEOTOP and ISMER, Université du Québec à Rimouski, Rimouski, Quebec, Canada*

Sea ice is a sensitive parameter characterized by a high variability in space and time that can be reconstructed from paleoclimatological archives. The most direct indication of past sea ice cover is found in marine sediments, which contain various tracers or proxies of environments characterized by sea ice. They include sedimentary tracers of particles entrained and dispersed by sea ice, biogenic remains associated with production under/within sea ice or with ice-free conditions, in addition to geochemical and isotopic tracers of brine formation linked to sea ice growth. Reconstructing the extent of past sea ice is, however, difficult because proxies are only indirectly related to sea ice and require the use of transfer functions having inherent uncertainties. In particular, we have to assume a correspondence between sea ice cover values from modern observations and the sea ice proxies from surface sediment samples, which is a source of bias since the time intervals represented by modern observations (here 1954–2000) and surface sediments ($10^0$–$10^3$ years) are not equivalent. Moreover, suitable sedimentary sequences for reconstructing sea ice are rare, making the spatial resolution of reconstructions very patchy. Nevertheless, although fragmentary in time and space and despite uncertainties, available reconstructions reveal very large amplitude changes of sea ice in response to natural forcing during the recent geological past. For example, during the early Holocene, about 8000 years ago, data from dinocyst assemblages suggest reduced sea ice cover as compared to present in some subarctic basins (Labrador Sea, Baffin Bay, and Hudson Bay), whereas enhanced sea ice cover is reconstructed along the eastern Greenland margin and in the western Arctic, showing a pattern not unlike the dipole anomaly that was observed during the 20th century.

Arctic Sea Ice Decline: Observations, Projections, Mechanisms, and Implications
Geophysical Monograph Series 180
Copyright 2008 by the American Geophysical Union.
10.1029/180GM04

## 1. INTRODUCTION

Sea ice is a parameter of the climate system that plays an important role because of its high albedo modulating the energy budget at the surface of the ocean and because it controls the exchanges of heat and gases at the interface between the ocean and the atmosphere [e.g., *Walsh*, 1983]. Arctic sea ice is also an important parameter with respect to the thermohaline circulation since it constitutes a reservoir of fresh water that is eventually exported through surface currents to the northern North Atlantic through Fram Strait and the Canadian Arctic Archipelago [e.g., *Barry et al.*, 1993; *Carmack*, 2000], whereas brines resulting from sea ice growth and evaporation contribute to enhance the salinity of the Deep North Atlantic Water masses [*Meincke et al.*, 1997]. Therefore, sea ice is a critical component of the Earth system which needs to be well constrained before any assessment of sea ice extent in the future based on coupled atmosphere-ocean modeling. Sea ice is, however, a complex parameter with respect to its dynamics and thermodynamics, and its parameterization may yield to diverse responses despite identical forcing. This is illustrated, for example, by the very large range of Arctic sea ice changes that are predicted for the end of the 21st century (see Intergovernmental Panel for Climate Change Fourth Assessment Report) [cf. also *Arzel et al.*, 2006; *Stroeve et al.*, 2007]. Moreover, as shown by *Stroeve et al.* [2007], almost all models simulate trends for the last decades that differ from observations and generally underestimate the actual decline in summer sea ice extent. Thus, sea ice modeling still remains a challenge for the scientific community. The examination of intervals characterized by the extreme climate of the recent geological past may provide a means for evaluating model capabilities for extrapolating sea ice under various boundary conditions [e.g., *Hewitt et al.*, 2001; *Smith et al.*, 2003; *Vavrus and Harrison*, 2003; *Renssen et al.*, 2005; *Goosse et al.*, 2007]. Near analogues of sea ice coverage for the next decades might well be found in the geological records that contain the archives for extreme sea ice cover at the surface of the Earth from total ice coverage to ice-free conditions. In this perspective, it is relevant to examine the recent geological history that has been marked by large-amplitude variations of the sea ice cover under natural forcing. However, sea ice is a sensitive parameter, which is characterized by a high variability in space and time that increases the difficulty of its reconstruction from paleoclimatological archives. The most direct indications of past sea ice cover are found in marine sediments, which contain various tracers or proxies of environments characterized by sea ice. They include sedimentary tracers of particles entrained and dispersed by sea ice, biogenic remains associated with production under sea ice or with ice-free conditions, in addition to geochemical and isotopic tracers of brine formation [*Hillaire-Marcel and de Vernal*, 2008; *Haley et al.*, 2008]. Here, we provide an overview of the approaches that can be used for the reconstruction of sea ice cover in the past, with emphasis given to proxies allowing quantification of sea ice coverage. Special attention is thus paid to the assemblages of

organic-walled dinoflagellate cysts (or dinocysts) that were used to develop transfer function for the reconstruction of sea ice cover [*de Vernal and Hillaire-Marcel*, 2000; *de Vernal et al.*, 2000, 2001, 2005a, 2005b]. We also refer to the isotopic composition ($\delta^{18}O$) of foraminifera that may provide complementary information on brine production related to sea ice formation [*Hillaire-Marcel and de Vernal*, 2008]. On the basis of these approaches, we provide examples of estimated changes in sea ice cover at the high latitudes of the Northern Hemisphere during the early Holocene, which is an interval characterized by higher summer insolation at high northern latitudes and by warmer climate than at present.

## 2. TRACERS AND PROXIES OF SEA ICE

### 2.1. Sedimentological Tracers

In marine sediments, which result from detrital terrigenous input and pelagic fluxes, there are particles related directly or indirectly to sea ice. Among tracers of sea ice that are independent from the biogenic production and purely relate to sedimentary processes, the grain size is most commonly used. The coarse mineral material (coarse silt and fine sand fractions) that is too large and too dense to be carried by wind or marine currents has to be associated with ice-rafted debris (IRD). Coarse material can be entrained by sea ice during ice formation on the shelf before being transported within the drifting ice that is eventually released by melting, notably at the sea ice margin [e.g., *Pfirman et al.*, 1990; *Hebbeln and Wefer*, 1991]. Coarse materials recovered in marine cores thus provide direct evidence for sea ice in the overlying surface water. However, the coarse material can result from drifted multiyear ice or icebergs having remote origin, as well as from annual ice having a more proximal source. The mineralogical or geochemical characteristics of IRD may help identifying the source rocks of detrital particles incorporated into the ice on the shallow shelves and are thus useful for interpretations in terms of ice drift patterns [e.g., *Bischof and Darby*, 1997; *Darby*, 2003; *Andrews and Eberl*, 2007].

In general, IRD accumulate where sea ice is melting and therefore characterize the areas close to distal margins of sea ice. Therefore, IRD usually correspond to seasonal sea ice, whereas the deep-sea sediment under perennial sea ice is generally deprived of coarse debris. The interpretation of IRD is not straightforward depending upon the context. For example, high-IRD content in sediments from the Arctic Ocean may indicate seasonal melt of sea ice and thus reduced ice concentration [e.g., *Darby et al.*, 2001; *de Vernal et al.*, 2005a], whereas IRD occurrences in North Atlantic sediments would rather reflect peaks in seasonal spreadings of sea ice [e.g., *Jennings et al.*, 2002; *Knudsen et al.*, 2004]. IRD may also relate to ablation at the outlet of glaciers, leading to the drift of icebergs that contain detrital particles and dust eventually released distally into the ocean [e.g., *Reeh*, 2004]. Thus, IRD in the open ocean certainly relate to ocean circulation pattern and temperature but not unequivocally to sea ice formation and extent.

## 2.2. Biogenic Tracers

Most sea ice proxies currently used in paleoceanography are related to biological activity, especially those related to planktic production in the photic zone, which is severely constrained by sea ice. Sea ice is indeed a determinant component of the ecosystem inasmuch as it influences the light penetration and photosynthesis. Thus, close relationships exist between sea ice cover and phytoplanktic populations. Sediments accumulating below the permanent multiyear pack ice are often deprived of biological remains, whereas sediments deposited below the ice marginal zone may contain abundant microfossils.

Among the microfossils used as sea ice proxy, diatoms are most useful because they include species specific to the ice marginal zone. In the circum-Antarctic the distribution of species and assemblages has been documented in relation with the extent of sea ice cover [e.g., *Armand et al.*, 2005; *Crosta et al.*, 2005]. On this basis, diatoms have been used qualitatively and quantitatively for assessing sea ice extent in the Southern Ocean during the Last Glacial Maximum 21,000 years ago [e.g., *Crosta et al.*, 1998; *Gersonde et al.*, 2005] and on longer time ranges [*Crosta et al.*, 2004]. In the Arctic and sub-Arctic, there are many studies using diatom assemblages, which led to qualitative reconstruction of changes in sea ice cover extent during the late Quaternary, in particular in the northernmost Pacific [e.g., *Sancetta*, 1981, 1983], the Nordic seas [*Koç et al.*, 1993], or the Russian Arctic seas [e.g., *Bauch and Polyakova*, 2000]. However, because diatoms are composed of opal, which is susceptible to dissolution in marine waters unsaturated with respect to dissolved silica, their preservation in Arctic Ocean and North Atlantic sediments is often very poor. Most studies dealing with past sea ice cover in the Arctic and subarctic seas based on diatoms only report on the abundance of diatoms and the presence of sea ice, but none provide yet any reconstruction of seasonal extent of sea ice.

Other proxies of sea ice include cysts of dinoflagellates (or dinocysts), which yield relatively diversified and abundant organic-walled microfossils in the sea ice marginal zone. The distribution of dinocysts appears dependent upon the duration of the ice-free season during which the life cycle, including sexual reproduction and cyst formation, has

to be accomplished. The development of modern dinocyst databases permitted the development of transfer functions for the quantitative reconstruction of changes in sea ice cover as expressed in number of months per year with more than 50% (in areal extent) of sea ice cover [e.g., *de Vernal and Hillaire-Marcel*, 2000; *de Vernal et al.*, 2001, 2005a, 2005b]. Time series of sea ice cover spanning the last glacial interval and the postglacial are thus available at the scale of the Arctic and subarctic seas. They are presented below after a critical examination of the approaches leading to past sea ice estimates.

In addition to diatoms and dinoflagellate cysts, the occurrence of a number of biological indicators, including coccoliths and foraminifera, can be used as proxies for sea ice–free conditions, at least seasonally. Molecular biomarkers such as alkenones, also referred to as $UK^{37}$, could be used as indicators of ice-free conditions since they rely on biogenic production of coccoliths [e.g., *Rosell-Melé and McClymont*, 2007]. Another biomarker named $IP_{25}$ produced by sea ice diatoms is currently under examination as a direct tracer of sea ice cover [cf. *Belt et al.*, 2007]. It seems to be a promising approach, but calibration still needs to be done prior to application on long time series.

Finally, another biological indication of past sea ice extent comes from the recovery of remains from marine mammals occurring in environments with seasonal sea ice. For example, the distribution of bones from bowheads, which follow the sea ice edge during their life span, was used to assess the seasonal sea ice opening in the Canadian Arctic Archipelago and Northwest Passage some 9000 years ago [cf. Dyke *et al.*, 1996; *Dyke and Savelle*, 2001].

## 2.3. Geochemical Tracers

The ice cores from continental ice caps can provide some indication on past sea ice because they may contain compounds related to marine brines in relation to sea ice–free conditions in adjacent basins. For example, sea salt variations in ice cores from Devon Island and Greenland have been interpreted as indices of sea ice cover changes in adjacent marine environment [e.g., *Mayewski and White*, 2002]. Such indications are, however, difficult to quantify in terms of concentration or density of sea ice cover. In a similar fashion, indirect inferences on sea ice conditions in the Arctic have been proposed based on mineralogical, geochemical and radiogenic isotope tracers in sediments that label sediment sources and transport mechanisms [e.g., *Darby*, 2003; *Haley et al.*, 2008]. However, here again, transcription of such data into sea ice concentration and/or density remains speculative. The isotopic composition ($\delta^{18}O$) of polar foraminifera may also provide indication of sea ice formation since the isotopic properties of the water are modified by brine formation associated with the freezing of seawater. The $\delta^{18}O$ of foraminifera also depends on salinity and temperature and cannot be interpreted directly by itself in terms of sea ice formation. Nevertheless, recent studies of isotopic properties of planktic foraminifera from the Arctic Ocean suggest that their light $\delta^{18}O$ values, far from isotopic equilibrium values expected under the cold Arctic conditions, are thought to relate to production rates of isotopically light brines resulting from sea ice formation [e.g., *Bauch et al.*, 1997; *Hillaire-Marcel et al.*, 2004].

# 3. QUANTIFICATION OF PAST SEA ICE COVER IN THE ARCTIC AND SUB-ARCTIC FROM DINOCYST ASSEMBLAGES

## 3.1. Dinocysts as Proxy of Sea Ice Cover

As mentioned above, at the scale of the Arctic and subarctic seas, most quantitative estimates of the past sea ice cover were derived from dinocyst assemblages. The relationship between sea ice cover and dinocyst assemblages is, however, not a simple function since none of the few dinoflagellates living in sea ice produces fossilizable cysts. Only two cyst-forming species are known to dwell in pack ice environment: *Polarella glacialis* and *Peridiniella catenata* [*Matthiessen et al.*, 2005]. However, their cysts are not recovered in sediment after usual laboratory procedures for the analyses of dinocyst assemblages. Thus, sediments from areas characterized by multiyear perennial pack ice are usually barren in dinocysts [e.g., *Rochon et al.*, 1999; *de Vernal et al.*, 2005a]. Nevertheless, there are a few dinocyst taxa that are known to occur in sediments from areas marked by seasonal sea ice. In Arctic and subarctic seas, *Islandinium* species are often abundant, and some morphotypes of the cyst of *Polykrikos* sp. occur exclusively in seasonally icecovered marine environments [e.g., *de Vernal et al.*, 2001, 2005a]. Many other taxa appear tolerant to sea ice cover and may occur in high proportions in sea ice environments. This is notably the case of the ubiquitous taxa *Operculodinium centrocarpum* and *Brigantedinium* spp. Other taxa such as *Pentapharsodinium dalei*, *Spiniferites elongatus-frigidus*, and *Impagidinium pallidum* have affinities for subarctic environments and often characterize areas with some winter sea ice.

The correspondence between dinocyst assemblages and sea surface conditions is not "straightforward" since the cyst stage only represents a fragmentary picture of the original dinoflagellate population and because biases related to lateral transport of particles or selective preservation cannot be discarded (for a review, see *de Vernal and Marret* [2007] and

*Zonneveld and Versteegh* [2008]). Nevertheless, the biogeography of dinocyst taxa and assemblages clearly illustrates close relationships with sea ice cover extent. The development of a large dinocyst database from the analyses of surface sediment samples (0–1 cm in box cores or multicores) indeed permitted the statistical evaluation of relationships between dinocysts and the sea ice cover (Figures 1 and 2). It also allowed the application of transfer functions for the reconstruction of past sea ice [*de Vernal and Hillaire-Marcel*, 2000; *de Vernal et al.*, 2001, 2005a, 2005b].

## 3.2. Accuracy and Limit of the Approach Based on Transfer Functions Using Dinocysts

There are various techniques of transfer functions allowing quantitative reconstructions of past climate or ocean parameters [e.g., *Guiot and de Vernal*, 2007]. Calibration techniques such as the *Imbrie and Kipp* [1971] method, the weighted averaging partial least squares regression or the artificial neural network approach were tested, but they were not used for reconstructions because the results depend strongly on the spatial extent of the reference database and because it seems advisable to avoid assumptions concerning the type of relationship (i.e., equations) between the dinocyst distribution and hydrographical parameters leading to hazardous extrapolations. We choose to apply the modern analogue technique (MAT) that is simply based on the comparison of past assemblages to modern ones in the reference database using a dissimilarity index. A detailed description of the MAT approach is given by *Guiot and de Vernal* [2007]. Details on the procedures used for sea ice reconstructions based on dinocyst assemblages are presented by *de Vernal et al.* [2001, 2005a, 2005b]. The results of validation exercises and reconstructions that are presented here are based on a database of 1189 reference sites that includes 64 dinocyst taxa. Of the 1189 sites, 584 are characterized by the occurrence of sea ice during the 1954 to 2000 A.D. interval, which is used as the "modern" reference.

The sea ice parameter compiled in the database is the mean annual sea ice occurrence with an areal concentration higher than 50%. The mean has been calculated on 1° by 1° (longitude-latitude) grid for the interval spanning 1954–2000 A.D. based on data provided by the National Snow and Ice Data Center in Boulder, Colorado. The reconstructions are expressed as the duration of sea ice in months per year, a parameter that is linearly correlated to the annual sea ice concentration [cf., e.g., *de Vernal et al.*, 2005a].

One difficulty in any approach based on the analyses of surface sediment samples compared to hydrographical observations lies in the fact that the two sets of data represent different time intervals. On one side, the upper centimeter

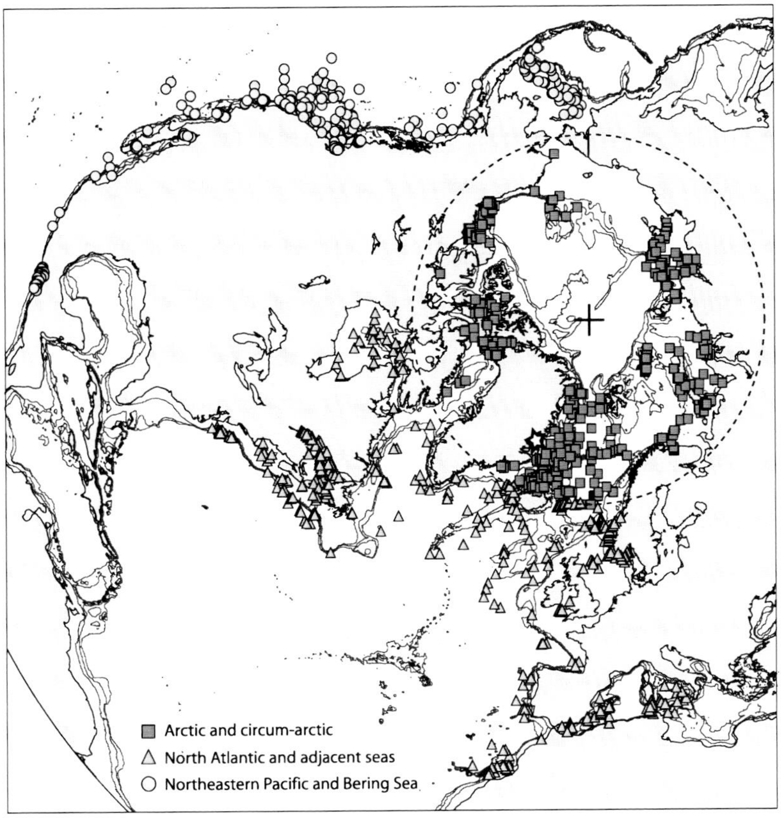

**Figure 1.** Map showing the location of surface sediment samples used to develop the reference dinocyst database. The dark gray squares correspond to Arctic sites, north of 66°N (see dashed line of the circle). The triangles and circles correspond to Atlantic and Pacific sites, respectively. The database includes assemblages for 584 sites with seasonal sea ice, about half of them corresponding to more than 6 months per year of sea ice. No sites under perennial sea ice are included inasmuch as the assemblages are barren in such conditions because of extremely low productivity. Isobaths correspond to 200, 1000, and 2000 m.

of the sediment may represent fluxes during a time range varying from about 10 to 1000 years depending upon biological mixing and accumulation rates. On the other side, the instrumental observations used as the "modern" reference only span the last few decades, at the most. There is therefore a discrepancy between the time intervals represented by the proxy, here the dinocyst assemblages, and the reference hydrographical values. Thus, when developing transfer functions based on calibration or similarity techniques, one uses an assumption concerning the correspondence between surface sediment samples and the modern hydrography that is potentially biased. In areas of relatively stable conditions over the last hundreds of years, this is not critical. However, in areas marked by a large interannual or interdecadal variability, this might be a problem and would yield noisy records. Moreover, in areas characterized by long-term trends and low sedimentation rates, we may expect some discrepancy

since the surface sediment sample would represent an older interval with a different state or mean than the one defined from the "modern" hydrographical values.

Despite fundamental limitations inherent to transfer functions as exposed above, the accuracy of the MAT applied to dinocyst assemblages permits reasonably adequate estimates of sea ice cover. The accuracy of the sea ice reconstructions has been evaluated from the leaving-one-out approach by estimating the modern sea ice cover from dinocyst assemblages (Figure 3). The coefficient of correlation ($r^2$) between observed and estimated values is 0.914, and the error of prediction establishes at ±1.1 months/year of sea ice on average (Figures 3a and 3b). Such accuracy seems appropriate given the interannual variability that characterizes the sea ice cover extent in the domain of seasonal sea ice. At the 584 sites of the database characterized by sea ice occurrence, the standard deviation (±1σ) around the 1954–2000 means

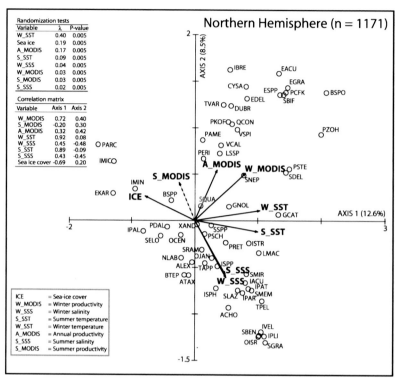

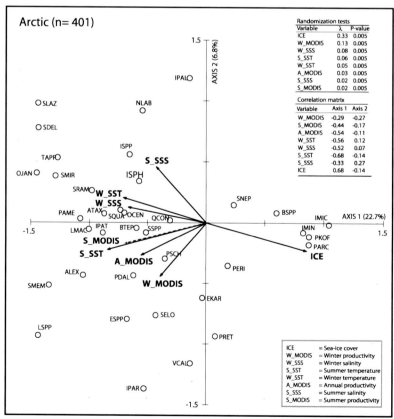

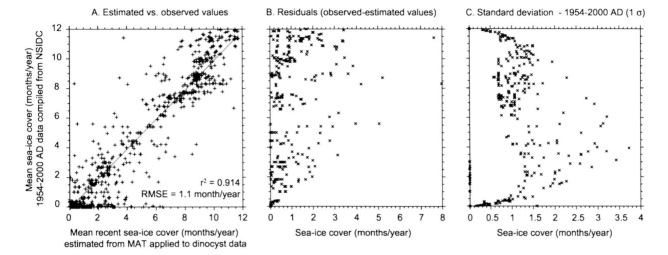

**Figure 3.** Illustration of the accuracy of the modern analogue technique (MAT) applied to dinocyst assemblages for estimating past sea ice cover extent: (a) estimated versus observed sea ice for the 1954–2000 A.D. time interval at reference sites; (b) distribution of the residuals, which are the differences between observed and estimated values, versus the mean sea ice cover for the 1954–2000 A.D. time interval; and (c) standard deviation (1σ) of the 1954–2000 A.D. sea ice cover extent at the reference sites versus the mean sea ice cover for the 1954–2000 A.D. time interval. All values are expressed in terms of months per years of sea ice cover with concentration greater than 50% at given sites defined on a 1° by 1° grid scale. In Figure 3a, the RMSE is the root mean square error, which corresponds to the standard deviation of the residual.

may reach 3.7 months/year but averages ±1.0 month/year (Figure 3c).

### 3.3. Variability and Secular Changes in Sea Ice Cover

The spatial distribution of the residuals (Figure 4), which are the differences between observed and estimated values of sea ice cover, may provide additional information on the reliability of reconstructions. Residuals are relatively high in ice marginal zones such as the eastern Greenland margins and the Barents Sea, which are also areas of high interannual variability as seen in observation data compiled for the 1954–2000 A.D. period (Figure 5). Therefore, we have to admit some uncertainty in reconstructions due to both the reliability of the transfer function and the inherent high variability of sea ice cover at the sea ice margin.

A closer examination of the geographical distribution of residuals shows interesting features. Along the eastern Greenland margins, within the limit of the mean sea ice edge, dinocyst data consistently yield underestimated sea ice cover values. This illustrates a bias that may result from a discrep-

ancy in the sea ice cover from the time interval included in the surface sediments (hundreds of years) and that of the last decades. On the average, dinocyst estimates that represent the last hundreds of years suggest less sea ice than during the last decades along the eastern Greenland margin (Figure 4). In the northern Barents Sea, dinocyst assemblages also yield underestimated sea ice cover values, but the assemblages south of the maximum sea ice limit provide overestimations. Such anomalies might reflect a more southward spread of sea ice in winter, and a reduced summer sea ice cover in the northern area of the eastern Barents Sea, during the previous centuries. If the above interpretation of the residuals is correct, the mean state of sea ice cover at secular scales could have been more zonal in the Nordic seas and with a less pronounced west to east gradient than at present. However, such an interpretation has to be considered with caution because of intrinsic limitations of the transfer function and because the MAT approach is an interpolation technique that might result in some smoothing. In any case, the spatial pattern of the residuals deserves some attention and points to the fact that secular time series with decadal resolution are needed to

---

**Figure 2.** (Opposite) Results of canonical correspondence analyses performed on the dinocyst assemblages (64 taxa) and surface ocean parameters (sea ice, temperature, salinity, and productivity) using the Canoco software [cf. *ter Braak and Smilauer*, 1998]. Sea ice stands as (top) a determinant parameter on the assemblage distribution at the scale of the Northern Hemisphere and as (bottom) the most determinant parameter at the scale of the Arctic data set (north of 66°N, see dark gray squares in Figure 1), which includes 401 sites and 37 dinocyst taxa. Details on the data treatments are reported elsewhere [*Radi and de Vernal*, 2008].

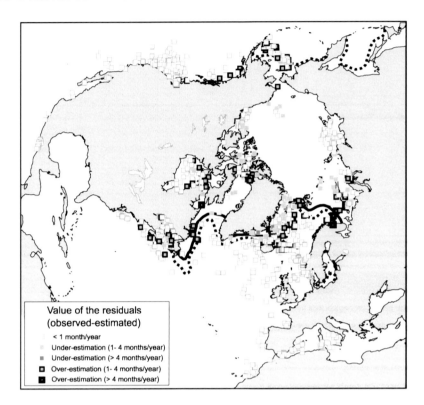

**Figure 4.** Spatial distribution of the residuals (observed minus estimated sea ice cover). The solid line and the dashed line correspond to the modern mean winter sea ice cover limit and the extreme sea ice limit, respectively.

document the natural variability and trends in sea ice cover extent beyond the mid-1900s.

The variability of sea ice over the last centuries is difficult to document from sediment cores because it requires particularly high sediment accumulation rates to achieve a suitable resolution in time. Nevertheless, a few cores permitted analyses with adequate resolution (Figure 6). Some historical records provide additional evidence for secular variations [e.g., *Lamb*, 1977; *Ogilvie*, 1984; *Hill*, 1998; *Hill et al.*, 2002]. The examples illustrated in Figure 6 are from the western Arctic and the northern Baffin Bay. Both reconstructions were made using the same approach and both show significant changes in sea ice on secular scale. However, the two records illustrate distinct trends. Whereas the Chukchi Sea data show a denser sea ice cover during the late 19th and 20th centuries, the northern Baffin Bay record suggests a decrease in sea ice cover during the 20th century after an interval of maximum areal extent at about 1750–1900 A.D. This maximum in the sea ice cover might well correspond to the episode of extremely cold winters, which have marked the second half of the last millennium in the subpolar North Atlantic and in western Europe and which is known as the "Little Ice Age" (LIA). The spreading of sea ice along the coasts of Iceland from 1600 to 1900 A.D. (Fig-

ure 7) [*Lamb*, 1977] is another example of the recording of the LIA by this parameter. However, as pointed out by *Jones and Mann* [2004], the LIA documented from historical or proxy data is far from a synchronous and uniform climate event over the Northern Hemisphere, and some discrepancies may be expected. Nonetheless, although sea ice records of the last millennium are still rare, the few data available indicate important secular changes in sea ice margins with some diachronous responses of sea ice extent from one region to another.

## 4. ISOTOPIC ($\delta^{18}O$) COMPOSITION OF MESOPELAGIC FORAMINIFER SHELLS: A PROXY FOR SEA ICE ACCRETION RATES?

Seasonal duration in the sea ice cover at given sites provides information on the sea ice distribution but not on its rate of formation. Other proxies are needed for estimating such rates that are dependent upon many parameters including air temperature, depth of the pycnocline, and stratification and salinity of the surface layer of the ocean, as well as wind strength and patterns. As briefly mentioned in section 2.3, some proxy for sea ice formation may be obtained in the isotopic ($\delta^{18}O$) composition of foraminifer shells, which pre-

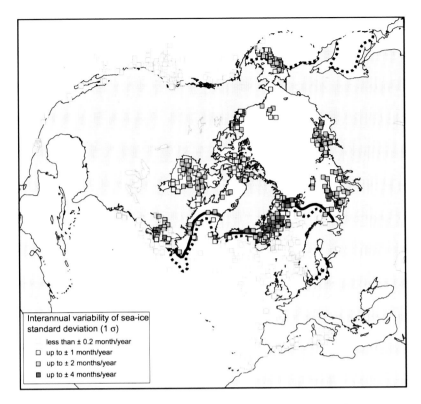

**Figure 5.** Spatial distribution of the standard deviation (1σ) of the 1954–2000 A.D. sea ice cover extent at the reference sites. The solid line and the dashed line correspond to the modern mean winter sea ice cover limit and the extreme sea ice limit, respectively.

serve the isotopic composition of the water mass in which foraminifera have formed their carbonate shell. The isotopic composition of seawater is thus a function of its salinity. It is regulated by its dilution with $^{18}O$-depleted meteoritic waters or by its concentration with some $^{18}O$ enrichment due to evaporative processes. However, it is also affected by the freezing of water that acts as a fractionation process for salt versus $^{18}O$ properties. First, sea ice usually forms in areas characterized by low salinity and low-$^{18}O$ content of surface waters. Second, if sea ice shows a very low salt content, it is slightly $^{18}O$ enriched in comparison with water, because of ice-water fractionation processes [e.g., *O'Neil*, 1968]. Both processes result in the production of $^{18}O$-depleted brines sinking deeper in the water column, whereas, in comparison, brines resulting from evaporative processes are $^{18}O$ enriched through Raleigh distillation processes. Sea ice formation thus has a peculiar effect on the isotopic composition of the water [e.g., *Tan and Strain*, 1980; *Bédard et al.*, 1981]. It is the only process that does not result in correlative changes of density or salinity and δ$^{18}O$. In all other situations, δ values and densities (σ$_\theta$) vary together, either in near colinearity when salinity is determinant or according to a polynomial relation when large δ and temperature gradients in the wa-

ter column are involved [cf. *Hillaire-Marcel et al.*, 2001a, 2001b], with

$$\delta^{18}O = \left[ \frac{\left(^{18}O/^{16}O\right)_{sample}}{\left(^{18}O/^{16}O\right)_{reference-value}} - 1 \right] \times 10^3$$

$$\sigma_\theta = \left[ \frac{\rho_{sample-water}}{\rho_{pure-water}} - 1 \right] \times 10^3,$$

where ρ stands for density.

In polar environments where sea ice may form, planktic foraminifer assemblages are characterized by monospecific assemblages composed of *Neogloboquadrina pachyderma* (in the sense of *Darling et al.* [2006]). In the Arctic Ocean, the calcareous shells of Np often record large oxygen isotope offsets from theoretical equilibrium values, assuming that carbonate precipitation occurs at approximately mid pycnocline depth between the cold and dilute surface water layer and the underlying warm and saline water mass, originating from the North Atlantic Ocean [cf. also *Aksu and Vilks*, 1988; *Kohfeld et al.*, 1996; *Volkmann and Mensch*,

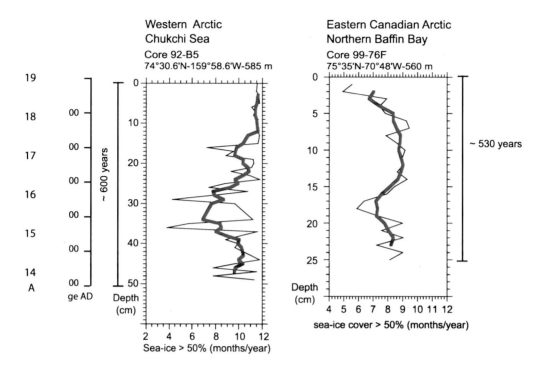

**Figure 6.** Examples of sea ice cover reconstruction for the last hundreds of years based on cores from the Chukchi Sea and northern Baffin Bay having exceptionally high sedimentation rates (for core location, see sites A and B in Figure 10 and Table 1). The sedimentation rates in both cores have been evaluated from $^{210}$Pb measurements. The data from core 92-B5 in the Chukchi Sea are from this study, and the data from core 99-76F in the northern Baffin Bay are reported by *Hamel et al.* [2002]. The thin lines correspond to the most probable values from sets of five analogues, and the thick lines correspond to a three-point running mean.

2001; *Simstich et al.*, 2003; *Hillaire-Marcel et al.*, 2004]. The offset in δ$^{18}$O values ranges from near −1‰ in the northern Greenland Sea to about −3‰ in the western Arctic Ocean [e.g., *Bauch et al.*, 1997; *Fisher et al.*, 2006]. It cannot be attributed to dilution with fresh water and lowering of salinity because of ecological characteristics of *N. pachyderma* that include minimum salinity requirements (≥34.5 [*Hilbrecht*, 1996]). It is, rather, attributed to brine extrusion occurring

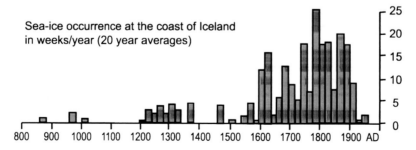

**Figure 7.** Historical record of sea ice occurrence off the Icelandic coasts (modified from *Lamb* [1977, Figure 17.13, p. 452]). Past records of sea ice extent in northwest North Atlantic are available from the accounts of voyages by Irish monk-explorers and by Vikings who established settlements in Iceland and in southwest Greenland at the end of the first millennium. Until 1200 A.D., sea ice was reported to occur exceptionally. All historical data suggest minimum sea ice extent around Iceland and off Greenland during the medieval warm episode and conditions favorable for navigation across the North Atlantic. Navigation reports and other historical archives show that the spreading of seasonal sea ice became more and more extensive in the northern North Atlantic after 1250 A.D. During the decade from 1340 to 1350 A.D., the increasing spread of Arctic sea ice around Greenland even forced the old sailing routes along the 65°N parallel to be abandoned, which caused the decline of the Viking settlements in southern Greenland.

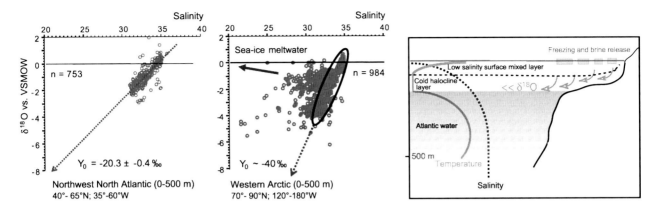

**Figure 8.** Relationship of $\delta^{18}O$ to salinity in surface and subsurface waters (0–500 m) of the (left) North Atlantic, the (middle) western Arctic, and (right) scheme of halocline and brine release due to sea ice formation from conceptual model of *Aagaard* [1981]. Arctic and North Atlantic data are from the Goddard Institute database (available at http://www.giss.nasa.gov/data/o18data/) [cf. *Schmidt*, 1999; *Bigg and Rohling*, 2000]. In the western Arctic and the northwest Atlantic, the $\delta^{18}O$ to salinity distribution defines linear relationships with an apparent freshwater end-member having an isotopic composition of about −40‰ and −20‰, respectively, which are incompatible with freshwater signatures of Arctic rivers (mean modern weighted values of approximately −17‰ [cf. *Hélie et al.*, 2006]) and thus provides evidence for the impact of sea ice brine distillation processes on isotopic composition of seawater.

during sea ice growth, when the freezing of low-salinity surface water produces isotopically light brines, either sinking rapidly in the water column or temporarily constrained in water vacuoles in the ice, whereas sea ice melting results in the addition of an isotopically heavy, but low-salinity water to the surface layer (see Figure 8). As already mentioned, the isotopic enrichment of sea ice with a separation factor near +3‰ under equilibrium conditions [*O'Neil*, 1968], although generally unattained because of kinetic and boundary layer effects when sea ice forms [cf. *Ekwurzel et al.*, 2001], may result in a further depletion in $^{18}O$ of the residual brines.

As a consequence of the above mentioned converging processes, sea ice formation provides the only situation leading to opposite trends of the salinity and $\delta$ values relationship in oceanic water, thus in biogenic carbonates. The interpretation of the $\delta^{18}O$ value of foraminifer shells remains difficult because it also results from thermodependent fractionation processes. Nevertheless, $\delta^{18}O$ data in biogenic carbonate from Arctic environments can be useful inasmuch as they allow inference about relative rates of sea ice formation.

## 5. EXAMPLE OF QUALITATIVE AND QUANTITATIVE SEA ICE RECONSTRUCTIONS

### 5.1. Western Arctic Holocene Time Series

In the Arctic Ocean areas presently characterized by perennial ice, the paleoceanographical records of the Holocene are rare and difficult to interpret because of extremely low

sedimentation rates [e.g., *Poore et al.*, 1999; *Hillaire-Marcel et al.*, 2004; *Not et al.*, 2008] in addition to limited biogenic contents of sediment. Nevertheless, a few cores close to the modern limit of the perennial ice in the western Arctic provided suitable biogenic content for both dinocyst analyses and stable isotope measurements in foraminifera. Figure 9 shows a Holocene record from the western Arctic, which is interesting despite its poor stratigraphical resolution, because it includes dinocyst-based reconstruction of sea ice cover and $\delta^{18}O$ in both planktic and benthic foraminifera [cf. *de Vernal et al.*, 2005a]. The record indicates very large amplitude changes during the Holocene, with extensive sea ice cover during the early Holocene until about 7000 years ago. The interval from 9000 to 7000 years B.P. is also marked by particularly low $\delta^{18}O$ values in *N. pachyderma*. The early Holocene isotopic data from planktic foraminifera in the western Arctic have been interpreted as the result of maximum inflow of warm North Atlantic water characterized by temperature higher than at present by up to 3°C in the subsurface to intermediate water layer [cf. *Hillaire-Marcel et al.*, 2004]. Such an interpretation implied a decoupling between freezing surface waters and relatively warm intermediate waters that could also be linked to increased heat accumulation under a capped layer [cf. *Mignot et al.*, 2007]. Not exclusive to the above mentioned interpretation, enhanced rates of sea ice formation in the Arctic with brine rejection to the top of the pycnocline also likely occurred as illustrated by the particularly low $\delta^{18}O$ values in all size classes of the analyzed planktic foraminifer shells. Thus, together, the

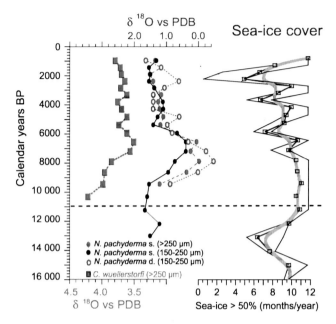

**Figure 9.** Holocene records from the Chukchi Sea in the western Arctic (core P1-92-AR-B15, Site 19 in Table 1 and Figure 10) showing reconstruction of sea ice cover based on dinocysts and isotopic composition of planktic and benthic foraminifera from different size fractions [cf. *de Vernal et al.*, 2005a; *Hillaire-Marcel et al.*, 2004]. The isotopic composition of foraminifer shells is reported against the commonly used standard Peedee belemnite (PDB), which is a calcite from a belemnite shell of the Cretaceous Pee Dee formation in North Carolina. Among planktic foraminifera, two morphotypes of *Neogloboquadrina pachyderma* occur: the sinistral (s.) and dextral (d.) forms. The isotope record is mostly Holocene in age because no foraminifer shells were recovered in the lower part of the sequence.

isotopic data and dinocyst reconstructions strongly suggest maximum sea ice formation in the western Arctic during the early Holocene. The remaining part of the record, after 7000 years B.P., would reflect a decreasing trend of sea ice cover and reduced rates of sea ice formation. Millennial oscillations might be recognized in the Chukchi Sea record, but the time resolution is inadequate for any reliable assessment.

*5.2. Early Holocene in the Subpolar North Atlantic and Adjacent Basins*

At the scale of the subpolar North Atlantic, many studies report on variations in the sea ice cover during the Holocene. Diatom analyses permitted evaluation of changes in the position of the sea ice edge in the Nordic seas [*Koç et al.*, 1993] and suggested variations in sea ice density off eastern Greenland [*Andersen et al.*, 2004]. The distribution of IRD

was also used to make inferences about sea ice changes at local scales [e.g., *Jennings et al.*, 2002] and to assess millennial cycles in the advection of sea ice across the northern North Atlantic [*Bond et al.*, 1997, 2001]. Clearly, sea ice recorded important variations during the Holocene, not only in the Arctic seas but also in the subpolar seas and along continental margins of the northwestern North Atlantic.

Quantitative Holocene sea ice reconstructions based on dinocysts are available in a number of cores from the northern North Atlantic and adjacent subpolar seas (see Figures 10–12 and Table 1). The results show regionalism in sea ice distribution even more pronounced than at present during the early Holocene. At some sites, data reveal a less extensive sea ice cover around 8000 years B.P. than during the late Holocene, whereas sites located along the Labrador and eastern Greenland margin recorded meanwhile a denser sea ice cover (Figures 11 and 12). Many sites show oscillations but without any clearly identifiable trend.

A reduced sea ice cover during the early Holocene is compatible with the higher summer insolation at high northern latitudes of the interval [*Berger*, 1978]. Despite lower winter insolation conditions, the thermal inertia of the ocean results in heat storage at the surface, thus in a delayed freezing in winter [e.g., *Renssen et al.*, 2005]. The minimum sea ice recorded at many sites during the early Holocene is thus probably a direct response to insolation. However, the changes from the early to late Holocene are not uniform and the spatial pattern of sea ice anomalies indicates that other mechanisms also played a role. Enhanced sea ice formation rates in the Arctic accompanied by increased Arctic sea ice export through Fram Strait could possibly explain the North Atlantic sea ice distribution of the early Holocene. If true, this would have important implications. On one hand, higher rates of ice growth in the Russian Arctic seas (East Siberian and Laptev seas) could have been induced by the lower surface salinity and enhanced stratification of the surface water layer, as a response to higher precipitation over the Russian Arctic during the interval [e.g., *Andreev and Klimanov*, 2000; *Stein et al.*, 2004], in addition to intensified cyclonic circulation resulting in convergence and pack ice accumulation in the western Arctic. On the other hand, high wind strength related to the intense cyclonic circulation could have resulted in a stronger Trans Polar Drift (TPD), thus resulting in enhanced sea ice export through Fram Strait. The scenario of higher rates of sea ice formation in the Russian Arctic seas during the early Holocene is compatible with isotopic data from the western Arctic cores (see section 5.1). It could also provide an example of the diachronous climate optimum recorded in circum-Arctic regions [cf. *Kaplan et al.*, 2003; *Kaufman et al.*, 2004]. The hypothesis of intensified cyclonic circulation and TPD strength during the early Holocene is consistent

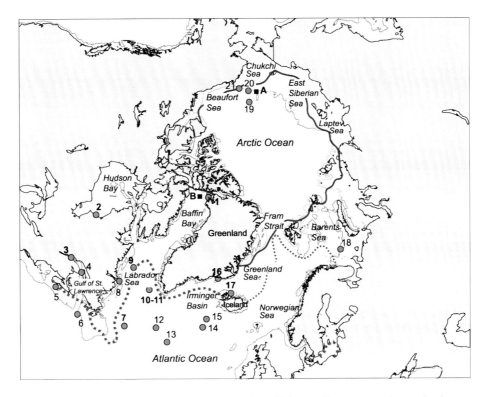

**Figure 10.** Map showing the location of sites (see Table 1) where Holocene dinocyst records permitted reconstruction of sea ice cover and evaluation of the difference between early Holocene (around 8000 years B.P.) and the late Holocene (the last millennium). The sites indicated in bold are those used to illustrate sea ice changes over the last 10,000 years. The thick line corresponds to the median of September sea ice extent (which is close to the limit of multiyear perennial sea ice cover) for 1979–2000 A.D. (limits from the National Snow and Ice Data Center (NSIDC) Web page http://nsidc. org/sotc/sea_ice.html). The dashed line corresponds approximately to the maximum extent of sea ice in winter.

with a dominant positive North Atlantic Oscillation (NAO) [see also *Darby et al.*, 2001; *Darby and Bischof*, 2004]. A NAO+ pattern could also explain a stronger penetration of the North Atlantic Current into the Arctic Ocean [*Duplessy et al.*, 2001; *Sarnthein et al.*, 2002; *Hald et al.*, 2004]. Such a scenario for the early Holocene very much resembles a situation with a prevailing positive phase of the multidecadal oscillation observed in records spanning the last 100 years [e.g., *Polyakov et al.*, 2004], whereas the intensified export of sea ice through Fram Strait would allow comparison with the Arctic dipole anomaly [cf. *Wu et al.*, 2006; *Watanabe et al.*, 2006].

## 6. CONCLUSION

Sea ice is a parameter in the climate system that is characterized by a high variability at all timescales and which has complex dynamics since it depends upon air temperature, salinity, and depth of the upper water layer and stratification and turbulence in the water column in addition to atmo-

spheric circulation and wind strength. Sea ice is thus a parameter that is difficult to quantify on a multiyear basis and which already experiences very large changes in response to the current anthropogenic warming. As a consequence, the development of approaches for reconstructing past sea ice cover is never a straightforward matter, especially since the establishment of transfer functions requires some calibration with "modern" values. One difficulty lies in the fact that the modern observations, which we need to use as reference, are not uniform [e.g., *Walsh and Chapman*, 2001; *Yi et al.*, 1999] and not necessarily representative of mean sea ice conditions over longer timescales. This is particularly critical because the time interval recorded within surface sediment samples used for analyses frequently ranges up to hundreds of years depending upon sedimentation rates and biological mixing of sediment. Such uncertainties add to limitations inherent to any transfer function using microfossils because of taphonomical processes. The discrepancies between modern observations and reconstructions from surface sediment samples illustrate the uncertainty in sea ice reconstruction

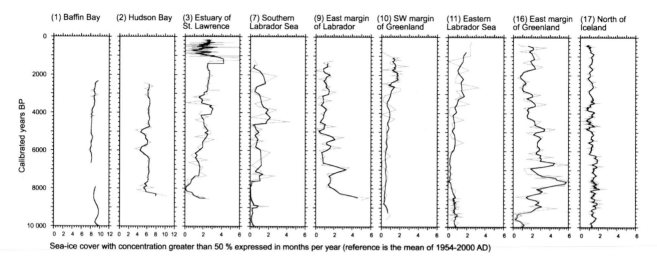

**Figure 11.** Sea ice variations at a few sites from the northwest North Atlantic and Arctic seas (see map in Figure 10 for core location and Table 1 for references).

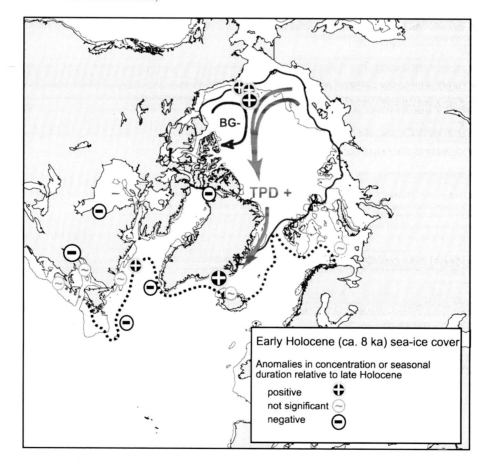

**Figure 12.** Map showing the pattern of sea ice anomalies of the early Holocene (at about 8000 years B.P.) compared to the recent or late Holocene (for references see Table 1). The thick line corresponds to the median of September sea ice extent (which is close to the limit of multiyear perennial sea ice cover) for 1979–2000 A.D. (limits from the NSIDC Web page http://nsidc.org/sotc/sea_ice.html). The dashed line corresponds approximately to the maximum extent of sea ice in winter.

**Table 1.** List of Cores Used in This Study[a]

| Cores | Site | Region | Longitude | Latitude | Water Depth (m) | 1954–2000 Mean Sea Ice >50% (months/year) | Standard Deviation (1σ) | References |
|---|---|---|---|---|---|---|---|---|
| P1-92-BC5 | A | Western Arctic | 159°58.6W | 74°30.6N | 585 | 11.56 | 0.84 | This study |
| 99-76F | B | Northern Baffin Bay | 70°48W | 75°35N | 560 | 8.12 | 1.02 | Hamel et al. [2002] |
| HU91-039-008P | 1 | Baffin Bay | 74° 19.9W | 77° 16.0N | 663 | 7.8 | 1.27 | Levac et al. [2001] |
| HU91-039-012P | 1 | Baffin Bay | 71° 51.5W | 76° 48.3N | 823 | 7.71 | 0.92 | Levac et al. [2001] |
| HU87-028-069P | 2 | Hudson Bay | 77° 57.58W | 55° 28.62N | 165 | 6.14 | 0.75 | Bilodeau et al. [1990] and de Vernal and Hillaire-Marcel [2006] |
| MD99-2220 | 3 | Estuary of St. Lawrence | 68°37.97W | 48°38.32W | 320 | 3 | -1.13 | This study |
| HU90-031-019P | 4 | Gulf of St. Lawrence | 63° 59.57W | 49° 17.44N | 322 | 0 | - | Simard and de Vernal [1998] |
| HU95-030-024P | 5 | Nova Scotia Shelf | 63°43W | 43°46N | 256 | 0 | - | Levac [2001] |
| MD95-2033 | 6 | Laurentian Fan | 55° 37.21W | 44° 39.87N | 1412 | 0 | - | de Vernal and Hillaire-Marcel [2006] |
| HU91-045-094P | 7 | Orphan Knoll | 45° 41.14W | 50° 12.26N | 3448 | 0 | - | de Vernal and Hillaire-Marcel [2000], Peyron and de Vernal [2001], and Solignac et al. [2004] |
| HU91-045-006TWC-P | 8 | Labrador shelf | 56° 27.06W | 54°42.26N | 530 |  |  | Levac and de Vernal [1997] |
| HU84-030-021TWC-P | 9 | Labrador Sea | 57° 30.42W | 58° 22.06N | 2853 | 4.22 | 1.50 | Hillaire-Marcel and de Vernal [1989] and de Vernal et al. [2001] |
| MD99-2227 | 10 | Labrador Sea | 48°22.32W | 58°12.64N | 3460 | 2 | 1.05 | de Vernal and Hillaire-Marcel [2006] |
| HU90-013-013P | 11 | Labrador Sea | 48° 22.40W | 58° 12.59N | 3379 | 0 | - | de Vernal and Hillaire-Marcel [2000], Peyron and de Vernal [2001], and Solignac et al. [2004] |
| HU91-045-085TWC | 12 | Charlie Gibbs fracture | 38° 38.25W | 53° 58.51N | 3603 | 0 | - | de Vernal and Hillaire-Marcel [2006] |
| HU91-045-080P | 13 | Charlie Gibbs fracture | 33° 31.78W | 53° 03.40N | 3024 | 0 | - | de Vernal and Hillaire-Marcel [2006] |
| MD99-2254 | 14 | Reykjanes Ridge | 30°39.86W | 56°47.78N | 2440 | 0 | - | Solignac et al. [2004] |
| HU91-045-072P | 15 | Reykjanes Ridge | 28° 44.32W | 58° 56.45N | 2237 | 0 | - | de Vernal and Hillaire-Marcel [2006] |
| JM96-1207 | 16 | East Greenland margin | 29°21W | 68°06N | 404 | 8.35 | 1.72 | Solignac et al. [2006] |
| MD99-2269 | 17 | Northern Iceland margin | 20°51.79W | 66°38.48N | 365 | 1.71 | 1.11 | Solignac et al. [2006] |
| PL-96-112 | 18 | Barents Sea | 42° 36.31E | 71° 44.18N | 286 | 2 | 0.76 | Voronina et al. [2001] |
| P1-92-AR-BC15 | 19 | Chukchi Sea | 160° 51.63W | 75° 44.03N | 2135 | 11.84 | 0.50 | Hillaire-Marcel et al. [2004] and de Vernal et al. [2005a] |
| P1-92-AR-P1 | 20 | Chukchi Sea | 162°44.6W | 75° 44N | 205 | 11.84 | 0.50 | de Vernal et al. [2005a] |
| HLY 05-1-05 | 21 | Chukchi Sea | 157°W | 72°41.68N | 415 | 11.46 | 0.73 | McKay et al. [2008] |

[a]The modern (1954–2000 A.D.) sea ice data are from the National Snow and Ice Data Center. Sites indicated in bold are those used in Figures 6, 9, and 11 to illustrate sea ice changes over the last 10,000 years.

based on micropaleontological approaches. They also point to the fact that the "modern" state defined from recent observation (1954–2000 A.D.) is probably not perfectly representative of the mean state of sea ice at the scale of the last hundreds of years. It would therefore be relevant to improve our knowledge of sea ice changes over the last hundreds of years, especially in areas such as the eastern Greenland margins where sea ice export from the Arctic Ocean determines thermohaline properties of the northern North Atlantic.

At the scale of the Holocene, a few sea ice time series from Arctic seas and the subpolar North Atlantic indicate significant fluctuations of sea ice extent. However, the trends from the early Holocene to the late Holocene appear different from one area to another, notably when comparing the western Arctic and the east Greenland coast on one side and the northern North Atlantic and subpolar epicontinental seas on the other side. This implies complex mechanisms involving the direct thermal effect of insolation, which was higher during the early Holocene, but also the rate of sea ice formation in the Russian Arctic in relation to freshwater budgets and wind strength, in addition to drift patterns across the Arctic. Improving the spatial coverage of past sea ice records during the last thousands of years would certainly help to develop a more comprehensive picture of sea ice variability on longer timescales and to help modelers with the challenge of simulating sea ice under natural forcing, prior to constraining its evolution in relation with the ongoing anthropogenically driven changes.

*Acknowledgments.* This study is a contribution to the "Polar Climate Stability Network" of the Canadian Foundation for Climate and Atmospheric Sciences (CFCAS). Support from the Natural Sciences and Engineering Research Council of Canada and the Fonds Québecois de Recherche sur la Nature et les Technologies (FQRNT) is also acknowledged.

# REFERENCES

Aagaard, K. (1981), On the deep circulation of the Arctic Ocean, *Deep Sea Res., Part A*, *28*, 251–268.

Aksu, A. E., and G. Vilks (1988), Stable isotopes in planktic and benthic foraminifera from Arctic Ocean surface sediments, *Can. J. Earth Sci.*, *25*, 701–709.

Andersen, C., N. Koç, A. Jennings, and J. T. Andrews (2004), Nonuniform response of the major surface currents in the Nordic seas to insolation forcing: Implications for the Holocene climate variability, *Paleoceanography*, *19*, PA2003, doi:10.1029/2002PA000873.

Andreev, A., and V. A. Klimanov (2000), Quantitative Holocene climatic reconstruction from Arctic Russia, *J. Paleolimnol.*, *24*, 81–91.

Andrews, J., and D. D. Eberl (2007), Quantitative mineralogy of surface sediments on the Iceland shelf, and application to downcore studies of Holocene ice-rafted sediments, *J. Sediment. Res.*, *77*, 469–479.

Armand, L., X. Crosta, O. Romero, and J. J. Pichon (2005), The biogeography of major diatom taxa in Southern Ocean sediments. 1. Sea-ice related species, *Palaeogeogr. Palaeoclimatol. Palaeoecol.*, *223*, 93–126.

Arzel, O., T. Fichefet, and H. Goosse (2006), Sea ice evolution over the 20th and 21st centuries as simulated by current AOGCMs, *Ocean Modell.*, *12*, 401–415.

Barry, R. G., M. C. Serreze, J. A. Maslanik, and R. H. Preller (1993), The Arctic sea ice-climate system: Observations and modeling, *Rev. Geophys.*, *31*, 397–422.

Bauch, D., J. Carstens, and G. Wefer (1997), Oxygen isotope composition of living *Neogloboquadrina pachyderma* (sin.) in the Arctic Ocean, *Earth Planet. Sci. Lett.*, *146*, 47–58.

Bauch, H. A., and Y. I. Polyakova (2000), Late Holocene variations in Arctic shelf hydrology and sea-ice regime: Evidence from north of the Lena Delta, *Int. J. Earth Sci.*, *89*, 569–577.

Bédard, P., C. Hillaire-Marcel, and P. Pagé (1981), $^{18}$O modelling of freshwater inputs in Baffin Bay and Canadian Arctic coastal waters, *Nature*, *293*, 287–289.

Belt, S. T., G. Masse, S. J. Rowland, M. Poulin, C. Michel, and B. LeBlanc (2007), A novel chemical fossil of palaeo sea ice: IP$_{25}$, *Org. Geochem.*, *38*, 16–27.

Berger, A. L. (1978), Long-term variations of daily insolation and Quaternary climatic changes, *J. Atmos. Sci.*, *35*, 2363–2367.

Bigg, G. R., and E. J. Rohling (2000), An oxygen isotope data set for marine water, *J. Geophys. Res.*, *105*(C4), 8527–8535.

Bilodeau, G., A. de Vernal, C. Hillaire-Marcel, and H. Josenhans (1990), Postglacial paleoenvironments of the Hudson Bay: Stratigraphic, microfaunal and palynological evidences, *Can. J. Earth Sci.*, *27*, 946–963.

Bischof, J. A., and D. A. Darby (1997) Mid to late Pleistocene ice drift in the western Arctic Ocean: Evidence for a different circulation in the past, *Science*, *277*, 74–78.

Bond, G., W. Showers, M. Cheseby, R. Lotti, P. Almasi, P. deMenocal, P. Priore, H. Cullen, I. Hajdas, and G. Bonani (1997), A pervasive millennial-scale cycle in North Atlantic Holocene and glacial climates, *Science*, *278*, 1257–1266.

Bond, G., B. Kromer, J. Beer, R. Muscheler, M. N. Evans, W. Showers, S. Hoffmann, R. Lotti-Bond, I. Hajdas, and G. Bonani (2001), Persistent solar influence on North Atlantic climate during the Holocene, *Science*, *294*, 2130–2136.

Carmack, E. C. (2000), The Arctic Ocean's freshwater budget: Sources, storage and export, in *The Freshwater Budget of the Arctic Ocean*, edited by E. L. Lewis et al., pp. 91–126, Kluwer Acad., Norwell, Mass.

Crosta, X., J. J. Pichon, and L. H. Burckle (1998), Application of modern analog technique to marine Antarctic diatoms: Reconstruction of maximum sea-ice extent at the Last Glacial Maximum, *Paleoceanography*, *13*, 284–297.

Crosta, X., A. Sturm, L. Armand, and J. J. Pichon (2004), Late Quaternary sea ice history in the Indian sector of the Southern Ocean as recorded by diatom assemblages, *Mar. Micropaleontol.*, *50*, 209–223.

Crosta, X., O. Romero, L. Armand, and J. J. Pichon (2005), The biogeography of major diatom taxa in Southern Ocean sediments: 2. Open ocean related species, *Palaeogeogr. Palaeoclimatol. Palaeoecol.*, *223*, 66–92.

Darby, D. A. (2003), Sources of sediment found in sea ice from the western Arctic Ocean, new insights into processes of entrainments and drift patterns, *J. Geophys. Res.*, *108*(C8), 3257, doi:10.1029/2002JC001350.

Darby, D. A., and J. Bischof (2004), A Holocene record of changing Arctic Ocean ice drift analogous to the effects of the Arctic Oscillation, *Paleoceanography*, *19*, PA1027, doi:10.1029/2003PA000961.

Darby, D., J. Bischof, G. Cutter, A. de Vernal, C. Hillaire-Marcel, G. Dwyer, J. McManus, L. Osterman, L. Polyak, and R. Poore (2001), New record shows pronounced changes in Arctic Ocean circulation and climate, *Eos Trans. AGU*, *82*(49), 601.

Darling, K. F., M. Kucera, D. Kroon, and C. M. Wade (2006), A resolution for the coiling direction paradox in *Neogloboquadrina pachyderma*, *Paleoceanography*, *21*, PA2011, doi:10.1029/2005PA001189.

de Vernal, A., and C. Hillaire-Marcel (2000), Sea-ice cover, sea-surface salinity and halo-/thermocline structure of the northwest North Atlantic: Modern versus full glacial conditions, *Quat. Sci. Rev.*, *19*, 65–85.

de Vernal, A., and C. Hillaire-Marcel (2006), Provincialism in trends and high frequency changes in the northwest North Atlantic during the Holocene, *Global Planet. Change*, *54*, 263–290.

de Vernal, A., and F. Marret (2007), Organic-walled dinoflagellates: Tracers of sea-surface conditions, in *Proxies in Late Cenozoic Paleoceanography*, edited by C. Hillaire-Marcel and A. de Vernal, pp. 371–408, Elsevier, New York.

de Vernal, A., C. Hillaire-Marcel, J.-L. Turon, and J. Matthiessen (2000), Reconstruction of sea-surface temperature, salinity, and sea-ice cover in the northern North Atlantic during the Last Glacial Maximum based on dinocyst assemblages, *Can. J. Earth Sci.*, *37*, 725–750.

de Vernal, A., et al. (2001), Dinoflagellate cyst assemblages as tracers of sea-surface conditions in the northern North Atlantic, Arctic and sub-Arctic seas: The new 'n = 677' data base and its application for quantitative palaeoceanographic reconstruction, *J. Quat. Sci.*, *16*, 681–699.

de Vernal, A., C. Hillaire-Marcel, and D. A. Darby (2005a), Variability of sea ice cover in the Chukchi Sea (western Arctic Ocean) during the Holocene, *Paleoceanography*, *20*, PA4018, doi:10.1029/2005PA001157.

de Vernal, A., et al. (2005b), Reconstruction of sea-surface conditions at middle to high latitudes of the Northern Hemisphere during the Last Glacial Maximum (LGM) based on dinoflagellate cyst assemblages, *Quat. Sci. Rev.*, *24*, 897–924.

Duplessy, J.-C., E. Ivanova, I. Murdmaa, M. Paterne, and L. Labeyrie (2001), Holocene paleoceanography of the northern Barents Sea and variations in the northward heat transport of the Atlantic Ocean, *Boreas*, *30*, 2–16.

Dyke, A. S., and J. M. Savelle (2001), Holocene history of the Bering Sea bowhead whale (*Balaena mysticetus*) in its Beaufort Sea summer grounds off southwestern Victoria Island, western Canadian Arctic, *Quat. Res.*, *55*, 371–379.

Dyke, A. S., J. Hooper, and J. M. Savelle (1996), A history of sea ice in the Canadian Arctic Archipelago based on postglacial remains of the bowhead whale (*Balaena mysticetus*), *Arctic*, *49*, 235–255.

Ekwurzel, B., P. Schlosser, R. A. Mortlock, R. G. Fairbanks, and J. H. Swift (2001), River runoff, sea ice meltwater, and Pacific water distribution and mean residence times in the Arctic Ocean, *J. Geophys. Res.*, *106*(C5), 9075–9092.

Fisher D., A. Dyke, F. Koerner, J. Bourgeois, C. Kinnard, C. Zdanowicz, A. de Vernal, C. Hillaire-Marcel, J. Savelle, and A. Rochon (2006), Natural variability of Arctic sea ice over the Holocene, *Eos Trans. AGU*, *87*(28), 273.

Gersonde, R., X. Crosta, A. Abelmann, and L. Armand (2005), Sea-surface temperature and sea ice distribution of the Southern Ocean at the EPILOG Last Glacial Maximum—A circum-Antarctic view based on siliceous microfossil records, *Quat. Sci. Rev.*, *24*, 869–896.

Goosse, H., E. Driesschaert, T. Fichefet, and M.-F. Loutre (2007), Information on the early Holocene climate constrains the summer sea ice projections for the 21st century, *Clim. Past*, *3*, 683–692.

Guiot, J., and A. de Vernal (2007), Transfer functions: Methods for quantitative paleoceanography based on microfossils, in *Proxies in Late Cenozoic Paleoceanography*, edited by C. Hillaire-Marcel and A. de Vernal, pp. 523–563, Elsevier, New York.

Hald, M., H. Ebbesen, M. Forwick, F. Godtliebsen, L. Khomenko, S. Korsun, L. Ringstad, L. Olsen, and T. O. Vorren (2004), Holocene paleoceanography and glacial history of the West Spitsbergen area, Euro-Arctic margin, *Quat. Sci. Rev.*, *23*, 2075–2088.

Haley, B. A., M. Frank, R. F. Spielhagen, and A. Eisenhauer (2008), Influence of brine formation on Arctic Ocean circulation over the past 15 million years, *Nat. Geosci.*, *1*, 68–72, doi:10.1038/ngeo.2007.5

Hamel, D., A. de Vernal, M. Gosselin, and C. Hillaire-Marcel (2002), Organic-walled microfossils and geochemical tracers: Sedimentary indicators of productivity changes in the North Water and northern Baffin Bay during the last centuries, *Deep Sea Res., Part II*, *49*, 5277–5295.

Hebbeln, D., and G. Wefer (1991), Effects of ice coverage and ice-rafted material on sedimentation in the Fram Strait, *Nature*, *350*, 409–411.

Hélie, J. F., J. Gibson, C. Hillaire-Marcel, and A. Myre (2006), Isotopic offsets of riverine supplies to the ocean vs. inland precipitation—Case studies of boreal and arctic rivers, *Eos Trans. AGU*, *87*(52), Fall Meet. Suppl., Abstract U43B-0864.

Hewitt, C. D., C. A. Senior, and J. F. B. Mitchell (2001), The impact of dynamic sea-ice on the climatology and climate sensitivity of a GCM: A study of past, present, and future climates, *Clim. Dyn.*, *17*, 655–668.

Hilbrecht, H. (1996), Extant planktic foraminifera and the physical environment in the Atlantic and Indian Oceans, *Mitt. Geol. Inst. Eidg. Tech. Hochsch. Univ. Zurich*, *300*, 93 pp.

Hill, B. T. (1998), Historical record of sea-ice and iceberg distribution around Newfoundland and Labrador, paper presented at the

ACSYS Workshop on Sea-Ice Charts of the Arctic, Arctic Clim. Syst. Stud., Seattle, Wash.

Hill, B. T., A. Ruffman, and K. Drinkwater (2002), Historical record of the incidence of sea ice on the Scotian Shelf and the Gulf of St. Lawrence from 1817 to 1962, paper presented at 16th IAHR International Symposium on Ice, Int. Assoc. of Hydraul. Eng. and Res., Dunedin, New Zealand.

Hillaire-Marcel, C., and A. de Vernal (1989), Isotopic and palynological records of the late Pleistocene in eastern Canada and adjacent basins, *Geogr. Phys. Quaternaire*, *43*, 263–290.

Hillaire-Marcel, C., and A. de Vernal (2008), Stable isotope clue to episodic sea ice formation in the glacial North Atlantic, *Earth Planet. Sci. Lett.*, *268*, 143–150.

Hillaire-Marcel, C., A. de Vernal, G. Bilodeau, and A. J. Weaver (2001a), Absence of deep water formation in the Labrador Sea during the last interglacial period, *Nature, 410*, 1073–1077.

Hillaire-Marcel, C., A. de Vernal, L. Candon, G. Bilodeau, and J. Stoner (2001b), Changes of potential density gradients in the northwestern North Atlantic during the last climatic cycle based on a multiproxy approach, in *The Oceans and Rapid Climate Change: Past, Present and Future, Geophys. Monogr. Ser.*, vol. 126, edited by D. Seidov, B. J. Haupt, and M. Maslin, pp. 83–100, AGU, Washington, D. C.

Hillaire-Marcel, C., A. de Vernal, L. Polyak, and D. Darby (2004), Size-dependent isotopic composition of planktic foraminifers from Chukchi Sea vs. NW Atlantic sediments—Implications for the Holocene paleoceanography of the western Arctic, *Quat. Sci. Rev.*, *23*, 245–260.

Imbrie, J., and N. G. Kipp (1971), A new micropaleontological method for quantitative paleoclimatology: Application to a late Pleistocene Caribbean core, in *The Late Cenozoic Glacial Ages*, edited by K. K. Turekian, pp. 71–181, Yale Univ. Press, New Haven, Conn.

Jennings, A. E., K. L. Knudsen, M. Hald, C. V. Hansen, and J. T. Andrews (2002), A mid-Holocene shift in Arctic sea-ice variability on the east Greenland shelf, *Holocene*, *12*, 49–58.

Jones, P. D., and M. E. Mann (2004), Climate over past millennia, *Rev. Geophys.*, *42*, RG2002, doi:10.1029/2003RG000143.

Kaplan, J. O., et al. (2003), Climate change and Arctic ecosystems: 2. Modeling, paleodata-model comparisons, and future projections, *J. Geophys. Res.*, *108*(D19), 8171, doi:10.1029/2002JD002559.

Kaufman, D. K., et al. (2004), Holocene thermal maximum in the western Arctic (0–180°W), *Quat. Sci. Rev.*, *23*, 529–560.

Knudsen, K.-L., J. Eiriksson, E. Jansen, H. Jiang, F. Rytter, and H. R. Gudmundsdottir (2004), Palaeoceanographic changes off north Iceland through the last 1200 years: Foraminifera, stable isotopes, diatoms and ice rafted debris, *Quat. Sci. Rev.*, *23*, 2231–2246.

Koç, N., E. Jansen, and H. Haflidason (1993), Paleoceanographic reconstruction of surface ocean conditions in the Greenland, Iceland, and Norwegian seas through the last 14,000 years based on diatoms, *Quat. Sci. Rev.*, *12*, 115–140.

Kohfeld, K. E., R. G. Fairbanks, S. L. Smith, and I. D. Walsh (1996), *Neogloboquadrina pachyderma* (sinistral coiling) as paleoceanographic tracers in polar oceans: Evidence from Northeast Wa-

ter Polynya plankton tows, sediment traps, and surface sediment, *Paleoceanography*, *11*, 679–699.

Lamb, H. H. (1977), *Climate: Present, Past and Future*, vol. 2, *Climate History and the Future*, 835 pp., Methuen, London,

Levac, E. (2001), High resolution Holocene palynological record from the Scotian Shelf, *Mar. Micropaleontol.*, *43*, 179–197.

Levac, E., and A. de Vernal (1997), Postglacial changes of terrestrial and marine environments along the Labrador coast: Palynological evidences from cores 91-045-005 and 91-045-006, Cartwright Saddle, *Can. J. Earth Sci.*, *34*, 1358–1365.

Levac, E., A. de Vernal, and W. Blake Jr. (2001), Holocene paleoceanography of the northernmost Baffin Bay: Palynological evidence, *J. Quat. Sci.*, *16*, 353–363.

Matthiessen, J., A. de Vernal, M. Head, Y. Okolodkov, K. Zonneveld, and R. Harland (2005), Modern organic-walled dinoflagellate cysts in Arctic marine environments and their (paleo-) environmental significance, *Palaontol. Z.*, *79*(1), 3–51

Mayewski, P. A., and F. White (2002), *The Ice Chronicles: The Quest to Understand Global Climate Change*, 233 pp., Univ. Press of N. Engl., Lebanon, N. H.

McKay, J., A. de Vernal, C. Hillaire-Marcel, C. Not, L. Polyak, and D. Darby (2008), Holocene fluctuations in Arctic sea-ice cover: Dinocyst-based reconstructions for the eastern Chukchi Sea, *Can. J. Earth Sci.*, in press.

Meincke, J, B. Rudels, and H. J. Friedrich (1997), The Arctic Ocean-Nordic Seas thermohaline system, *J. Mar. Sci.*, *54*, 283–299.

Mignot, J., A. Levermann, S. Nawrath, and S. Rahmstorf (2007), The role of northern sea ice cover for the weakening of the thermohaline circulation under global warming, *J. Clim.*, *20*, 4160–4171.

Not, C., C. Hillaire-Marcel, B. Ghaleb, L. Polyak, and D. Darby (2008), Stratigraphic implication of [210]Pb-distribution in very low sedimentation rate sediments from the Mendeleev Ridge (Arctic Ocean), *Can. J. Earth Sci.*, in press.

Ogilvie, A. (1984), The past climate and sea-ice record from Iceland, part I: Data to A.D. 1780, *Clim. Change*, *6*, 131–152.

O'Neil, J. R. (1968), Hydrogen and oxygen isotope fractionation between ice and water, *J. Phys. Chem.*, *72*, 3683–3684.

Peyron, O., and A. de Vernal (2001), Application of artificial neural network (ANN) to high-latitude dinocyst assemblages for the reconstruction of past sea-surface conditions in Arctic and sub-Arctic seas, *J. Quat. Sci.*, *16*, 699–711.

Pfirman, S., M. E. Lange, I. Wollenburg, and P. Schlosser (1990), Sea-ice characteristics and the role of sediment inclusions in deep-sea deposition: Arctic-Antarctic comparison, in *Geological History of the Polar Oceans: Arctic Versus Antarctic*, edited by U. Bleil and J. Thiede, pp. 187–211, Kluwer Acad., Norwell, Mass.

Polyakov, I. V., G. V. Alekseev, A. Timokhov, U. S. Bhatt, R. L. Colony, H. L. Simmons, D. Walsh, J. E. Walsh, and V. F. Zakharov (2004), Variability of the Intermediate Atlantic Water of the Arctic Ocean over the last 100 years, *J. Clim.*, *17*, 4489–4497.

Poore, R. Z., L. Osterman, W. B. Curry, and R. L. Phillips (1999), Late Pleistocene and Holocene meltwater events in the western Arctic Ocean, *Geology*, *27*, 759–762.

Radi, T., and A. de Vernal (2008), Dinocysts as proxy of primary productivity in mid-high latitudes of the Northern

Hemisphere, *Mar. Micropaleontol.*, *68*, 84–114, doi:10.1016/j.marmicro.2008.01.012.

Reeh, N. (2004), Holocene climate and fjord glaciations in northeast Greenland: Implications for IRD deposition in the North Atlantic, *Sediment. Geol.*, *165*, 333–342.

Renssen, H., H. Goosse, T. Fichefet, V. Brovkin, E. Driesschaert, and F. Wolk (2005), Simulating the Holocene climate evolution at northern high latitudes using a coupled atmosphere-sea ice-ocean-vegetation model, *Clim. Dyn.*, *24*, 23–43.

Rochon, A., A. de Vernal, J. L. Turon, J. Matthiessen, and M. J. Head (1999), *Distribution of Dinoflagellate Cyst Assemblages in Surface Sediments From the North Atlantic Ocean and Adjacent Basins and Quantitative Reconstruction of Sea-Surface Parameters*, AASP Found. Contrib. Ser., vol. 35, 146 pp., Am. Assoc. of Stratigr. Palynol., College Station, Tex.

Rosell-Melé, A., and E. L. McClymont (2007), Biomarkers as paleoceanographic proxies, in *Proxies in Late Cenozoic Paleoceanography*, edited by C. Hillaire-Marcel and A. de Vernal, pp. 459–509, Elsevier, New York.

Sancetta, C. (1981), Oceanographic and ecologic significance of diatoms in surface sediments of the Bering and Okhotsk seas, *Deep Sea Res., Part A*, *28*, 789–817.

Sancetta, C. (1983), Effect of Pleistocene glaciation upon oceanographic characteristics of the North Pacific Ocean and Bering Sea, *Deep Sea Res., Part A*, *30*, 851–869.

Sarnthein, M., S. Van Kreveld, H. Erlenkeuser, P. M. Grootes, M. Kucera, U. Pflaumann, and M. Schulz (2002), Centennial-to-millennial-scale periodicities of Holocene climate and sediment injections off the western Barents shelf, 75°N, *Boreas*, *32*, 447–461.

Schmidt, G. A. (1999), Forward modeling of carbonate proxy data from planktic foraminifera using oxygen isotope tracers in a global ocean model, *Paleoceanography 14*, 482–497.

Simard, A., and A. de Vernal (1998), distribution des kystes de *Alexandrium excavatum* dans les sédiments récents et postglaciaires des marges est-canadiennes, *Geogr. Phys. Quaternaire*, *52*, 361–371.

Simstich, J., M. Sarnthein, and H. Erlenkeuser (2003), Paired δ$^{18}$O signals of *Neogloboquadrina pachyderma* (s) and *Turborotalita quinqueloba* show thermal stratification structure in Nordic Seas, *Mar. Micropaleontol.*, *48*, 107–125.

Smith, L. M., G. H. Miller, B. Otto-Bliesner, and S.-I. Shin (2003), Sensitivity of the Northern Hemisphere climate system to extreme changes in Holocene Arctic sea ice, *Quat. Sci. Rev.*, *22*, 645–658.

Solignac, S., A. de Vernal, and C. Hillaire-Marcel (2004), Holocene sea-surface conditions in the North Atlantic—Contrasted trends and regimes between the eastern and western sectors (Labrador Sea vs. Iceland Basin), *Quat. Sci. Rev.*, *23*, 319–334.

Solignac, S., J. Giraudeau, and A. de Vernal (2006), Holocene sea surface conditions in the western North Atlantic: Spatial

and temporal heterogeneities, Paleoceanography, *21*, PA2004, doi:10.1029/2005PA001175.

Stein, R., et al. (2004), Arctic (palaeo) river discharge and environmental change: Evidence from the Holocene Kara Sea sedimentary record, *Quat. Sci. Rev.*, *23*, 1485–1511.

Stroeve, J., M. M. Holland, W. Meier, T. Scambos, and M. Serreze (2007), Arctic sea ice decline: Faster than forecast, *Geophys. Res. Lett.*, *34*, L09501, doi:10.1029/2007GL029703.

Tan, F. C., and P. M. Strain (1980), The distribution of sea ice meltwater in the eastern Canadian Arctic, *J. Geophys. Res.*, *85*(C4), 1925–1932.

ter Braak, C. J. F., and P. Smilauer (1998), *Canoco Reference Manual and User's Guide to Canoco for Windows, Software for Canonical Community Ordination (Version 4)*, 351 pp., Cent. for Biometry, Wageningen, Netherlands.

Vavrus, S., and S. P. Harrison (2003), The impact of sea ice dynamics on the Arctic climate system, *Clim. Dyn.*, *20*, 741–757.

Volkmann, R., and M. Mensch (2001), Stable isotope composition (δ$^{18}$O, δ$^{13}$C) of living planktic foraminifers in the outer Laptev Sea and the Fram Strait, *Mar. Micropaleontol.*, *42*, 163–188.

Voronina, E., L. Polyak, A. de Vernal, and O. Peyron (2001), Holocene variations of sea-surface conditions in the southeastern Barents Sea, reconstructed from dinoflagellate cyst assemblages, *J. Quat. Sci.*, *16*, 717–727.

Walsh, J. E. (1983), Role of sea ice in climate variability: Theories and evidence, *Atmos. Ocean*, *21*, 229–242.

Walsh, J. E., and W. L. Chapman (2001), Twentieth-century sea-ice variations from observational data, *Ann. Glaciol.*, *33*, 444–448.

Watanabe, E., J. Wang, A. Sumi, and H. Hasumi (2006), Arctic dipole anomaly and its contribution to sea ice export from the Arctic Ocean in the 20th century, *Geophys. Res. Lett.*, *33*, L23703, doi:10.1029/2006GL028112.

Wu, B., J. Wang, and J. E. Walsh (2006), Dipole anomaly in the winter Arctic atmosphere and its association with sea ice motion, *J. Clim.*, *19*, 210–225, doi:10.1175/JCLI3619.1

Yi, D., L. A. Mysak, and S. A. Venegas (1999), Decadal-to-interdecadal fluctuations of Arctic sea-ice cover and the atmospheric circulation during 1954–1994, *Atmos. Ocean*, *37*, 389–415.

Zonneveld, K. A. F., G. Versteegh, and M. Kodrans-Nsiah (2008), Preservation and organic chemistry of Late Cenozoic organic-walled dinoflagellate cysts: A review, *Mar. Micropaleontol.*, *68*, 179–197, doi:10.1016/j.marmicro.2008.01.015.

A. de Vernal, C. Hillaire-Marcel, T. Radi, and S. Solignac, GEOTOP, Université du Québec à Montréal, CP 8888, succ. Centre-Ville, Montréal, Québec H3C 3P8, Canada. (devernal.anne@uqam.ca)

A. Rochon, GEOTOP and ISMER, Université du Québec à Rimouski, Rimouski, Quebec G5L 3A1, Canada.

# Arctic Cloud Properties and Radiative Forcing From Observations and Their Role in Sea Ice Decline Predicted by the NCAR CCSM3 Model During the 21st Century

Irina V. Gorodetskaya[1]

*Laboratoire de Glaciologie et Géophysique de l'Environnement, Centre National de la Recherche Scientifique, Université Joseph Fourier, Grenoble, France*

L.-Bruno Tremblay

*Department of Atmospheric and Oceanic Sciences, McGill University, Montreal, Quebec, Canada*

Arctic sea ice is sensitive to changes in surface radiative fluxes. Clouds influence shortwave radiation primarily through their high albedo and longwave radiation by changing atmospheric emissivity and determining the height (temperature) of the layer of the highest emission. We review Arctic cloud properties affecting radiative fluxes, estimate sea ice effect on the top-of-atmosphere albedo, and discuss cloud response and contribution to the Arctic sea ice decline during the 21st century predicted by the National Center for Atmospheric Research Community Climate System Model, version 3 (CCSM3). Over perennial sea ice, clouds decrease incoming shortwave flux at the surface compared to clear skies from zero in winter to ~100 W m$^{-2}$ during the summer. On average over the Arctic Ocean, sea ice retreat decreases the shortwave radiation reflected at the top of the atmosphere within the same range for all-sky conditions. In addition, Arctic clouds warm the surface increasing the annual mean downwelling longwave flux by ~40 W m$^{-2}$. During the 21st century, CCSM3 predicts a drastic sea ice decline accompanied by larger cloud cover and liquid water content, which increase both cloud cooling and warming effects at the surface. The surface albedo decrease caused by sea ice retreat is partly compensated but not canceled by stronger shortwave cloud cooling. Warming of the near-surface atmosphere is an additional factor increasing the downwelling longwave flux at the surface. The ultimate effect of cloud changes in this model is facilitating the sea ice decline.

[1]Formerly at Lamont-Doherty Earth Observatory and Department of Earth and Environmental Sciences, Columbia University, Palisades, New York, USA.

Arctic Sea Ice Decline: Observations, Projections, Mechanisms, and Implications
Geophysical Monograph Series 180
Copyright 2008 by the American Geophysical Union.
10.1029/180GM05

## 1. INTRODUCTION

During the last 30 years, satellite observations have shown a large reduction in the Arctic sea ice cover [*Overpeck et al.*, 2005; *Stroeve et al.*, 2005; *Serreze et al.*, 2007]. Significant future reduction in the extent and thickness of the Arctic sea ice is predicted by coupled models participating in the Fourth Assessment Report of Intergovernmental Panel on Climate Change (IPCC AR4) [*Zhang and Walsh*, 2006]. When forced with Special Report on Emission Scenarios (SRES) A1B scenario, in which atmospheric $CO_2$ concentration doubles by 2100, half of the model population simulates an ice-free Arctic Ocean in late summer by the end of the 21st century [*Arzel et al.*, 2006]. In reality, the sea ice decrease has accelerated during the past several years, and seasonally ice-free Arctic can happen even faster than the most pessimistic forecast [*Serreze et al.*, 2007; *Stroeve et al.*, 2005]. According to satellite data, September 2007 average Arctic sea ice extent was already as low as $4.28 \times 10^6$ km$^2$, which is 23% lower than the previous record minimum for this month in 2005 and 39% below the long-term average from 1979 to 2000 (National Snow and Ice Data Center press release of 1 October 2007, available at http://nsidc.org/news/index.html).

Studies reviewed by *Serreze et al.* [2007] showed that a combination of the ocean, ice, and atmosphere changes have produced feedbacks, which worked together toward the rapid sea ice reduction. The ice-albedo feedback is considered to be a major process accelerating the sea ice reduction. Analyzing the 21st century SRES A1B simulations of the National Center for Atmospheric Research Community Climate System Model, version 3 (NCAR CCSM3), *Holland et al.* [2006] found that the Arctic sea ice retreat accelerates because of the larger open water area and increased absorption of solar radiation, triggered by increased oceanic heat transport into the Arctic. Similarly, *Winton* [2006] named both the ice albedo feedback and ocean heat flux as two primary drivers for sea ice removal in one of IPCC standard experiments, where $CO_2$ increases by 1%/year to quadrupling. The increase in the oceanic heat flux has been supported by observations, which showed warming of the Atlantic layer residing at intermediate depths within the Arctic Ocean [*Polyakov et al.*, 2004] and the increase in both transport and temperature of the waters entering the Arctic from the Atlantic [*Schauer et al.*, 2004; *Swift et al.*, 1997]. The exchange between relatively warm and saline water entering from the Atlantic and the sea ice ultimately depends on the strength of the Arctic halocline, which has significant spatial, seasonal, and interannual variability [*Björk et al.*, 2002; *Martinson and Steele*, 2001; *Rudels et al.*, 2004]. In the Bering Strait, observations (from 1990 to 2004) also suggest an increase in ocean heat flux in the Chukchi Sea due to both increased volume flux and temperature [*Woodgate et al.*, 2006]. While the time series is relatively short, the 2004 ocean heat flux is the highest on record, and the 2001–2004 heat input could melt an equivalent of 640 km$^3$ of sea ice.

Another possible contribution to the Arctic surface warming is the increase in the downwelling longwave (LW) radiation associated with increased cloud cover and water vapor. Satel-lite data analysis by *Francis et al.* [2005] indicates that the sea ice edge annual anomalies are most highly correlated with the fluctuations in the downwelling LW flux, more than with the changes in wind advection, sensible heat transport, or short-wave (SW) radiation. Negative correlation was found between the sea ice edge anomalies and downwelling SW flux, indicating that stronger ice retreat occurred when the incoming SW flux at the surface was smaller. Observations in the Canadian Arctic showed that the earlier snowmelt onset and decrease in the annual mean albedo have been possibly driven by the increase in winter incoming LW radiation [*Weston et al.*, 2007]. According to the modeling study by *Zhang et al.* [1996], it is the downwelling LW flux that controls interannual variability in the timing of melt onset. Sea ice thinning in response to increased cloudiness was demonstrated by several studies using thermodynamic sea ice models of different complexity [*Ebert and Curry*, 1993; *Shine and Crane*, 1984]. Satellite observation have shown a strong positive trend in spring cloudiness over the Arctic Ocean since 1980 [*Schweiger*, 2004; *Wang and Key*, 2003]. Cloudier atmosphere during spring could also contribute to the recent sea ice retreat in addition to the oceanic heat increase discussed above.

Surface cloud radiative forcing is defined as the difference in the downwelling and/or net radiative fluxes at the surface during cloudy conditions compared to clear skies. Clouds have competing effects on the surface radiative budget: they cool the surface by reflecting the SW radiative flux and warm the surface by increasing the downwelling LW radiation. The uniqueness of the Arctic cloud effects on radiation is in the strong seasonality of their SW and LW forcing. Cloud ability to cool the surface in the polar regions strongly depends on the surface albedo. Clouds reflecting solar radiation over the sea ice covered with fresh snow have zero or very little effect on the surface net SW flux because of the high surface albedo. When sea ice is covered with melt ponds during the summer or over the ice-free ocean areas, clouds greatly reduce the amount of the SW flux absorbed at the surface. Cloud ability to warm the surface depends only on the marcophysical and microphysical cloud properties, which influence the downwelling LW flux. The upwelling LW flux, in turn, responds to changes in surface temperature. During the winter, cloud presence is most of the times associated with warmer surface temperatures [*Walsh and Chapman*, 1998]. During the summer melt period, when the surface temperature is constrained by 0°C, increased downwelling LW flux is used to melt the snow and ice.

There is a high uncertainty in the cloud response to already occurring and future sea ice changes. Cloud response to the reduction in sea ice thickness and area can have two opposite effects on the surface radiative budget. First is the "umbrella" effect: melting of ice promotes greater cloud formation, which reduces the amount of SW radiation absorbed by the ocean. This diminishes the ice-albedo feedback and mitigates the increase in surface temperature. Second is the "greenhouse" effect: increased cloudiness and atmospheric temperatures accompanying the sea ice reduction in the Arctic enhance LW cloud warming of the surface. This extra energy causes earlier and/or greater snowmelt, which, in turn, triggers the ice-albedo feedback by reducing the surface albedo.

In the present chapter we use satellite and ground-based observations to estimate and compare the effects of clouds on the shortwave and longwave radiation. Further, we use the NCAR CCSM3 coupled global climate model to examine the change of sea ice and clouds during the 20th and 21st centuries. The chapter is structured as follows: Section 2 describes observations and the model. Section 3 reviews the studies based on observations subdivided into two parts: section 3.1 discusses the effects of sea ice and clouds on the SW fluxes at the top of atmosphere using satellite data, and section 3.2 reviews Arctic cloud radiative properties using ground-based observations. Section 4 presents the CCSM3 model simulations of sea ice, clouds, and radiation during the 20th and 21st centuries. The summary and the discussion of the study are given in section 5.

## 2. DATA AND METHODOLOGY

### 2.1. Observational Data

A suite of accurate up-to-date data of Arctic cloud properties, radiative fluxes, and sea ice are used in this study. Shortwave radiative fluxes and albedo spectrally integrated over the 0.2–5.0 µm band at the top of the atmosphere (TOA) derived from the Earth Radiation Budget Experiment (ERBE) data are used to analyze sea ice effects on the TOA shortwave radiation for all-sky conditions. The program combines the ERBS, NOAA 9, and NOAA 10 satellite measurements for the period from November 1984 to February 1990 [Barkstrom et al., 1989; Barkstrom and Smith, 1986]. We use the narrow field of view product with a spatial resolution of 2.5° × 2.5°. Global error for ERBE monthly SW fluxes is 5.5 W m$^{-2}$ [Wielicki et al., 1995]. Larger errors in the polar regions (up to 20 W m$^{-2}$ for all-sky fluxes and 50 W m$^{-2}$ for clear-sky fluxes) are caused by inaccuracies in surface scene and cloud identification over sea ice (a review of the ERBE data errors is given by Gorodetskaya et al. [2006]). Large uncertainties in clear-sky identification over ice surfaces make separate analyses of the data for clear-sky conditions and cloudy-sky conditions unreliable [Li and Leighton, 1991]. For all-sky data, the errors contribute to the scatter, but they are substantially smaller than the changes in the SW fluxes

associated with seasonal variations in sea ice concentrations. Significant improvements in the radiative flux retrievals in the polar regions are achieved in the new product from the Clouds and the Earth's Radiant Energy System (CERES) program, which is currently becoming available after a series of validations [Kato et al., 2006]. However, the TOA albedo from the CERES data differs significantly from the ERBE [Bender et al., 2006] and requires more comparison with observations prior to be used in sensitivity studies. The choice in the present study is given to the widely used ERBE data set, taking into account its shortcomings.

Sea ice concentration (SIC) data used for the analysis of sea ice effects on the TOA shortwave fluxes are from the UK Met Office Hadley Centre's sea ice and sea surface temperature data set (HadISST1) available from 1870 to the present on a 1° latitude-longitude grid [Rayner et al., 2003]. Beginning in 1978, the data are derived from Special Sensor Microwave/Imager and the Scanning Multichannel Microwave Radiometer [Gloersen et al., 1992]. The microwave radiance data have a monthly averaged SIC error of about 7%, increasing up to 11% during the melt season [Gloersen et al., 1992]. The biases are greatly reduced in the HadISST1 homogenization process using other satellite and in situ sea ice concentration and sea ice extent data [Rayner et al., 2003].

Cloud and surface radiative flux ground-based data are obtained during the Surface Heat Budget of the Arctic Ocean (SHEBA) program conducted on an ice floe that drifted more than 1400 km in the Beaufort and Chukchi seas between 74°N and 81°N latitudes and between 165°W and 140°W longitudes from late October 1997 to mid-October 1998 [Uttal et al., 2002]. The major meteorological and surface energy budget measurements were conducted on multiyear pack ice with summertime melt ponds and occasional nearby leads [Perovich et al., 2002; Tschudi et al., 2001]. Upwelling and downwelling fluxes were measured using broadband radiometers [Persson et al., 2002]. The uncertainty in the downwelling and upwelling LW flux was estimated at ±2.5 W m$^{-2}$ and at ±4 W m$^{-2}$ for the net LW radiation. The estimated uncertainty in the downward and upward SW flux is ±3% with a bias from –5 to +1 W m$^{-2}$ for the downward and 0 to –6 W m$^{-2}$ for the upward SW flux [Persson et al., 2002]. Cloud presence and base height were derived from a combination of lidar and radar measurements [Intrieri et al., 2002b]. Column-integrated liquid water path (LWP) was derived from brightness temperatures measured by microwave radiometer with 25 g m$^{-2}$ accuracy [Intrieri et al., 2002b; Westwater et al., 2001].

### 2.2. NCAR CCSM3 Model

The output from the 20th and 21st century simulations of the NCAR CCSM3 global coupled model is examined to improve our understanding of the role of clouds in the Arctic sea ice decline. The model consists of components simulating the Earth's atmosphere, ocean, land surface, and sea ice connected by a flux coupler [*Collins et al.*, 2006]. This paper shows results from the configuration used for climate change simulations with a T85 grid for the atmosphere and land and a 1° grid for the ocean and sea ice. Cloud amount is diagnosed by the relative humidity, atmospheric stability, and convective mass fluxes [*Boville et al.*, 2006]. Cloud ice and liquid phase condensates are predicted separately [*Rasch and Kristijansson*, 1998; *Zhang et al.*, 2003]. This links the radiative properties of the clouds with their formation and dissipation. However, after advection, convective detrainment, and sedimentation, the cloud phase is recalculated as a function of cloud temperature. Cloud liquid and ice are assumed to coexist within a temperature range of −10°C and −40°C with the ice fraction linearly increasing with decreasing temperature [*Boville et al.*, 2006]. Clouds are all liquid above −10°C, and all ice below −40°C. Compared with observations, the model produces too much atmospheric moisture in the polar regions and too little in the tropics and subtropics, suggesting that the poleward moisture flux is excessive [*Collins et al.*, 2006]. As consequence of the excessive moisture advection and allowing cloud liquid water to exist at low temperatures, the model Arctic clouds contain large amount of cloud liquid water [*Gorodetskaya et al.*, 2008].

The sea ice in the CCSM3 is represented by a dynamic-thermodynamic model that includes a subgrid-scale ice thickness distribution, energy conserving thermodynamics, and elastic-viscous-plastic dynamics [*Briegleb et al.*, 2004]. The surface albedo for the visible and near infrared bands is a function of ice and snow thickness and surface temperature.

In this paper, the model output during 1950–1999 is obtained from the "Climate of the 20th century experiment" 20C3M scenario simulations conducted from 1870 to present. The model output for 2000–2100 time period is from the SRES A1B experiment, where simulations are initialized with conditions from the end of the 20C3M simulations and run to 2100 under imposed SRES A1B conditions. The SRES A1B scenario assumes moderate population growth and rapid economic growth according to the estimates in the end of 1990s with atmospheric $CO_2$ concentration increase to 720 ppm by 2100 (doubling the 1990 amount) [*Houghton et al.*, 2001].

### 2.3. Methods

Satellite data analysis of sea ice effects on the TOA short-wave fluxes is performed during the ERBE time period from November 1984 to February 1990 on 2.5° grid. Cloud properties influencing radiative forcing are obtained from point

ground-based observations during 1 year from 1997 to 1998. The CCSM3 model output is analyzed on monthly mean and annual mean timescales on the 1° grid for the time period from 1950 to 2100. Although the data are available on different time periods and spatial resolutions, they are used for sensitivity estimates rather than for providing the absolute magnitudes. Observations give a picture of the sea ice/cloud/radiation relationships for modern climate. These relationships are then applied to understand the role of cloud and radiation changes on sea ice cover during the 21st century as predicted by the CCSM3 model.

## 3. SEA ICE AND CLOUD EFFECTS ON RADIATION: OBSERVATIONS

### 3.1. Effect of Sea Ice on the Top-of-Atmosphere Albedo

Top-of-atmosphere albedo represents a fraction of the incoming solar energy reflected from the surface and atmosphere system. Clouds shield the surface from solar radiation mitigating the effect of melting sea ice on the TOA albedo. While the surface albedo decreases drastically when sea ice cover declines (exposing dark ocean surface) the TOA albedo changes are much smaller if the skies are cloudy. The magnitude of the ice-albedo feedback is defined by the changes in the TOA albedo, rather than in the surface albedo, in response to the sea ice changes. Thus, cloud changes can modify the magnitude of the ice-albedo feedback. To show the role of clouds in mitigating the ice-albedo feedback, we estimate sea ice effect on the SW radiation reflected at the TOA for average cloud conditions.

In the Arctic, the ice-free ocean areas are almost always associated with overcast skies, which reduces the difference between the TOA albedo over the open ocean and sea ice. Figure 1 shows spatial distribution of the all-sky TOA albedo over the Arctic and adjacent ocean area during March and July, together with the climatological sea ice extent for these months. During July the sea ice surface has the lowest albedo because of ubiquitous melt ponds, while the sea ice extent reaches the minimum area only in September. In March, Arctic sea ice extent is at maximum, and insolation starts to increase after the polar night (incoming SW flux at the TOA averaged over the ocean north of 70°N reaches 100 W m$^{-2}$ in March). The monthly mean TOA albedo over sea ice in March lies between 60 and 70% (Figure 1a), reducing to 50–60% during the summer (Figure 1b). During the summer, the ice-free ocean, including the area of the ocean occupied by ice during the winter, has TOA albedo ranging between 40 and 50% (Figure 1b). Thus, while the difference between the TOA albedo over the ice-free ocean and over the sea ice is reduced by high cloud amount, seasonal

## a) March

## b) July

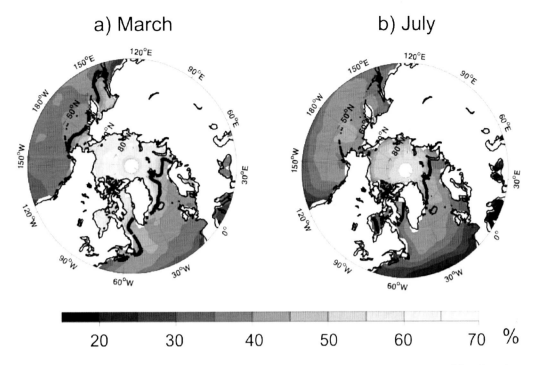

20    30    40    50    60    70 %

**Figure 1.** Spatial distribution of monthly mean top-of-atmosphere (TOA) albedo (percent, gray scale) and sea ice extent (thick black line) over the ocean north of 40°N during (a) March and (b) July. The TOA albedo values for each month are based on the ERBE data during 1985–1989 on 2.5° × 2.5° grid. The sea ice extent is defined as the area of the ocean with sea ice concentrations (SICs) of at least 10%. SICs are from the HadISST1 data, subset to the ERBE time period and regridded to the ERBE resolution. White area around the North Pole contains no data.

changes in sea ice cause about 10–20% change in the all-sky TOA albedo.

Sea ice effect on the TOA albedo is summarized in Figure 2, which shows the TOA albedo and the surface albedo as a function of sea ice concentration. The radiative effectiveness of sea ice is defined as RE = TOA albedo (SIC = 100%) − TOA albedo (SIC = 0%), following *Yamanouchi and Charlock* [1997]. The average RE of the Arctic sea ice with respect to the TOA albedo is 0.22 for all-sky conditions [*Gorodetskaya et al.*, 2006]. Arctic sea ice experiences extensive melt during the summer, which decreases the sea ice surface albedo down to about 38%, while during winter, freshly fallen snow increases surface albedo up to 84% according to ground-based observations [*Curry et al.*, 2001]. Surface albedo over the open ocean are calculated as a function of the solar zenith angle increasing from 3% during summer to almost 30% during winter. These surface albedo values define an envelope of the surface albedo ranges shown by thin lines in Figure 2: the upper line representing winter/fall/spring surface albedo and the lower line representing the melt season.

Monthly mean TOA albedo averaged for all-sky conditions is above the range of the surface albedo at low SICs

and within the surface albedo range at high SICs. While the solar zenith angle causes a significant change in the open ocean surface albedo, the effects of clouds overwhelm these changes and increase the mean TOA albedo over the open ocean well above the maximum observed ice-free surface albedo. In contrast, over the 100% SICs the combination of sea ice and cloud effects results in the mean TOA albedo lying within the extremes of observed surface albedo values (Figure 2).

Figure 3 shows seasonal changes in the downwelling and reflected SW flux at the TOA, and the sea ice RE, calculated as the product of the downwelling SW flux averaged over the Arctic Ocean and sea ice RE with respect to the TOA albedo for each corresponding month [*Gorodetskaya et al.*, 2006]. The largest impact of sea ice on the SW flux occurs during the summer: for an incident solar flux of about 450 W m$^{-2}$ reaching the TOA in the polar latitudes in June (averaged over the ocean north of 70°N), local reduction of SIC from 100% to 0% results in about 100 W m$^{-2}$ decrease in reflected SW radiation at the top of the atmosphere. Sea ice effect on SW flux is relatively small during the fall and early spring reducing to zero during the winter months proportionate to the incoming SW flux. Thus, variations in monthly

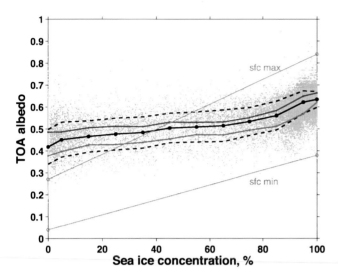

**Figure 2.** Monthly mean top-of-atmosphere (TOA) albedo (grey dots) against Arctic sea ice concentration (SIC). Black dots connected by a line represent area-weighted TOA albedo averages for 0% and 100% SIC, as well as for each 10% SIC bin based on all data regardless of season from the grid boxes where sea ice was present for at least 1 month during the ERBE time period (November 1984 to February 1990). The dashed lines are standard deviations. The upper thick solid line represents the averages for winter (December–February); the lower thick solid line represents averages for summer (June–August). The thin lines denoted by "sfc max" and "sfc min" connect maximum and minimum observed surface albedo values for the open ocean and sea ice, showing the TOA albedo envelope during clear skies.

mean sea ice induce noticeable changes in the TOA albedo, yet much smaller than the associated changes in the surface albedo because of the compensating effect of clouds.

*3.2. Cloud Properties and Surface Radiative Forcing*

In the previous section we showed that the sea ice effect on the top-of-atmosphere albedo is strongly compensated by clouds. Thus, changes in cloud shortwave radiative forcing should be taken into account when calculating the magnitude of the ice-albedo feedback. In turn, the effect of clouds on the SW flux absorbed at the surface (net SW radiation defined as the difference between the incoming and reflected SW fluxes at the surface) depends on the surface albedo: for the same cloud albedo, the cloud radiative forcing is greater over the surface with a smaller albedo, while over the highly reflective surface, clouds have little effect on the surface net SW flux. In addition, clouds have a warming effect on the surface by increasing the downwelling LW radiation. Thus, the surface radiative budget is strongly influenced by com-

plex interactions between the sea ice and clouds. In this section we discuss separately cloud effects on the surface SW and LW radiative fluxes using ground-based observations obtained during the year of the SHEBA program [*Intrieri et al.*, 2002a, 2002b; *Persson et al.*, 2002].

The annual mean SW cloud forcing with respect to the net flux is $-10 \pm 0.5$ W m$^{-2}$; that is, clouds decrease the net SW flux by about 10 W m$^{-2}$. The annual mean LW cloud forcing with respect to the net flux is $38 \pm 3$ W m$^{-2}$ [*Intrieri et al.*, 2002a]. However, cloud radiative forcing has a strong seasonal variability. Variations in the cloud effects on the net SW flux at the surface depend on cloud transmittance, the solar zenith angle (insolation), and also on the surface albedo. The cloud effect on the net LW flux at the surface is dominated by the downwelling component, which is dependent on cloud properties, such as cloud base temperature, cloud liquid and ice content, and cloud particle radius. Classical definition of cloud radiative forcing as a function of cloud presence (cloudy versus clear sky) or cloud fraction is insufficient in the Arctic, where a wide range of cloud radiative forcing exists for the same cloud fraction.

At SHEBA location (74°–81°N) polar night lasts from November to February. The effect of cloud on the net SW

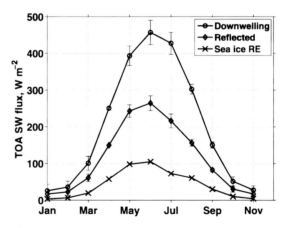

**Figure 3.** Seasonal cycles of the top-of-atmosphere (TOA) downwelling shortwave (SW) radiative flux, reflected SW radiative flux (both averaged over the ocean north of 70°N), and sea ice radiative effectiveness (RE) with respect to the SW flux, calculated as the product of the downwelling SW flux averaged over the ocean area north of 70°N by the sea ice RE with respect to the TOA albedo for each month, defined as the difference between the TOA albedo averaged over 100% and 0% sea ice concentration. Monthly sea ice RE with respect to the TOA albedo are 0.10, 0.13, and 0.18 for December, January, and February; 0.19, 0.23, and 0.25 for March, April, and May; 0.23, 0.17, and 0.20 for June, July, and August; and 0.20, 0.20, and 0.15 for September, October, and November [*Gorodetskaya et al.*, 2006].

radiation outside of the melt season is negligible because of the small amount of solar radiation reaching the surface and high surface albedo (Figure 4). Figure 5 from *Intrieri et al.* [2002a] shows annual cycles of the downwelling, upwelling, and net radiative fluxes and cloud radiative forcing with respect to each flux. During the melt season, from early July until mid-August, clouds reduce the downward solar flux reaching the surface by about 100 W m$^{-2}$ (Figure 5a). At the same time, cloud effect on the surface net SW flux strongly depends on the surface albedo. The cloud SW forcing with respect to the surface net SW flux is small until the end of May when sea ice is covered by snow with albedo of about 0.85 (Figure 4). The importance of cloud cooling increases dramatically with the progression of melt, when the surface albedo is lowered by snow melting and formation of melt ponds. Figure 4 shows a large drop in the surface albedo during June–July over the sea ice surface without even including the melt ponds. While downwelling SW flux reaches 200

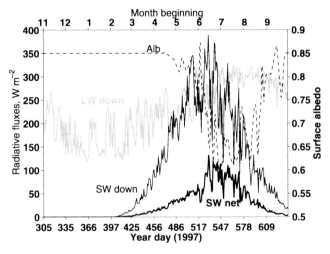

**Figure 4.** Annual cycle of downwelling longwave flux (LW down, thick grey line), downwelling shortwave flux (SW down, thin black line), net shortwave flux (SW net, thick black line), and surface albedo (Alb, dashed line), based on SHEBA observations from 1 November 1997 to 30 September 1998. The daily radiative fluxes form a composite data set in which 82% of the observations are from the Atmospheric Surface Flux Group (ASFG) radiometers, 6% are from the Atmospheric Radiation Measurement (ARM) program radiometers, 7% are from the SHEBA Project Office (SPO) radiometers, and 5% are from climatology [*Lindsay*, 2003]. Albedo data are provided by H. Huwald (personal communication, 2004). Surface albedo is fixed at 0.85 during winter months. Year day (1997) zero is defined as 1 January 1997, 0000 LT. Year days for 1 December, 1 March, 1 June, and 1 September are 335, 425, 517, and 609, correspondingly.

W m$^{-2}$ in May, the net SW flux is below 50 W m$^{-2}$ until the onset of melt in June when the net SW flux doubles (Figure 4). The largest cloud effect on the surface net SW flux occurs in the beginning of July (50 W m$^{-2}$, Figure 5a), when surface albedo is the lowest.

The cloud forcing with respect to both the downwelling and net LW flux is large throughout the year with the annual range of 45 W m$^{-2}$ (from the minimum of 15 W m$^{-2}$ in December to the maximum of 60 W m$^{-2}$ in September) (Figure 5b). During the winter, the downwelling LW flux is the largest source of energy for the surface, and changes in the winter surface air temperature are closely tied to changes in the LW radiation budget [*Walsh and Chapman*, 1998]. The LW cloud forcing is dominated by the downwelling component, while the effect of clouds on the upwelling LW flux is small [*Intrieri et al.*, 2002a].

Cloud surface longwave forcing depends mostly on the cloud emissivity (or opacity) and cloud emitting temperature [*Chen et al.*, 2006]. Because of the frequent presence of liquid water in the Arctic clouds, their opacity is mostly determined by the cloud liquid water path (cloud liquid water content integrated over the entire atmospheric column). Cloud longwave forcing is highly sensitive to the cloud LWP for low LWP values, while the impact of changes in LWP on cloud longwave forcing is practically zero for LWP values greater than 30 g m$^{-2}$. This property of clouds is known as the longwave saturation effect; that is, a cloud with large LWP emits as a blackbody, and thus any further increase in its LWP has no effect on the cloud longwave forcing.

The cloud emitting temperature, in turn, is most often determined by the cloud base temperature. Large values of the downwelling LW flux correspond mostly to clouds with base temperatures warmer than –15°C (Figure 6a) [*Shupe and Intrieri*, 2004]. Some clouds with colder cloud base temperature are almost indistinguishable from clear-sky emission. The strongest longwave cloud forcing was also found for clouds residing at heights lower than 0.5 km (Figure 6b) [*Shupe and Intrieri*, 2004]. Cloud base height, however, has mostly an indirect effect on cloud longwave forcing by influencing the cloud base temperature. *Chen et al.* [2006] showed that the clouds with bases between 1 and 3 km have decreasing longwave forcing with increasing cloud base height, while clouds with a base height lower than about 0.6 km have increasing longwave forcing with increasing height. This relationship shows if the cloud resides in or above the temperature inversion layer (typically about 1 km). The cloud base temperature effect on the LW cloud forcing is enhanced when clouds reside at or near the peak of the near-surface temperature inversion layer. It has to be taken into account that when a cloud resides below the temperature inversion layer, the maximum emitting

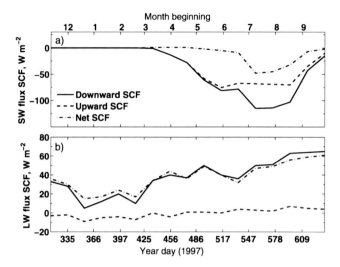

**Figure 5.** Annual cycle of (a) shortwave surface cloud forcing (SW flux SCF) and (b) longwave surface cloud forcing (LW flux SCF) for the downwelling (solid line), upwelling (dashed), and the net flux (dash-dot). Year day (1997) zero is defined as 1 January 1997, 0000 LT. Year days for 1 December, 1 March, 1 June, and 1 September are 335, 425, 517, and 609, correspondingly (adapted from *Intrieri et al.* [2002a]).

temperatures can be located within the cloud rather than at the cloud base [*Stramler*, 2006].

Presence of liquid in Arctic clouds substantially increases their impact on both the SW and LW fluxes [*Shupe and Intrieri*, 2004; *Zuidema et al.*, 2005]. The annual mean LW (SW) cloud forcing with respect to the net surface fluxes was estimated at 52 (−21) W m$^{-2}$ for liquid-containing clouds compared to 16 (−3) W m$^{-2}$ for ice-only clouds [*Shupe and Intrieri*, 2004]. The time mean cloud optical thickness of liquid clouds is estimated at 10.1 ± 7.8 corresponding to LWP of 37 g m$^{-2}$. This by far exceeds the mean optical thickness for ice-only clouds of 0.2 [*Zuidema et al.*, 2005]. Similarly, in mixed phase clouds, the cloud liquid content dictates cloud optical thickness [*Zuidema et al.*, 2005].

For both LW and SW radiative fluxes, cloud optical thickness has a threshold value above which further increase in the liquid water path has no influence on SW or LW cloud forcing. Clouds become saturated in the LW at lower values of LWP than in the SW; that is, as LWP increases, the cloud SW cooling effect continues to increase after the LW warming effect reaches saturation (clouds emit as blackbodies regardless of further increase in LWP) [*Shupe and Intrieri*, 2004]. To compute cloud SW forcing as a function of cloud LWP over a reflective surface is difficult as cloud SW forcing also depends on the downwelling solar radiation and surface albedo. Monthly mean LWP at SHEBA is about 10–20 g m$^{-2}$ during the winter and increases to 90 g m$^{-2}$ by the end of summer [*Gorodetskaya et al.*, 2008]. Low sensitivity

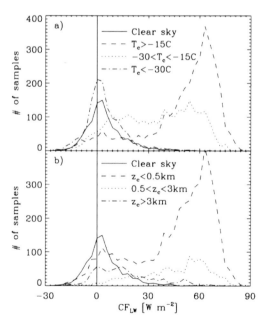

**Figure 6.** Distribution of the longwave cloud forcing (CF$_{LW}$) defined as difference between net surface LW fluxes for all-sky and clear-sky conditions distinguished by (a) cloud base temperature and (b) cloud base height. The clear-sky distribution (an uncertainty estimate for these calculations) is also shown in both Figures 6a and 6b. From *Shupe and Intrieri* [2004], with permission of the American Meteorological Society.

of cloud LW forcing to LWP has been found during the late spring and summer when LWP is large, while high sensitivity of cloud LW forcing to changes in cloud LWP is found in winter and spring when cloud LWP values are small [*Chen et al.*, 2006]. As will be discussed in section 4, the magnitude of the cloud LWP during the present climate has a substantial effect on the cloud SW and LW forcing changes during the 21st century.

Figure 7 reproduced from *Zuidema et al.* [2005] illustrates the change in the cloud forcing with respect to the net radiative fluxes as a function of cloud optical depth based on low-level mixed phase clouds observed at constant height during 4 days in May 1998 over the SHEBA site. It shows that cloud LW forcing dominates the total cloud forcing for

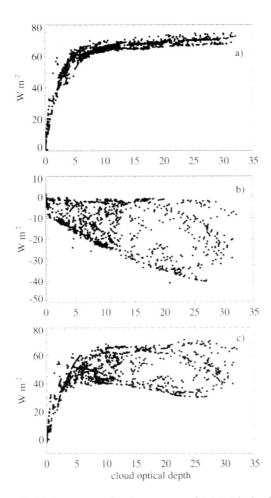

**Figure 7.** (a) Longwave, (b) shortwave, and (c) total cloud surface forcing with respect to the net radiative fluxes as a function of cloud optical depth. Dotted lines denote optical depths of 3 and 6, based on observations during the SHEBA from 1 to 10 May 1998. From *Zuidema et al.* [2005], with permission of the American Meteorological Society.

small cloud optical thickness (less than 3). After passing a threshold of 6, further increase in the cloud optical thickness has no influence on the LW radiation, and net cloud forcing is mostly determined by the shortwave component. Cloud effect on the net SW radiation depends on surface reflectivity causing large spread in cloud SW and total forcing, in particular for large cloud optical thickness.

The relative magnitude of the cloud SW and LW forcing discussed above has a large effect on the surface temperature and sea ice thickness. Over perennial sea ice, cloud warming effect overpasses the cooling effect during most of the year. During the winter polar night and transition seasons, cloudy conditions are associated with increased downwelling LW flux and warmer surface air temperatures over sea ice [*Walsh and Chapman*, 1998]. During the summer melt period, surface temperatures hover around 0°C, while the intensity of net surface melt is strongly influenced by the surface albedo and timing of melt onset: earlier melt onset increases the amount of solar radiation absorbed during the entire melt season [*Perovich et al.*, 2007]. *Perovich et al.* [2002] explained the melt onset during SHEBA by the decrease in surface albedo due to a rain event which occurred at the end of May. On the other hand, in late spring warm air masses enter the Arctic from lower latitudes. This increases both the cloud liquid water content and effective emitting temperature [*Stramler*, 2006]. Increased downwelling LW flux during spring can provide energy for initiating the surface melt [*Zhang et al.*, 1996; *Weston et al.*, 2007]. Several studies using sea ice thermodynamic models found a strong relationship between changes in cloud radiative forcing and sea ice thickness. *Curry et al.* [1993] showed that a large increase in sea ice thickness occurs in response to reduction in annual mean cloudiness. According to *Shine and Crane* [1984], surface albedo can reverse the effects of cloudiness increase during the summer. They found that cloud increase outside of summer months leads to sea ice thinning, while if clouds increase during July and August, their radiative cooling will be greater than their longwave forcing, slowing down sea ice melt and thus leading to overall thicker ice if the forcing persists for years. In the next section, we will discuss changes in the Arctic cloud forcing predicted for the 21st century and their role in the sea ice cover changes as simulated by the NCAR CCSM3 coupled global climate model.

## 4. ROLE OF CLOUDS IN SEA ICE CHANGES DURING THE 21ST CENTURY

Coupled models participating in the IPCC AR4 assessment show large differences in future Arctic sea ice thickness and extent when forced with $CO_2$ emissions as specified by

SRES A1B scenario [*Arzel et al.*, 2006]. The CCSM3 model simulates close to the observed sea ice changes during the 20th century and predicts the most drastic sea ice loss during the 21st century when compared to other models. Already by 2040, CCSM3 predicts that majority of the Arctic basin will be ice free in September. *Holland et al.* [2006] linked the drastic sea ice decline in the CCSM3 model to increased oceanic heat flux, which triggers the ice-albedo feedback by melting large areas of sea ice. Below we demonstrate that cloud changes also play an important role in sea ice decline. The results are based on one realization (run 1) of a group of seven ensemble members from the SRES A1B simulations.

Figure 8 shows the annual mean sea ice concentration averaged north of 70°N simulated by CCSM3 from 1950 to 2100. The model simulates SIC decrease from about 50% in 1950 to 45% in 2000 and 25% by 2100. The sea ice decline is accompanied by significant warming of the atmospheric boundary layer and a strong increase in the cloud liquid water content: the model predicts a 5 K increase in the boundary layer air temperature and about 50 g m$^{-2}$ increase in the cloud liquid water path during the 21st century (Figure 9a). As discussed in the previous sections, the increase in cloud liquid water path increases cloud optical thickness and thus increases both the cloud SW cooling and LW warming. At the same time, cloud effect on the downwelling LW flux has an additional strong contributing factor: increase in the near-surface air temperature.

Figure 9b disentangles the surface albedo and cloud effects on the shortwave radiation and compares cloud effects on the shortwave and longwave radiative fluxes. The thin black line

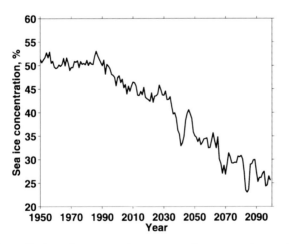

**Figure 8.** Annual mean sea ice concentration averaged over the ocean north of 70°N from the NCAR CCSM3 model. Data during 1950–1999 are from the Climate of the 20th Century Experiment (20C3M) simulations run 1, and data during 2000–2100 are from the SRES A1B simulations run 1, performed as part of IPCC AR4.

represents the difference between the annual mean values of reflected radiation from 1950 to 2100 and the 1990–1999 average denoted in the equations below by $\langle\ \rangle_{90}$ (in Figure 9b values of the reflected flux are multiplied by minus showing anomalies in absorbed radiation due to surface albedo):

$$\Delta SW^{\uparrow} = SW^{\uparrow} - \langle SW^{\uparrow}\rangle_{90} = SW^{\downarrow}\alpha_s - \langle SW^{\downarrow}\alpha_s\rangle_{90}, \quad (1)$$

where $SW^{\uparrow}$ is reflected SW flux, $SW^{\downarrow}$ is downwelling SW flux, and $\alpha_s$ is surface albedo.

The thick black line in Figure 9b represents the anomalies in the net SW flux with respect to the 1990–1999 average:

$$\begin{aligned}
\Delta SW^{net} &= SW^{net} - \langle SW^{net}\rangle_{90} \\
&= SW^{\downarrow}(1-\alpha_s) - \langle SW^{\downarrow}(1-\alpha_s)\rangle_{90} \\
&= SW^{\downarrow} - SW^{\downarrow}\alpha_s - \langle SW^{\downarrow}\rangle_{90} + \langle SW^{\downarrow}\alpha_s\rangle_{90}. \quad (2)
\end{aligned}$$

The difference between the anomalies in the net absorbed SW flux ($\Delta SW^{net}$) and the anomalies in the absorbed radiation due to surface albedo ($-\Delta SW^{\uparrow}$), that is the difference between thin and thick black lines in Figure 9b, shows by how much clouds offset the increase in the absorbed SW radiation at the surface:

$$\Delta SW^{net} - (-\Delta SW^{\uparrow}) = SW^{\downarrow} - \langle SW\rangle^{\downarrow}_{90}. \quad (3)$$

The sea ice decline lowers the surface albedo, which decreases the surface reflected SW flux. If no changes in clouds occurred, the surface would gain about 42 W m$^{-2}$ more during the summer months in 2100 compared to the 1990s because of the surface albedo drop (thin black line in Figure 9b). At the same time, less shortwave flux reaches the surface because of the larger cloud liquid water path. This cloud effect offsets but does not cancel the increase in the net SW flux due to the strong drop in the surface albedo. Increase in cloud reflectivity reduces the actual gain in the net SW by 2100 compared to the 1990s to about 18 W m$^{-2}$ (thick black line in Figure 9b). At the same time, larger cloud liquid water path increases the cloud longwave emissivity, while warmer atmospheric temperatures increase the effective emitting temperature. This leads to a significant increase in the downwelling longwave flux (grey line in Figure 9b). The yearly average increase in the downwelling LW flux by 2100 compared to the 1990s is as high as 34 W m$^{-2}$, which is almost twice the magnitude of the net SW flux increase during the summer.

Figure 10 compares seasonal cycles in cloud LWP and boundary layer atmospheric temperature during the first and last decade of the 21st century, and Figure 11 shows

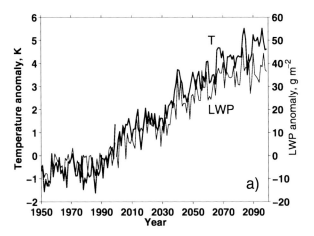
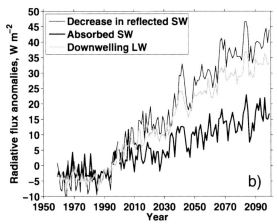

**Figure 9.** Anomalies relative to 1990–1999 in (a) air temperature (T) averaged for 850 and 925 mbar pressure levels (thick line) and cloud liquid water path (LWP) (thin line) and (b) May–August mean decrease in the surface reflected shortwave flux (thin black line), May–August mean surface absorbed shortwave flux (thick black line), and annual mean downwelling longwave flux (grey line) simulated by the NCAR CCSM3 model averaged over the ocean north of 70°N. Data during 1950–1999 are from the 20C3M simulations run 1, and data during 2000–2100 are from the SRES A1B scenario simulations run 1, performed as part of IPCC AR4.

the effect of these cloud properties changes on the surface radiative fluxes. CCSM3 simulates high amount of liquid water in Arctic clouds during the entire year compared to ground-based observations performed during SHEBA from 1997 to 1998 [*Gorodetskaya et al.*, 2008]. Its monthly average LWP exceeds the threshold value for cloud saturation in both SW and LW radiation (about 30 g m$^{-2}$), especially during summer months (Figure 10a). This is limiting the impact of increasing cloud LWP on SW cloud forcing. The largest increase by year 2100 in LWP is predicted during the win-

ter months, when clouds contain smaller amounts of liquid compared to summer (Figure 10a). As discussed in the previous section, downwelling LW flux is strongly influenced by increasing LWP for thin clouds. This allows the increased LWP during winter months to have a noticeable effect on the downwelling LW flux (Figure 11a).

In addition to increase in cloud LWP, the atmospheric layer from 1000 to 850 mbar, where low stratus clouds usually reside, warms up significantly during the entire year but especially strongly during the winter (Figure 10b). Warming

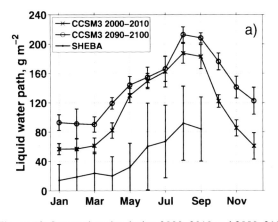
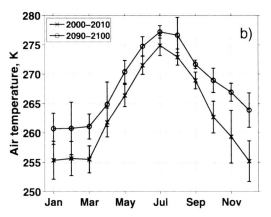

**Figure 10.** Seasonal cycles during 2000–2010 and 2090–2100 of (a) cloud liquid water path (LWP) and (b) air temperature averaged over 1000, 925, and 850 mbar pressure levels simulated by the NCAR CCSM3 model forced with SRES A1B scenario and averaged over the ocean north of 70°N. The error bars for the model output are standard deviations based on monthly means. SHEBA data are monthly averaged LWPs based on daily values available from October 1997 to September 1998. SHEBA error bars are standard deviations based on daily values.

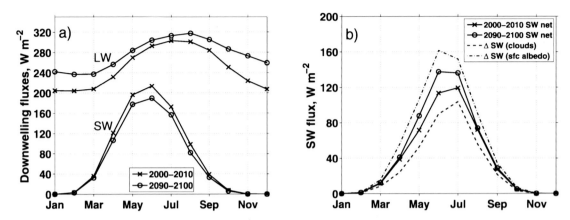

**Figure 11.** Seasonal cycles during 2000–2010 and 2090–2100 of (a) surface downwelling longwave (LW) and shortwave (SW) radiative fluxes and (b) all-sky net (incoming minus reflected) surface SW flux simulated by the NCAR CCSM3 model forced with SRES A1B scenario and averaged over the ocean north of 70°N. Decrease in surface incoming SW due to changes in clouds is shown as "Δ SW (clouds)." Increase in the net SW due to surface albedo changes is shown as "Δ SW (sfc albedo)."

of the low troposphere contributes to the large increase in downwelling LW flux in winter and is the main factor responsible for increased downwelling LW flux during the summer when clouds are optically thick (Figure 11a).

Figure 11b shows seasonal cycles of the Arctic Ocean surface net SW fluxes during the first and last decades of the 21st century together with the contribution of the cloud and surface albedo changes to the difference between the two

**Table 1.** Differences in Cloud Fraction, Cloud Liquid Water Path, Surface Downwelling Longwave and Shortwave Fluxes, and Net Shortwave Flux Between the Last and First Decades of the 21st Century Averaged Over the Ocean North of 70°N Predicted by the NCAR CCSM3 Model SRES A1B Scenario for Each Month[a]

| Month | Cloud (%) | LWP (g m⁻²) | DLF (W m⁻²) | DSF (W m⁻²) | Net SW (W m⁻²) |
|---|---|---|---|---|---|
| December | 9 | 62 | 52 | 0 | 0 |
| January | 7 | 36 | 38 | 0 | 0 |
| February | 6* | 34 | 33 | 0 | 0 |
| DJF | 7 | 44 | 41 | 0 | 0 |
| | | | | | |
| March | 2.5 | 29 | 30 | −3 | 0.8 |
| April | 5 | 37 | 25 | −15 | 2.2 |
| May | 2.5 | 14 | 14 | −19 | 17 |
| MAM | 3 | 27 | 23 | −12 | 6 |
| | | | | | |
| June | 4 | 5* | 11 | −24 | 24 |
| July | 3 | 4* | 10 | −16 | 17 |
| August | 5 | 25 | 17 | −16 | 2 |
| JJA | 4 | 11 | 13 | −19 | 14 |
| | | | | | |
| September | 3 | 26 | 21 | −6 | 2 |
| October | 5 | 54 | 36 | −1.5 | 0 |
| November | 9 | 56 | 49 | 0 | 0 |
| SON | 6 | 45 | 35 | −2.5 | 0.8 |

[a] Abbreviations are LWP, liquid water path; DLF and DSF, surface downwelling longwave and shortwave fluxes, respectively; and SW, shortwave flux. Seasonal means are for December–February (DJF), March–May (MAM), June–August (JJA), and September–November (SON). All differences are significant at 95% level, except for those marked with asterisks.

decades. In CCSM3, increased cloud cooling of the surface in the end of the 21st century only partly compensates for the large surface albedo decrease. The largest increase in the net SW flux by 24 W m$^{-2}$ is found in June. At the same time, a large increase in the LW cloud forcing is simulated during the entire year with a maximum of 52 W m$^{-2}$ in December and minimum of 10 W m$^{-2}$ in July (Figure 11a).

Cloud changes are often described in terms of cloud fraction. The CCSM3 model indeed predicts an increase in both cloud fraction and cloud LWP during the 21st century. However, magnitudes of the increase differ depending on the month, and in some months comparable increase in cloud fraction can be accompanied by different increase in cloud LWP and as a consequence different changes in downwelling LW and SW fluxes (Table 1). Changes in the net SW flux, shown in Table 1, also strongly depend on the insolation and surface albedo.

## 5. SUMMARY AND DISCUSSION

Drastic sea ice retreat has been observed in the Arctic during the last decade of the 20th century and beginning of the 21st century. Most global climate models forced with today's trends in atmospheric greenhouse gas concentrations predict drastic sea ice decline in the Arctic by the end of the 21st century. The response of the Arctic climate system to initial warming is not linear and involves multiple feedbacks able to accelerate or diminish the surface warming. The sea ice-albedo feedback is considered one of the main factors accelerating sea ice disappearance by increasing the amount of absorbed solar radiation at the surface [see *Winton*, this volume]. Increased cloud formation is thought to mitigate the Arctic warming by replacing the highly reflective sea ice surface. However, clouds also contribute to the surface warming by increasing the downwelling longwave flux, thus enhancing the greenhouse effect. In the present chapter, we examined satellite and ground-based observations in attempt to disentangle cloud effects on the shortwave and longwave radiative fluxes. We also discussed changes in sea ice, clouds and surface radiative fluxes predicted by the coupled global climate model NCAR CCSM3 during the 21st century.

Over the Arctic perennial sea ice, clouds have a net warming effect on the surface during most of the year by increasing the downwelling longwave flux. The magnitude of the downwelling longwave flux strongly depends on cloud properties, such as cloud liquid water content and cloud base temperature. Clouds in the Arctic are usually mixed phase with frequent presence of liquid during winter and continuous large amounts of liquid during the summer. During winter, when cloud events are occasional and clouds are thin,

even a small increase in the cloud liquid water content significantly increases the downwelling longwave flux. Cloudy skies in winter are always associated with warmer surface temperatures because of the efficient energy transfer from the relatively warm cloudy atmosphere to the surface by increasing the downwelling LW flux. CCSM3 simulations for the 21st century predict an increase in the cloud liquid water content together with warming of the atmospheric boundary layer. The largest increase in the cloud liquid water content and near-surface atmospheric temperature is predicted during the winter with a large impact on the downwelling longwave flux. Thermodynamic model studies showed that this is the time when cloud changes have the largest impact on the sea ice thickness [*Curry et al.*, 1993; *Shine and Crane*, 1984].

In summer, clouds are present practically continuously and contain large amounts of liquid. Above certain threshold, clouds emit as blackbodies, and further increase in the cloud liquid content has no effect on the cloud longwave forcing. For this reason, increase in the cloud base temperature plays a more important role for downwelling longwave flux during the summer. Cloud cooling effect also becomes significant during the summer because of both high cloud optical thickness and thus high cloud albedo and the lowered surface albedo over melting sea ice. Clouds reduce monthly mean shortwave radiation reaching the surface by up to 100 W m$^{-2}$ on average over the Arctic Ocean during the summer. At the same time, an increase in the shortwave radiation absorbed by the surface-atmosphere column up to the same magnitude can occur when sea ice gives way to the open ocean for all-sky conditions. Simulations with the CCSM3 model showed that during the 21st century, clouds significantly diminish but do not cancel the effect of reduced surface albedo on the surface-absorbed shortwave flux.

The relative role of the shortwave and longwave cloud forcing during the 21st century depends strongly on how the model simulates cloud properties. Mixed phase cloud parameterization in the CCSM3 model allows liquid water to be present at temperatures between −10 and −40°C. Together with excessive poleward moisture flux, this leads to very high liquid water content in the Arctic clouds, which is overestimated compared to the SHEBA ground-based observations. There are not enough Arctic-wide observations to say if this overestimation is significant. Part of the model's overestimation of the cloud LWP comes from the fact that CCSM3 values are averaged over the ocean and sea ice areas north of 70°N, while SHEBA measurements are from perennial ice area. In order to improve model simulation of cloud properties, there is a need for more year-round ground-based observations of clouds and radiative fluxes for various sea ice conditions. Correctly predicting seasonality of cloud changes plays an important role. Overestimation of cloud LWP during summer decreases changes in both the shortwave and the longwave cloud forcing. During winter, CCSM3 cloud LWP is comparable to the observed values when cloud LWP changes are largest and the sensitivity of the cloud longwave forcing to LWP is high.

The NCAR CCSM3 model predicts an increase in cloud fraction, cloud liquid water content, and near-surface atmosphere warming in the Arctic during the 21st century. These atmospheric changes provide additional energy to the surface; thus increase in cloudiness facilitates rather than reduces the surface warming accompanying sea ice decline. The ultimate sea ice changes depend on a combination of factors. The present study showed that clouds directly respond to initial Arctic warming and are among the key factors accelerating the Arctic sea ice decline. Given the larger magnitude of the longwave flux increase induced by clouds compared to the absorbed shortwave flux, as predicted by CCSM3, the cloud surface warming may be a more dominant driver of sea ice loss than the sea ice albedo feedback.

*Acknowledgments.* We are grateful to Matthew Shupe for his continuous help with the SHEBA data, Martin Vancoppenolle and Nikolay Koldunov for their positive and critical input, and Gerhard Krinner and Annette Rinke for helpful discussions. Our great appreciation goes to everyone involved in the SHEBA fieldwork and subsequent data processing. We acknowledge the NCAR CCSM3 group for providing the model output for analysis, the Program for Climate Model Diagnosis and Intercomparison (PCMDI) for collecting and archiving the model data, the JSC/CLIVAR Working Group on Coupled Modeling (WGCM) and their Coupled Model Intercomparison Project (CMIP) and Climate Simulation Panel for organizing the model data analysis activity, and the IPCC WG1 TSU for technical support. The IPCC Data Archive at Lawrence Livermore National Laboratory is supported by the Office of Science, U.S. Department of Energy. We thank the National Snow and Ice Data Center, the Hadley Centre for Climate Prediction and Research, and the Earth Radiation Budget Experiment team for providing satellite data. We appreciate the efforts of the two anonymous reviewers and the Editor for both the positive feedback and constructive criticism that helped to improve the manuscript. I.G. was supported by NASA Fellowship ESSF0400000163 and Agence Nationale de la Recherche (France) grant OTP 232333. B.T. was supported by the Natural Sciences and Engineering Research Council of Canada Discovery Grant Program and by the National Science Foundation under grant OPP-0230325 from the Office of Polar Programs and grant ARC-05-20496 from the Arctic Science Program.

## REFERENCES

Arzel, O., T. Fichefet, and H. Goosse (2006), Sea ice evolution over the 20th and 21st centuries as simulated by current AOGCMs, *Ocean Modell.*, *12*, 401–415.

Barkstrom, B. R., and G. L. Smith (1986), The Earth Radiation Budget Experiment: Science and implementation, *Rev. Geophys.*, *24*, 379–390.

Barkstrom, B. R., E. Harrison, G. Smith, R. Green, J. Kibler, R. Cess, and the ERBE Science Team (1989), Earth Radiation Budget Experiment (ERBE) archival and April 1985 results, *Bull. Am. Meteorol. Soc.*, *70*, 1254–1262.

Bender, F. A.-M., H. Rodhe, R. J. Charlson, A. M. L. Ekman, and N. Loeb (2006), 22 views of the global albedo—Comparison between 20 GCMs and two satellites, *Tellus, Ser. A*, *58*, 320–330.

Björk, G., P. Söderkvist, P. Winsor, A. Nikolopoulos, and M. Steele (2002), Return of the cold halocline layer to the Amundsen Basin of the Arctic Ocean: Implications for the sea ice mass balance, *Geophys. Res. Lett.*, *29*(11), 1513, doi:10.1029/2001GL014157.

Boville, B. A., P. J. Rasch, J. J. Hack, and J. R. McCaa (2006), Representation of clouds and precipitation processes in the Community Atmosphere Model version 3 (CAM3), *J. Clim.*, *19*, 2184–2198.

Briegleb, B. P., C. M. Bitz, E. C. Hunke, W. H. Lipscomb, M. M. Holland, J. L. Schramm, and R. E. Moritz (2004), Scientific description of the sea ice component in the Community Climate System Model, version three, *Tech. Rep. NCAR/TN-463+STR*, Natl. Cent. for Atmos. Res., Boulder, Colo.

Chen, Y., F. Aires, J. A. Francis, and J. R. Miller (2006), Observed relationships between Arctic longwave cloud forcing and cloud parameters using a neural network, *J. Clim.*, *19*, 4087–4104.

Collins, W. D., et al. (2006), The Community Climate System Model version 3 (CCSM3), *J. Clim.*, *19*, 2122–2143.

Curry, J. A., J. L. Schramm, and E. E. Ebert (1993), Impact of clouds on the surface radiation balance of the Arctic Ocean, *Meteorol. Atmos. Phys.*, *51*, 197–217.

Curry, J. A., J. L. Schramm, D. K. Perovich, and J. O. Pinto (2001), Applications of SHEBA/FIRE data to evaluation of snow/ice albedo parameterizations, *J. Geophys. Res.*, *106*, 15,345–15,355.

Ebert, E. E., and J. A. Curry (1993), An intermediate one-dimensional thermodynamic sea ice model for investigating ice-atmosphere interactions, *J. Geophys. Res.*, *98*, 10,085–10,109.

Francis, J. A., E. Hunter, J. R. Key, and X. Wang (2005), Clues to variability in Arctic minimum sea ice extent, *Geophys. Res. Lett.*, *32*, L21501, doi:10.1029/2005GL024376.

Gloersen, P., W. J. Campbell, D. J. Cavalieri, J. C. Comiso, C. L. Parkinson, and H. J. Zwally (1992), Arctic and Antarctic sea ice, 1978—1987: Satellite passive microwave observations and analysis, *NASA Spec. Publ.*, *SP-511*, 290 pp.

Gorodetskaya, I. V., M. A. Cane, L.-B. Tremblay, and A. Kaplan (2006), The effects of sea ice and land snow concentrations on planetary albedo from the Earth Radiation Budget Experiment, *Atmos. Ocean*, *44*, 195–205.

Gorodetskaya, I. V., L.-B. Tremblay, B. Liepert, M. A. Cane, and R. I. Cullather (2008), The influence of cloud and surface properties on the Arctic Ocean shortwave radiation in coupled models, *J. Clim.*, *21*, 866–882.

Holland, M. M., C. M. Bitz, and B. Tremblay (2006), Future abrupt reductions in the summer Arctic sea ice, *Geophys. Res. Lett.*, *33*, L23503, doi:10.1029/2006GL028024.

Houghton, J. T., Y. Ding, D. J. Griggs, M. Noguer, P. J. vander Linden, X. Dai, K. Maskell, and C. A. Johnson (Eds.) (2001), *Climate Change 2001: The Scientific Basis: Contribution of Working Group I to the Third Assessment Report of the Intergovernmental Panel on Climate Change*, 881 pp., Cambridge Univ. Press, Cambridge, U. K.

Intrieri, J. M., C. W. Fairall, M. D. Shupe, P. O. G. Persson, E. L. Andreas, P. S. Guest, and R. E. Moritz (2002a), An annual cycle of Arctic surface cloud forcing at SHEBA, *J. Geophys. Res.*, *107*(C10), 8039, doi:10.1029/2000JC000439.

Intrieri, J. M., M. D. Shupe, T. Uttal, and B. J. McCarty (2002b), An annual cycle of Arctic cloud characteristics observed by radar and lidar at SHEBA, *J. Geophys. Res.*, *107*(C10), 8030, doi:10.1029/2000JC000423.

Kato, S., N. G. Loeb, P. Minnis, J. A. Francis, T. P. Charlock, D. A. Rutan, E. E. Clothiaux, and S. Sun-Mack (2006), Seasonal and interannual variations of top-of-atmosphere irradiance and cloud cover over polar regions derived from the CERES data set, *Geophys. Res. Lett.*, *33*, L19804, doi:10.1029/2006GL026685.

Key, E. L., P. J. Minnett, and R. A. Jones (2004), Cloud distributions over the coastal Arctic Ocean: Surface-based and satellite observations, *Atmos. Res.*, *72*, 57–88.

Li, Z., and H. G. Leighton (1991), Scene identification and its effect on cloud radiative forcing in the Arctic, *J. Geophys. Res.*, *96*, 9175–9188.

Lindsay, R. W. (2003), Changes in the modeled ice thickness distribution near the Surface Heat Budget of the Arctic Ocean (SHEBA) drifting ice camp, *J. Geophys. Res.*, *108*(C6), 3194, doi:10.1029/2001JC000805.

Martinson, D. G., and M. Steele (2001), Future of the Arctic sea ice cover: Implications of an Antarctic analog, *Geophys. Res. Lett.*, *28*, 307–310.

Overpeck, J., M. Sturm, J. Francis, and D. K. Perovich (2005), Arctic system on trajectory to new, seasonally ice-free state, *Eos Trans. AGU*, *86*(34), 309.

Perovich, D. K., T. C. Grenfell, B. Light, and P. V. Hobbs (2002), Seasonal evolution of the albedo of multiyear Arctic sea ice, *J. Geophys. Res.*, *107*(C10), 8044, doi:10.1029/2000JC000438.

Perovich, D. K., S. V. Nghiem, T. Markus, and A. Schweigher (2007), Seasonal evolution and interannual variability of the local solar energy absorbed by the Arctic sea ice–ocean system, *J. Geophys. Res.*, *112*, C03005, doi:10.1029/2006JC003558.

Persson, P. O. G., C. W. Fairall, E. L. Andreas, P. S. Guest, and D. K. Perovich (2002), Measurements near the Atmospheric Surface Flux Group tower at SHEBA: Near-surface conditions and surface energy budget, *J. Geophys. Res.*, *107*(C10), 8045, doi:10.1029/2000JC000705.

Polyakov, I. V., G. V. Alekseev, L. A. Timokhov, U. S. Bhatt, R. L. Colony, H. L. Simmons, D. Walsh, J. E. Walsh, and V. F. Zakharov (2004), Variability of the intermediate Atlantic water of the Arctic Ocean over the last 100 years, *J. Clim.*, *17*, 4485–4497.

Rasch, P. J., and J. E. Kristijansson (1998), A comparison of the CCM3 model climate using diagnosed and predicted condensate parameterization, *J. Clim.*, *11*, 1587–1614.

Rayner, N. A., D. E. Parker, E. B. Horton, C. K. Folland, L. V. Alexander, D. P. Rowell, E. C. Kent, and A. Kaplan (2003), Global analyses of sea surface temperature, sea ice, and night marine air temperature since the late nineteenth century, *J. Geophys. Res.*, *108*(D14), 4407, doi:10.1029/2002JD002670.

Rudels, B., E. P. Jones, U. Schauer, and P. Eriksson (2004), Atlantic sources of the Arctic Ocean surface and halocline waters, *Polar Res.*, *23*, 181–208.

Schauer, U., E. Fahrbach, S. Osterhus, and G. Rohardt (2004), Arctic warming through the Fram Strait: Oceanic heat transport from 3 years of measurements, *J. Geophys. Res.*, *109*, C06026, doi:10.1029/2003JC001823.

Schweiger, A. J. (2004), Changes in seasonal cloud cover over the Arctic seas from satellite and surface observations, *Geophys. Res. Lett.*, *31*, L12207, doi:10.1029/2004GL020067.

Serreze, M. C., M. M. Holland, and J. Stroeve (2007), Perspectives on the Arctic's shrinking sea-ice cover, *Science*, *315*, 1533–1536.

Shine, K. P., and R. G. Crane (1984), The sensitivity of a one-dimensional thermodynamic sea ice model to changes in cloudiness, *J. Geophys. Res.*, *89*, 10,615–10,622.

Shupe, M. D., and J. M. Intrieri (2004), Cloud radiative forcing of the Arctic surface: The influence of cloud properties, surface albedo, and solar zenith angle, *J. Clim.*, *17*, 616–628.

Stramler, K. L. (2006), The influence of synoptic atmospheric motions on the Arctic energy budget, Ph.D. thesis, Columbia Univ., New York.

Stroeve, J. C., M. C. Serreze, F. Fetterer, T. Arbetter, W. Meier, J. Maslanik, and K. Knowles (2005), Tracking the Arctic's shrinking ice cover: Another extreme September minimum in 2004, *Geophys. Res. Lett.*, *32*, L04501, doi:10.1029/2004GL021810.

Swift, J. H., E. P. Jones, K. Aagaard, E. C. Carmack, M. Hingston, R. W. MacDonald, F. A. McLaughlin, and R. G. Perkin (1997), Waters of the Makarov and Canada basins, *Deep Sea Res., Part I*, *48*, 1503–1529.

Tschudi, M. A., J. A. Curry, and J. A. Maslanik (2001), Airborne observations of summertime surface features and their effect on surface albedo during FIRE/SHEBA, *J. Geophys. Res.*, *106*, 15,335–15,344.

Uttal, T., et al. (2002), The surface heat budget of the Arctic, *Bull. Am. Meteorol. Soc.*, *83*, 255–275.

Walsh, J. E., and W. L. Chapman (1998), Arctic cloud-radiation-temperature associations in observational data and atmospheric reanalyses, *J. Clim.*, *11*, 3030–3045.

Weston, S. T., W. G. Bailey, L. J. B. McArthur, and O. Hertzman (2007), Interannual solar and net radiation trends in the Canadian Arctic, *J. Geophys. Res.*, *112*, D10105, doi:10.1029/2006JD008000.

Westwater, E. R., Y. Han, M. D. Shupe, and S. Y. Matrosov (2001), Analysis of integrated cloud liquid and precipitable water vapor retrievals from microwave radiometers during the Surface Heat Budget of the Arctic Ocean project, *J. Geophys. Res.*, *106*, 32,019–32,030.

Wielicki, B. A., R. D. Cess, M. D. King, D. A. Randall, and E. F. Harrison (1995), Mission to planet Earth: Role of clouds and radiation in climate, *Bull. Am. Meteorol. Soc.*, *76*, 2125–2153.

Winton, M. (2006), Does the Arctic sea ice have a tipping point?, *Geophys. Res. Lett.*, *33*, L23504, doi:10.1029/2006GL028017.

Winton, M. (2008), Sea ice–albedo feedback and nonlinear Arctic climate change, this volume.

Woodgate, R. A., K. Aagaard, and T. J. Weingartner (2006), Interannual changes in the Bering Strait fluxes in volume, heat and freshwater between 1991 and 2004, *Geophys. Res. Lett.*, *33*, L15609, doi:10.1029/2006GL026931.

Yamanouchi, T., and T. P. Charlock (1997), Effects of clouds, ice sheet, and sea ice on the Earth radiation budget in the Antarctic, *J. Geophys. Res.*, *102*, 6953–6970.

Zhang, M., W. Lin, C. S. Bretherton, J. J. Hack, and P. J. Rasch (2003), A modified formulation of fractional stratiform condensation rate in the NCAR community atmospheric model (CAM2), *J. Geophys. Res.*, *108*(D1), 4035, doi:10.1029/2002JD002523.

Zhang, T., K. Stamnes, and S. A. Bowling (1996), Impact of clouds on surface radiative fluxes and snowmelt in the Arctic and subarctic, *J. Clim.*, *9*, 2110–2123.

Zhang, X., and J. E. Walsh (2006), Toward a seasonally ice-covered Arctic Ocean: Scenarios from the IPCC AR4 model simulations, *J. Clim.*, *19*, 1730–1747.

Zuidema, P., et al. (2005), An Arctic springtime mixed-phase cloudy boundary layer observed during SHEBA, *J. Atmos. Sci.*, *62*, 160–176.

I. V. Gorodetskaya, Laboratoire de Glaciologie et Géophysique de l'Environnement, 54 rue Moliere, Domaine Universitaire, B.P. 96, F-38402 Saint-Martin d'Hères CEDEX, France. (Irina.Gorodetskaya@lgge.obs.ujf-grenoble.fr)

L.-B. Tremblay, Department of Atmospheric and Oceanic Sciences, McGill University, Montreal, QC H3A 2K6, Canada.

# Some Aspects of Uncertainty in Predicting Sea Ice Thinning

Cecilia M. Bitz

*Department of Atmospheric Sciences, University of Washington, Seattle, Washington, USA*

A high proportion of the uncertainty in the decline of Arctic sea ice thickness in recent global climate models can be explained by the uncertainty in the ice thickness in the late 20th century. Experiments with one model indicate that this sensitivity to the mean state remains even when ice-albedo feedback is eliminated from the model. The magnitude of ice-albedo feedback is quantified and found to be too small to be a major source of uncertainty in thickness decline in climate models. Instead, it is shown that the sea ice growth-thickness feedback in combination with large biases in the sea ice thickness during the 20th century can easily give rise to very large uncertainty in future thickness decline. Reducing biases in the surface fluxes and better tuning the surface albedo would improve uncertainty in both present and future prediction.

## 1. INTRODUCTION

Large and rapid changes in the Arctic sea ice in the past few decades have attracted attention to future sea ice predictions in global climate models. Models are consulted to see if future changes will continue at the current pace or if they will accelerate or decelerate [e.g., *Holland et al.*, 2006b]. But uncertainty (i.e., spread) in the 21st century sea ice predictions in the models used for the Intergovernmental Panel on Climate Change Fourth Assessment Report is considerable [*Arzel et al.*, 2006; *Zhang and Walsh*, 2006]. Understanding the cause for this uncertainty could help scientists interpret the results of current models, and may help reduce the uncertainty in developing future models.

Arctic Sea Ice Decline: Observations, Projections, Mechanisms, and Implications
Geophysical Monograph Series 180
Copyright 2008 by the American Geophysical Union.
10.1029/180GM06

Some have claimed that a model's success at simulating the observed mean climatology and recent trends is a metric of model reliability for future forecasts [e.g., *Stroeve et al.*, 2007]. Indeed, it has been shown that the mean state of sea ice strongly influences trends in the volume of Arctic sea ice in a given model. *Gregory et al.* [2002] found this to be true in a global climate model when they noted that sea ice volume declined more rapidly in the early 21st century simulation, when the sea ice was thicker and more extensive, and the rate of decline slowed long before the ice disappeared. The rate change could not be explained by a difference in forcing. This result may seem surprising if one expects that ice-albedo feedback increases as sea ice thins and therefore might cause sea ice decay to accelerate in the 21st century (e.g., see *Holland et al.* [2006a] and counter arguments by *Winton* [this volume]).

*Bitz and Roe* [2004] explained *Gregory et al.*'s [2002] results in terms of a strong negative feedback that depends inversely on sea ice thickness. When subject to an increase in downwelling longwave radiation, perennial sea ice melts faster during the melt season, but it also tends to grow faster

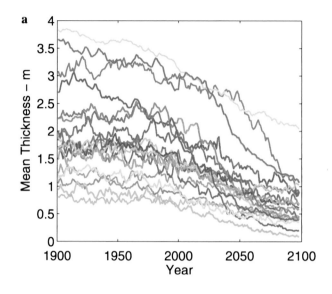

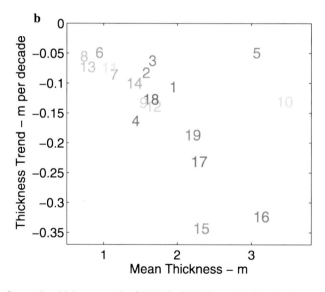

**Plate 1** (a) Time series of mean ice thickness north of 70°N in CMIP3 models for the 20th century and SRES A1B future scenario. (b) Scatterplot of the thickness trend from 2010 to 2050 versus the mean from 1950 to 2000. The numbers corresponds to the order the models are listed in Table 1. The IAB FGOALS model was excluded because it has more than 10 m thick ice in the Arctic in the 20th century, and the IPSL CM4 model was excluded because the run archived at CMIP3 had an erroneous discontinuity in the aerosol forcing (S. Denvil, personal communication, 2008).

where $\bar{\alpha}_i$ is the albedo averaged over all ice thickness categories, $\alpha_o$ is the albedo of open water, and $A_i$ is the fraction of the grid cell covered by sea ice. In the fixed-albedo runs, the grid cell average albedo is fixed to

$$\alpha_{\text{fix}} = [\bar{\alpha}_i][A_i] + [\alpha_o](1 - [A_i]), \qquad (2)$$

where the brackets denote the climatological monthly mean annual cycle that is taken from the normal control run.

Even though the grid cell average albedo is held fixed, the ice fraction may depart from the climatological mean according to the evolution of the climate in the model (particularly when $CO_2$ is raised to 710 ppm). When this occurs, $\bar{\alpha}_i$ or $\alpha_o$ is adjusted to maintain $\alpha = \alpha_{\text{fix}}$, as illustrated in Figure 1.

The sea ice thickness in the model exhibits considerable low-frequency variability. As previously discussed by *Bitz et al.* [1996], the variability is a strong function of the mean

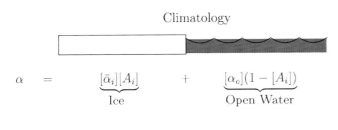

**Figure 1.** Illustration of how the grid cell average albedo is fixed in a run with evolving ice fraction. The shading of the ice and ocean in the illustration is meant to crudely indicate the relative reflectivity. If the ice-covered portion of a grid cell in a run with the albedo fixed should fall below (rise above) the climatological ice fraction, the albedo of the open water (ice covered) portion is increased (decreased) to compensate. For simplicity, the ice-thickness distribution is represented by a single rectangle to indicate the total ice-covered fraction of the grid cell.

thickness, such that thicker ice is far more variable (see Figure 2). Therefore the control runs are at least 150 years long, and all averages were taken for the last 100 years. The perturbed runs, which are thinner, were run for at least 80 years, and averages were taken for the last 50 years (thinner ice also equilibrates faster [see *Bitz and Roe*, 2004]).

The procedure for fixing the albedo succeeded to the extent that the fixed-albedo control reproduces the total surface albedo (sum of all four components) of the normal control when averaged June–August with an average random error of 0.0095 (the standard deviation across grid cells with sea ice) and a systematic error of 0.006 (the average difference across grid cells with sea ice). The average total surface albedo over the same area and months is 0.5. Hence compared to the total, the random error is less than 2%, and the systematic error is about 1.2%. When comparing the total surface albedo in the fixed-albedo perturbed run to the fixed-albedo control, the errors are even lower, with a random error of 0.2% and a systematic error of 0.7%.

The sea ice thickness in the fixed-albedo control has nearly the same pattern of thickness in the normal control, but it tends to be 10–30 cm thicker in the Arctic Ocean (see Figure 2, with an across–grid cell average random error of 33 cm and systematic error of 19 cm in 100-year averages from each control. This thickness difference is almost 10% of the model's mean ice thickness in the Arctic Ocean. Because the percentage of systematic error in the surface albedo is so much smaller, it would appear that variations in the surface albedo have a nonlinear influence on sea ice thickness.

The goal of this study is to investigate the influence of the mean thickness on the 21st century sea ice retreat. The difference in the mean state of the fixed-albedo control and the normal control is therefore an issue that must be considered. In the next section, changes from doubling $CO_2$ are computed with respect to the control that corresponds to the experiment: Normal perturbed is compared to normal control and fixed-albedo perturbed is compared to fixed-albedo control.

### 3.3. Ice-Albedo Feedback Quantified

The annual mean ice thickness and March and September extents from the normal control are shown in Plate 2a. As in many climate models [see, e.g., *Holland and Bitz* 2003], the thickness pattern is biased. The thickest ice appears in the Chukchi Sea and in the center of the Arctic Ocean rather than next to the Canadian Archipelago. These biases result primarily from errors in the surface circulation in the model [*Bitz et al.*, 2001]. The thickness biases in CCSM3 are significantly reduced in integrations at T85 [*DeWeaver and Bitz*, 2006]. Thus the ice thickness patterns in CCSM3 runs

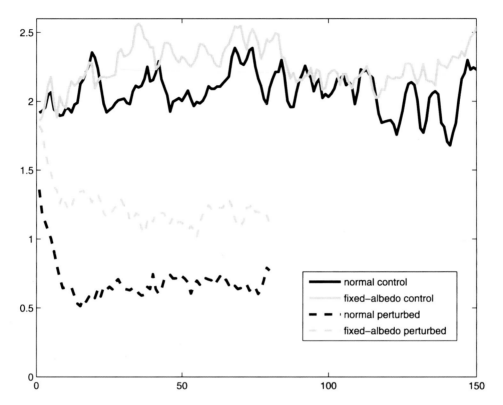

**Figure 2.** Time series of mean ice thickness north of 70°N in CCSM3 experiments (see Table 2).

in CMIP3, which used T85 resolution, are much better. In addition, the transient forcing during the 20th century leads to a somewhat thinner Arctic by the end of the 20th century compared to the fixed forcing 1990s control. In spite of the biases in Plate 2a, the experiments are, nonetheless, useful for evaluating general relationships among ice-albedo feedback, the mean state, and the response to anthropogenic forcing.

The change in sea ice thickness that results from doubling $CO_2$ with freely varying albedo is shown in Plate 2b, and the change with fixed albedo is shown in Plate 2c. It is clear that the pattern of thickness change is a strong function of the control thickness. As in the across-model analysis with the CMIP3 models, the thickness changes most where ice is thickest in the control. Although the overall magnitude of change is less without ice-albedo feedback, the functional dependence on the control thickness appears broadly the same.

The influence of ice-albedo feedback can be made more explicit by dividing the two thickness change maps in Plates 2b and 2c, as shown in Plate 3a. In the parlance of feedback analysis from electrical engineering, this quantity is called the "gain":

$$G = \Delta h / \Delta h_0, \qquad (3)$$

where $\Delta h$ is the thickness change in the normal perturbed case and $\Delta h_o$ is the thickness change from the same perturbation but in the absence of some feedback (or feedbacks), which here is the ice-albedo feedback. $\Delta h_o$ can also be thought of as the thickness change of a "reference system" [*Roe and Baker*, 2007], which comprises all the feedbacks in the system except the feedback that gives rise to the gain.

The gain from ice-albedo feedback ranges from about 1.1 to 1.5 in most of the Arctic, with an average of 1.26 north of 70°N. The gain tends to be larger near the location of the ice edge in summer, at the interface of the perennial and seasonal ice. The gain also appears somewhat noisy in spite of the long time periods that were used to compute the means.

A feedback factor $f$ can also be defined such that

$$\Delta h = \frac{\Delta h_o}{1 - f}, \qquad (4)$$

where $f$ is related to $G$ by $f = 1 - G^{-1}$. Plate 3b shows $f$ for CCSM3. Where $G > 1$, the feedback factor is positive,

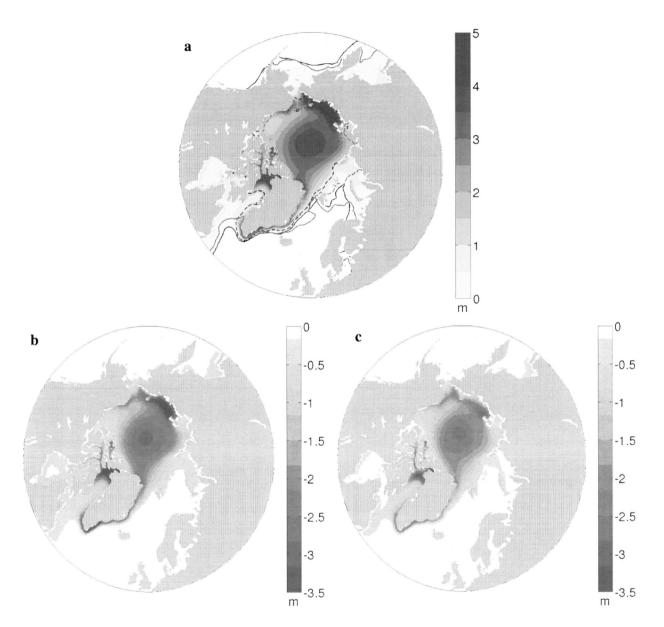

**Plate 2** (a) Annual mean sea ice thickness in the normal control with contour of 15% ice concentration for March (solid) and September (dashed) for the control (black) and 1979–2000 passive microwave (green) [*Comiso*, 1995]. (b) Annual mean thickness difference between perturbed and control. (c) Annual mean thickness difference between fixed-albedo perturbed and control.

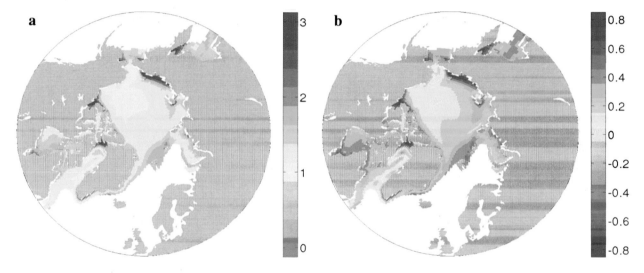

**Plate 3** (a) Gain to ice thickness from ice-albedo feedback, computed from the ratio of $\Delta h$ (Plate 2b) to $\Delta h_o$ (Plate 2c). (b) Feedback factor on ice thickness from ice albedo feedback.

which is the case for all but a few tiny areas. In most of the Arctic, $f$ varies between about 0.1 and 0.3, with an average of 0.21 north of 70°N. The error in my estimate of $f$ that results from variability in the mean thickness north of 70°N is 0.02. The closer $f$ is to one, the closer the system is to experiencing a runaway feedback. Although $f$ indicates ice-albedo feedback is positive, it is not very big.

## 4. INFLUENCE OF ICE-ALBEDO FEEDBACK ON UNCERTAINTY IN FUTURE THICKNESS

In a recent landmark paper, *Roe and Baker* [2007] showed that much of the uncertainty in climate model predictions of future global mean warming could be estimated analytically from the uncertainty in climate feedbacks. I shall borrow heavily from their work to show that, in contrast, uncertainty in ice-albedo feedback has very little influence on the uncertainty of 21st century ice thickness trends among CMIP3 models.

In the equation of climate sensitivity for global temperature change ($\Delta T$):

$$\Delta T = \frac{\Delta T_o}{1 - f}, \tag{5}$$

$\Delta T_o$ derives from assuming a blackbody planet, in which blackbody radiation emitted by the planet stabilizes the climate by cooling (or warming) the planet when the planet exceeds (or falls below) its equilibrium temperature and $f$ is the feedback factor (this time for $\Delta T$ not $\Delta h$). Thus the most basic negative feedback process for stabilizing temperature is normally excluded from $f$, and, instead, its influence is contained in the reference climate sensitivity $\Delta T_o$. For the blackbody planet assumption, $\Delta T_o$ can be estimated from the first term in a Taylor's series expansion:

$$\Delta T_o = \left[\frac{\partial}{\partial T}\sigma T^4\right]^{-1}_{T_E} \Delta R, \tag{6}$$

where $\sigma T^4$ is the Steffan-Boltzmann law, $T_E$ is the effective radiative temperature of the planet, and $\Delta R \approx 3.2$ W m$^{-2}$ is an estimate of the change in the top of atmosphere outgoing longwave radiation when $CO_2$ is doubled. Because $\sigma T^4$ is well approximated by a tangent line, the first term in the Taylor's series is a good approximation. In other words, the temperature dependence of $\Delta T_o$ is easily neglected, and equation (6) gives $\Delta T_o = 1.2°$C, which is fairly accurate for a wide range of temperatures.

*Roe and Baker* [2007] note that $f$ in equation (5) is on average about 0.65 for recent climate models. Hence the feedback factor that affects global temperature change is about 3 times larger than the one I estimated for the influence of ice-albedo feedback on sea ice thickness.

A number of studies have attempted to estimate what portion of $f$ in equation (5) results from the influence of ice-albedo feedback on global mean temperature, not to be confused with the influence of ice-albedo feedback on sea ice thickness. *Bony et al.* [2006] summarize estimates of ice-albedo feedback in three studies and find a mean ice-albedo feedback factor of about 0.12 with a range of about –0.03 to 0.4. (Note that I have used a different definition for feedback than Bony et al., so I have had to convert their estimates to match my definition.) For reference from CCSM3, I find the feedback factor on global mean temperature for all feedbacks is 0.57, and for ice-albedo feedback alone it is 0.24. (There is no reason to expect that ice-albedo feedback should have the same feedback factor for global mean temperature as it does for ice thickness.)

*Roe and Baker* [2007] also pointed out that it is possible to compute the uncertainty in $\Delta T$ that results from uncertainty in $f$. Specifically, the uncertainty can be related to the probability density function $D_T(\Delta T)$ that the global temperature change is $\Delta T$. And the distribution in $\Delta T$ can be related to a distribution in $f$ by

$$D_T(\Delta T) = D_f(f)\frac{df}{d\Delta T}. \tag{7}$$

*Roe and Baker* [2007] assumed a normal distribution for $f$, with mean $\bar{f}$ and variance $\sigma_f^2$,

$$D_f(f) = \frac{1}{\sqrt{2\pi}\sigma_f}\exp\left[-\frac{(f-\bar{f})^2}{2\sigma_f^2}\right], \tag{8}$$

and then computed the resulting distribution for $\Delta T$,

$$D_T(\Delta T) = \frac{1}{\sqrt{2\pi}\sigma_f}\frac{\Delta T_o}{\Delta T^2}\exp\left[-\frac{(1-\bar{f}-\Delta T_o/\Delta T)^2}{2\sigma_f^2}\right]. \tag{9}$$

With the same basic relation between ice thickness and feedback as with global mean temperature and feedback, by analogy the distribution for $\Delta h$ is

$$D_h(\Delta h) = \frac{1}{\sqrt{2\pi}\sigma_h}\frac{\Delta h_o}{\Delta h^2}\exp\left[-\frac{(1-\bar{f}-\Delta h_o/\Delta h)^2}{2\sigma_f^2}\right]. \tag{10}$$

Here, I use this equation to represent the distribution of the thickness change averaged north of 70°N owing to ice-albedo feedback, so all variables in equation (10) are considered averaged north of 70°N as well.

Figure 3 shows examples of distributions from equations (9) and (10) that arise from doubling $CO_2$. The parameters used for the distribution of $\Delta T$ ($\Delta T_o = 1.2^\circ C$, $\bar{f} = 0.65$, and $\sigma_f = 0.13$) are derived from recent climate models as discussed by *Roe and Baker* [2007]. The parameters used for the distributions of $\Delta h$ are $\bar{f} = 0.21$, estimated from CCSM3 (see section 3.3); $\Delta h_o = -1$ m, chosen to give a peak in the distributions at a little over 1 m; and $\sigma_f = 0.1$ for the narrowest and $\sigma_f = 0.21$ for the slightly broader distribution.

I do not know the correct values for $\bar{f}$ and $\sigma_f$ that represent the mean and uncertainty of the influence of ice-albedo feedback on ice thickness from current models. My estimate of uncertainty in $f$ from just one model is bound to be much smaller than the range of $f$ across models. Presumably, the main factors that give rise to different values of $f$ across models are difference between open ocean and sea ice albedos and how the ocean-ice heat flux is partitioned between lateral and basal melt. However, Figure 3 shows that even a very large $\sigma_f$ gives a narrow distribution for $\Delta h$. With an uncertainty of 100% of $\bar{f}$ ($\sigma_f = 0.21$), ice-albedo feedback still only has a rather modest influence on uncertainty in $\Delta h$ because $f$ is so small. In contrast, the feedbacks that influence global mean temperature give an $f$ that is more than 3 times larger. Even a small uncertainty in $f$ is important as $f$ approaches 1 because $1 - f$ appears in the denominator of equations (4) and (5). Thus a more important issue is whether I have underestimated $\bar{f}$. I have let $\bar{f}$ ($\sigma_f = 0.21$), which is the feedback factor I computed for CCSM3 in the previous section. This is unlikely to be an underestimate

of the true $\bar{f}$ for the CMIP3 models because CCSM3 has among the highest Arctic climate sensitivity of any CMIP3 model [*Bitz et al.*, 2008].

## 5. INFLUENCE OF THE MEAN STATE ON UNCERTAINTY IN FUTURE THICKNESS

Because the present-day thickness north of 70°N in the CMIP3 model differs by more than a factor of 3 (see Figure 1), an estimate of the uncertainty caused by errors in the mean state is in order. Sea ice is stabilized primarily by the inverse relation between net sea ice growth and thickness, which *Bitz and Roe* [2004] called the growth-thickness feedback process. On an annual mean basis and provided the climate conditions are not too anomalous, sea ice experiences net melt (growth) when the ice exceeds (falls below) its equilibrium thickness. This leads to an adjustment process, which was described by *Untersteiner* [1961, 1964], that yields an equilibrium thickness. The growth adjustment can be considered analogous to the blackbody-radiative adjustment process that causes the planet to reach an equilibrium temperature.

When the climate is perturbed, such as by increasing $CO_2$, this adjustment process acts to damp the response somewhat. However, for sea ice, the damping is a strong function of thickness itself. In other words, $\Delta h_o$ is a strong function of $h$, while as explained above $\Delta T_o$ is nearly a constant. *Bitz and Roe* [2004] calculated the dependence of $\Delta h_o$ on $h$ for an idealized coupled atmosphere and ice slab without ice-albedo feedback using the formulation given by *Thorndike* [1992]:

$$\widehat{\Delta h_o} = -\frac{(kn_w + Bh)^2}{Bn_w k(-A/n_w + D/2)}\left[-\frac{1}{n_w} - \frac{1}{n_s} + \frac{hB/n_w}{kn_w + Bh}\right]\Delta A,$$
$$(11)$$

where the parameters and variables are defined in Table 3. The hat over $\Delta h_o$ is added to emphasize that this idealized model lacks many processes that I had lumped into $\Delta h_o$ above. The term in brackets in equation (11) has a fairly weak thickness dependence, so its $h$ can be replaced with a constant $\bar{h}$, and the leading dependence on $h$ is parabolic:

$$\widehat{\Delta h_o} \approx -(q + rh)^2, \quad (12)$$

where $q$ and $r$ are independent of $h$.

*Bitz and Roe* [2004] considered how ice export might alter equation (12). *Hibler and Hutchings* [2002] (updated by *Hibler et al.* [2006]) argue that export increases with thickness for ice thickness between about 0 and 4 m. *Bitz and Roe* [2004] reasoned that this sensitivity of export on the mean state is likely to enhance the sensitivity of $\Delta h_o$ to $h$ among

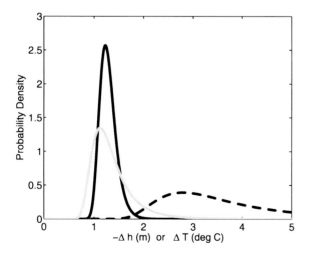

**Figure 3.** Distributions from equations (9) and (10) for $\Delta h$ (with $\Delta h_o = -1$ m, $\bar{f} = 0.21$, and $\sigma_f = 0.1$ (black line) and $\sigma f = 0.21$ (grey line)) and $\Delta T$ (dashed line, with $\Delta T_o = -1.2^\circ C$, $\bar{f} = 0.65$, and $\sigma_f = 0.13$). The distributions for $\Delta h$ are much narrower because $\bar{f}$ is much smaller for $h$ than for $T$.

**Table 3.** Definitions for the Idealized Analytic Model

| Variable | Description | Unit of Measure |
|---|---|---|
| $h$ | annual mean ice thickness | variable |
| $\widehat{\Delta h_o}$ | thickness change for idealized model | variable |
| $\Delta A$ | radiative forcing | 22.6 W m$^{-2}$ |
| $A$ | $\sigma T_f^4$ with $T_f = 273$ K | 320 W m$^{-2}$ |
| $B$ | $4\sigma T_f^3$ | 4.6 W m$^{-2}$ |
| $k$ | thermal conductivity | 2 W m$^{-1}$ K$^{-1}$ |
| $n_{w,s}$ | optical depth for winter or summer | 2.5 or 3.25 |
| $D$ | atmospheric heat transport | 100 W m$^{-2}$ |

models with ice in motion, which is the case for nearly all CMIP3 models. Neglecting the influence of the mean state on export here gives a conservative estimate for the the uncertainty in thickness change because of uncertainty in the mean state.

Given a distribution for $h$, $D_h(h)$, the distribution of $\widehat{\Delta h_o}$ is

$$D_{h_o}(\widehat{\Delta h_o}) = D_h(h)\frac{dh}{d\widehat{\Delta h_o}}. \qquad (13)$$

If $h$ is assumed to be normally distributed with mean $\bar{h}$ and variance $\sigma_h^2$, then using equation (12) and equation (8) with $f$ replaced by $h$ gives

$$D_{h_o}(\widehat{\Delta h_o}) \approx \frac{1}{2r\sigma_h\sqrt{-2\pi\widehat{\Delta h_o}}}\exp\left[-\frac{\left[\left(\sqrt{-\widehat{\Delta h_o}}-q\right)\bigg/r-\bar{h}\right]^2}{2\sigma_h^2}\right].$$

$$(14)$$

Now if $\widehat{\Delta h_o}$ is the only source of uncertainty in $\Delta h$, then

$$D_h(\Delta h) = D_{h_o}(\widehat{\Delta h_o})\frac{d\widehat{\Delta h_o}}{d\Delta h} = D_{h_o}(\Delta h(1-f))(1-f). \qquad (15)$$

Figure 4 shows examples of distributions from equations (14) and (15) that arise from doubling $CO_2$ but without any uncertainty in $f$ (hence $\sigma_f = 0$). Again, I use these equations to represent the distribution of thickness change averaged north of 70°N, so I have taken averages north of 70°N that give $\bar{h} = 1.8$ m and $\sigma_h = 0.77$ m from the CMIP3 models for 1950–2000 (see Figure 1) and $f = 0.21$ from CCSM3 (see section 3.3). The distribution for $\Delta h$ is influenced by ice-albedo feedback such that it is broader and the thickness change is larger than for $\Delta h_o$.

It is possible to compute the distribution of $\Delta h$ with uncertainty in both $\Delta h_o$ (via $h$) and $f$ by computing the ratio of distributions. The result is the Mellin convolution

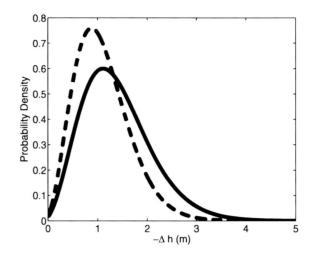

**Figure 4.** Distributions from equations (14) and (15) for $\Delta h_o$ (dashed line) and $\Delta h$ (solid line) with $\bar{h} = 1.8$ m, $\sigma_h = 0.77$ m, $f = 0.21$, and $\sigma_f = 0$.

$$D_h(\Delta h) = \int_{-\infty}^{1} D_f(f)D_{h_o}(\Delta h(1-f))(1-f)df \qquad (16)$$

[see, e.g., *Springer*, 1979]. A numerical solution to this integral is shown in Figure 5 with the same parameters as in the previous paragraph except $\sigma_f = 0.21$. Uncertainty in $f$ increases slightly the probability of greater thickness change at the expense of decreasing the probability of the peak.

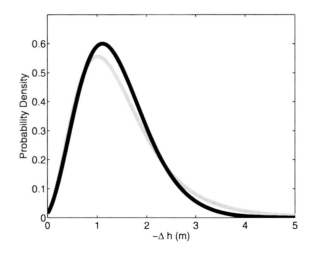

**Figure 5.** Distributions from equation (16) for $\Delta h$ with $\sigma_f = 0.21$ (grey line) and with $\sigma_f = 0$ (black line, which is identical to the black line in Figure 4). Both lines have $\bar{h} = 1.8$ m, $\sigma_h = 0.77$ m, and $f = 0.21$

## 6. DISCUSSION

I have estimated the uncertainty from two primary thermodynamic feedbacks: ice-albedo feedback and the growth-thickness feedback. No doubt there are also feedbacks between the ice and ocean that vary from model to model, and these feedbacks may also depend on the mean state. Because I have not accounted for them, I have focused on ice thickness north of 70°N, where I expect far less influence from the ocean than in the subpolar seas. I have also not tried to quantify the uncertainty in how models treat sea ice dynamics and ice export. My estimates of the distribution widths should be thought of as a lower limit.

For simplicity (and by analogy to the work of *Roe and Baker* [2007]), the distributions here are meant to represent the climate in equilibrium after doubling $CO_2$. In the future scenario shown in Figure 1, the trends of sea ice thinning in the early 21st century span almost an order of magnitude. I have assumed that the uncertainty in equilibrium ice thickness change from doubling $CO_2$ would be similar.

Earlier I noted that the CMIP3 models with thicker ice in the late 20th century thin at a faster rate in the early 21st century (see Plate 1a). As a result the uncertainty in mean ice thickness among CMIP3 models tends to decline in time over the two centuries. Figure 6 recasts estimates of the probability density functions from the section 5 in terms of thickness

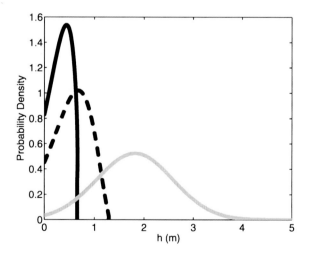

**Figure 6.** Distributions recast as a function of ice thickness rather than change in ice thickness, showing that the distribution narrows as the ice thins. The grey line is the initial assumption of a normal distribution of ice thickness with $\bar{h} = 1.8$ m, $\sigma_h = 0.77$ m. The other two lines are distributions after doubling $CO_2$ without ice-albedo feedback ($f = 0$, dashed line) and with ice-albedo feedback ($f = 0.21$ and $\sigma_f = 0$, black line).

rather than thickness change to illustrate this narrowing of uncertainty in time. The distribution in thickness becomes more sharply peaked after doubling $CO_2$, especially when the gain from ice-albedo feedback is included.

In the experiments described here, $CO_2$ was increased from 355 to 710 ppm, so the albedo effect is evaluated for a perturbation that transforms most of the perennial ice to seasonal ice in the Arctic Ocean in CCSM3. This forcing is roughly equivalent to the total anthropogenic forcing in the first half of the 21st century of the SRES A1B scenario. Experiments were run with $CO_2$ raised to just 550 ppm as well (not shown), which gave nearly the same estimate for the ice albedo feedback factor on ice thickness $f$ in CCSM3. Because $f$ depends little on the magnitude of the perturbation, I expect $f$ would not vary much during a transient integration either.

## 7. CONCLUSIONS

The average sea ice thickness north of 70°N in CMIP3 models ranges from less than 1 m to more than 3 m in the late 20th century. The rate of sea ice thinning in the 21st century in these models is a strong function of the late 20th century thickness, such that models with above-average thickness also thin faster than average. The average ice thickness north of 70°N across the CMIP3 models is highly correlated with the September ice extent and therefore strongly influences marine ecosystems and early winter surface temperatures. Because sea ice thickness change depends sensitively on the mean state, error in a model's climatology gives rise to error in future predictions of ice thinning and extent.

I have shown that uncertainty in the strength of ice albedo feedback is probably not a major source of uncertainty for ice thinning in future predictions. This result stems from the fact that the ice-albedo feedback factor on ice thickness $f$ is rather small. I estimated $f$ in a global climate model by holding the surface albedo of sea ice and ocean fixed while doubling $CO_2$. Ice-albedo feedback causes sea ice to thin about 26% more compared to a model run without ice-albedo feedback. A gain of 26% corresponds to a feedback factor of only $f = 0.21 \pm 0.02$, where the error here is an estimate of uncertainty in this one model (which is bound to be much smaller than the range of $f$ across models).

Such a small value for $f$ can only give rise to a fairly narrow estimate for thickness change provided the range of $f$ across models is the sole source of uncertainty. Even if the uncertainty of $f$ across models is as high as 100% (ranging from 0 to 0.42), it causes little uncertainty in the ice thickness change.

Instead, the uncertainty in the mean state has a much larger influence on uncertainty in the thickness change. I have argued that the principal cause is the growth-thickness feedback, which is regulated by the conduction of heat through the ice. This feedback controls the adjustment to equilibrium and is strongly thickness-dependent. Heat conduction depends roughly on the inverse of thickness, or $1/h$. When surface fluxes are perturbed, the ice thickness adjusts until the conduction of heat through the ice achieves surface energy balance. The thickness need not adjust very much for thin ice owing to the $1/h$ dependence. Consequently, when the thickness is biased, the thickness change in response to a perturbation is also biased. A bias of $\pm 0.77$m, as from the CMIP3 models, gives rise to an uncertainty of more than $\pm 1$m for the thickness change because of doubling $CO_2$.

I have not explained why the models have so much spread in the mean state. Another paper in this monograph argues that a large portion of the error can be explained by the summertime atmospheric energy fluxes and the surface albedo in particular [DeWeaver et al., this volume]. If this is the case, then modelers should do a better job reducing biases in the atmosphere and tuning the surface albedo to reduce the spread in model uncertainty for present and future prediction.

Acknowledgments. I am grateful for support from the National Science Foundation through grants ATM0304662 and OPP0454843. I thank Gerard Roe and Eric DeWeaver for helpful conversations and two anonymous reviewers whose comments greatly improved this paper. I thank the modeling groups, the Program for Climate Model Diagnosis and Intercomparison (PCMDI) and the WCRP's Working Group on Coupled Modeling (WGCM), for their roles in making available the WCRP CMIP3 multimodel data set. Support of this data set is provided by the Office of Science, U.S. Department of Energy.

## REFERENCES

Arzel, O., T. Fichefet, and H. Goosse (2006), Sea ice evolution over the 20th and 21st centuries as simulated by current AOGCMs, Ocean. Modell., 12, 401–415.

Bitz, C. M., and W. H. Lipscomb (1999), An energy-conserving thermodynamic model of sea ice, J. Geophys. Res., 104, 15,669–15,677.

Bitz, C. M., and G. H. Roe (2004), A mechanism for the high rate of sea ice thinning in the Arctic ocean, J. Clim., 17, 3623–3632.

Bitz, C. M., D. S. Battisti, R. E. Moritz, and J. A. Beesley (1996), Low-frequency variability in the Arctic atmosphere, sea ice, and upper-ocean system, J. Clim., 9, 394–408.

Bitz, C. M., M. M. Holland, A. J. Weaver, and M. Eby (2001), Simulating the ice-thickness distribution in a coupled climate model, J. Geophys. Res., 106, 2441–2463.

Bitz, C. M., J. K. Ridley, M. M. Holland, and H. Cattle (2008), Global climate models and 20th and 21st century Arctic climate change, in Arctic Climate Change—The ACSYS Decade and Beyond, edited by P. Lemke, Springer, Heidelberg, Germany, in press.

Bony, S., et al. (2006), How well do we undersand and evaluate climate change feedback processes?, J. Clim., 19, 3445–3482, doi:10.1175/JCLI3819.1.

Collins, W. D., et al. (2006), The Community Climate System Model Version 3 (CCSM3), J. Clim., 19, 2122–2143.

Comiso, J. C. (1995), SSM/I sea ice concentrations using the bootstrap algorithm, NASA Tech. Rep., RP 1380, 40 pp.

DeWeaver, E., and C. M. Bitz (2006), Atmospheric circulation and Arctic sea ice in CCSM3 at medium and high resolution, J. Clim., 19, 2415–2436.

DeWeaver, E., E. C. Hunke, and M. M. Holland (2008), Sensitivity of Arctic sea ice thickness to intermodel variations in the surface energy budget simulation, this volume.

Gregory, J. M., P. A. Stott, D. J. Cresswell, N. A. Rayner, C. Gordon, and D. M. H. Sexton (2002), Recent and future changes in Arctic sea ice simulated by the HadCM3 AOGCM, Geophys. Res. Lett., 29(24), 2175, doi:10.1029/2001GL014575.

Hibler, W. D., and J. K. Hutchings (2002), Multiple equilibrium Arctic ice cover states induced by ice mechanics, paper presented at Ice in the Environment: Proceedings of the 16th IAHR International Symposium on Ice, Int. Assoc. of Hydraul. Eng. and Res., Dunedin, New Zealand.

Hibler, W. D., J. K. Hutchings, and C. F. Ip (2006), Sea ice arching and multiple flow states of Arctic pack ice, Ann. Glaciol., 44, 339–344.

Holland, M. M., and C. M. Bitz (2003), Polar amplification of climate change in coupled models, Clim. Dyn., 21, 221–232.

Holland, M. M., C. Bitz, and A. Weaver (2001), The influence of sea ice physics on simulations of climate change, J. Geophys. Res., 106, 2441–2464.

Holland, M. M., C. M. Bitz, E. C. Hunke, W. H. Lipscomb, and J. L. Schramm (2006a), Influence of the sea ice thickness distribution on polar climate in CCSM3, J. Clim., 19, 2398–2414.

Holland, M. M., C. M. Bitz, and B. Tremblay (2006b), Future abrupt reductions in the summer Arctic sea ice, Geophys. Res. Lett., 33, L23503, doi:10.1029/2006GL028024.

Hunke, E. C., and J. K. Dukowicz (2002), The elastic-viscous-plastic sea ice dynamics model in general orthogonal curvilinear coordinates on a sphere—Incorporation of metric terms, Mon. Weather Rev., 130, 1848–1865.

Lipscomb, W. H. (2001), Remapping the thickness distribution in sea ice models, J. Geophys. Res., 106, 13,989–14,000.

Meehl, G. A., W. M. Washington, B. D. Santer, W. D. Collins, J. M. Arblaster, A. Hu, D. M. Lawrence, H. Teng, L. E. Buja, and W. G. Strand (2006), Climate change projections for the twenty-first century and climate change commitment in the CCSM3, J. Clim., 19, 2597–2616.

Roe, G. H., and M. B. Baker (2007), Why is climate sensitivity so unpredictable?, Science, 218, 629–632, doi:10.1126/science. 1144735.

Springer, M. D. (1979), The Algebra of Random Variables, 470 pp., John Wiley, New York.

Stroeve, J., M. M. Holland, W. Meier, T. Scambos, and M. Serreze (2007), Arctic sea ice decline: Faster than forecast, Geophys. Res. Lett., 34, L09501, doi:10.1029/2007GL029703.

Thorndike, A. S. (1992), A toy model linking atmospheric thermal radiation and sea ice growth, *J. Geophys. Res.*, *97*, 9401–9410.

Untersteiner, N. (1961), On the mass and heat budget of Arctic sea ice, *Arch. Meteorol. Geophys. Bioklimatol., Ser. A*, *12*, 151–182.

Untersteiner, N. (1964), Calculations of temperature regime and heat budget of sea ice in the central Arctic, *J. Geophys. Res.*, *69*, 4755–4766.

Winton, M. (2008), Sea ice–albedo feedback and nonlinear Arctic climate change, this volume.

Zhang, X., and J. E. Walsh (2006), Toward a seasonally ice-covered Arctic Ocean: Scenarios from the IPCC AR4 model simulations, *J. Clim.*, *19*, 1730–1747.

_____

C. M. Bitz, Department of Atmospheric Sciences, University of Washington, MS 351640, Seattle, WA 98195-1640, USA. (bitz@atmos.washington.edu)

# Sensitivity of Arctic Sea Ice Thickness to Intermodel Variations in the Surface Energy Budget

Eric T. DeWeaver

*Center for Climatic Research, University of Wisconsin-Madison, Madison, Wisconsin, USA*

Elizabeth C. Hunke

*Los Alamos National Laboratory, Los Alamos, New Mexico, USA*

Marika M. Holland

*National Center for Atmospheric Research, Boulder, Colorado, USA*

Sea ice simulations from an ensemble of climate models show large differences in the mean thickness of perennial Arctic sea ice. To understand the large thickness spread, we assess the sensitivity of thickness to the ensemble spread of the surface energy budget. Intermodel thickness and energy flux variations are related through a diagnostic calculation of thickness from surface temperature and energy fluxes. The calculation shows that an ensemble range of 60 W m$^{-2}$ in energy fluxes, as simulated by climate models, results in an approximate range of 1–5 m in ice thickness. The ensemble mean value of the melt season energy flux, together with a budget residual term that represents the effects of ocean heat exchange and ice divergence, are the key factors that determine the range in ice thickness owing to the flux spread. The ensemble spread in summertime energy flux is strongly related to the spread in surface albedo, while differences in longwave radiative forcing, due in part to cloud simulation errors, play a smaller role.

## 1. INTRODUCTION

The recent dramatic decline in September Arctic sea ice extent has led to increased interest in climate model projections of future Arctic sea ice conditions. However, future sea ice projections are subject to substantial uncertainty, as can be seen by comparing results from different climate models.

One metric of this uncertainty is the range of simulated values for the mean thickness of present-day perennial Arctic sea ice: 1 to 6 m (excluding one outlier) over the simulation ensemble of "20th century climate in coupled models" (20C3M) [e.g., *Randall et al.*, 2007]. Improved understanding of the factors determining this range and the requirements for reducing it through model improvement would be of some interest, particularly given the use of climate model projections for Arctic-related policy decisions [e.g., *Amstrup et al.*, this volume].

Simulation assessments prepared for policy makers [e.g., *Kattsov et al.*, 2005; *Randall et al.*, 2007; *DeWeaver*, 2007] typically account for sea ice simulation uncertainty by describing the delicate physical processes which determine ice

*Arctic Sea Ice Decline: Observations, Projections, Mechanisms, and Implications*
Geophysical Monograph Series 180
Copyright 2008 by the American Geophysical Union.
10.1029/180GM07

thickness and concentration, such as cloud radiative forcing, snow cover, melt pond fraction, upper ocean stability, surface wind stress, etc. While consideration of these inputs and their uncertainties is clearly necessary, their impact on thickness depends on the intrinsic sensitivity of thickness to errors in inputs. Furthermore, the intrinsic sensitivity determines the extent to which incremental improvement in the inputs will lead to better thickness simulations.

In this chapter, we examine the intrinsic sensitivity of sea ice thickness to errors in surface energy fluxes using the 20C3M simulation archive and a diagnostic thermodynamic equation. Our goal is to describe the sensitivity in the simplest possible terms, foregoing as much model-specific complexity as possible to achieve an understanding in terms of basic physical principles applicable to all climate models. In this simple analysis, the behavior of thickness sensitivity can be understood as a consequence of the inverse relationship between thickness and summertime net energy gain, so that the thickness error associated with an error in, say, insolation is strongly dependent on the true value of the summertime energy gain. This result is analogous to the global temperature sensitivity to $CO_2$ doubling found by *Roe and Baker* [2007] [see also *Bitz*, this volume], in which the inverse relationship between temperature sensitivity and net climate feedback leads to a long-tailed distribution of temperature sensitivity, or a poorly constrained upper bound on global warming.

Further motivation for this research comes from the issue of model credibility. The credibility of climate model projections of Arctic sea ice decline is often judged by the models' ability to simulate present-day sea ice conditions. While success in present-day simulation is a logical criterion, it is possible that successful simulations have been achieved through nonphysical "tuning" which masks severe model physics deficiencies. In that case, simulation success confers a false confidence in future projections, an undesirable outcome when projections are used as input for policy decisions. As discussed by *Eisenman et al.* [2007] (hereinafter referred to as EUW) [see also *DeWeaver et al.*, 2008] (hereinafter referred to as DHH), nonphysical tuning could be manifested as an incompatibility between the intermodel ranges of thickness and surface energy fluxes. According to EUW's calculations, the ensemble spread of 40 W m$^{-2}$ in downwelling longwave fluxes found in the 20C3M ensemble should produce an ensemble sea ice thickness spread which greatly exceeds the actual spread. They offer this discrepancy as evidence that nonphysical tuning of sea ice albedo has been used to artificially compress the spread of thickness values. In our analysis, the ensemble ranges of thickness and surface energy flux are inherently compatible, without the need for nonphysical tuning. Of course, the simplifying

assumptions made in the diagnostic equation preclude a definitive judgment of state-of-the-art climate models. Yet it is reassuring that the simple calculation can capture the sensitivity of the 20C3M ensemble in an approximate sense.

The remainder of the chapter is divided into four sections. Section 2 lays out the theoretical framework of the diagnostic analysis and shows how energy balance and surface flux continuity can be used to obtain an equation for sea ice thickness, which is used in section 3 to reconcile the discrepancy between flux and thickness spread found by EUW. Section 4 uses the equation diagnostically to relate the spread of thickness in the 20C3M ensemble to the corresponding spread in surface fluxes and shows that the flux-derived thickness spread is consistent with the actual spread. Surface fluxes for the ensemble are compared with available observations, and the spread in the summertime energy balance is found to be strongly related to albedo and surface temperature. Conclusions follow in section 5.

## 2. THEORETICAL FRAMEWORK

The diagnosis presented here and by EUW is based on a variant of *Thorndike*'s [1992] (hereinafter referred to as T92) simple model, shown schematically in Figure 1. The ice is represented as a motionless slab with thickness $h$, surface temperature $T$ (in Celsius degrees), and bottom temperature $0°C$, taken as the approximate freezing temperature of seawater. The year is partitioned into ice growth and ice melt seasons of length $\tau_G$ and $\tau_M$, respectively, with no insolation during the growth season and a constant mean insolation during the melt season (a square-wave seasonal cycle of insolation). $T = 0$ during the melt season and $T$ during ice growth satisfies a heat conduction equation in which the net surface heat loss via longwave radiation is balanced by upward conductive heat flux through the ice.

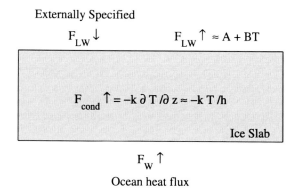

**Figure 1.** Schematic depiction of sea ice and energy fluxes for equations (1), (2), and (3) in sections 2 and 3.

Averaged over the growth season, the heat conduction equation is written as

$$W - BT_G = kT_G/h. \quad (1)$$

Here $h$ and $T_G$ are the growth season mean ice thickness and surface temperature, $k$ is the thermal conductivity of sea ice (2 W m$^{-1}$ C$^{-1}$), and the conductive heat flux $F_{cond}$ is approximated as $-k \partial T_G/\partial z \approx -kT_G/h$, where the seasonal average of $(T/h)$ is approximated by the ratio of the seasonal averages of $T$ and $h$ (variables and values are also given in Table 1).

**Table 1.** Definitions of Variables Used in Text

| Variable | Definition | Constant Value If Any |
|---|---|---|
| $A$ | upwelling longwave radiation at 0°C | 320 W m$^{-2}$ |
| $B$ | derivative of Stefan-Boltzmann equation at 0°C | 4.6 W m$^{-2}$ K$^{-1}$ |
| $F_{LW}, F_{SW}$ and $F_{SL}$ | surface net longwave, shortwave, and turbulent fluxes | |
| $F\downarrow, F\uparrow$ | upwelling and downwelling fluxes | |
| $h$ | mean sea ice thickness | |
| $h_D$ | sea ice thickness calculated from surface fluxes by (4) | |
| $k$ | thermal conductivity of sea ice | 2 W m$^{-1}$ C$^{-1}$ |
| $L$ | latent heat of fusion | $3.34 \times 10^5$ J kg$^{-1}$ |
| $N$ | ice growth season longwave atmospheric optical depth | 3 |
| $R$ | annual mean surface energy budget residual in W m$^{-2}$ | |
| $S$ | net surface heat flux during the melt season | |
| $\hat{S}$ | $S + R(\tau_M + \tau_G)/\tau_M$ | |
| $T_G$ | mean surface temperature for ice growth season | |
| $\vec{V}$ | sea ice velocity | |
| $W$ | net surface heat flux during ice growth season | |
| $\alpha$ | surface albedo | |
| $\gamma$ | $\rho L \nabla \cdot \vec{V}$ in (6) | |
| $\rho$ | sea ice density | 917 kg m$^{-3}$ |
| $\sigma$ | Stefan-Boltzmann constant | $5.67 \times 10^{-8}$ W m$^{-2}$ K$^{-4}$ |
| $\tau_G$ | length of ice growth season | 8 months |
| $\tau_M$ | length of ice melt season | 4 months |

$W - BT_G$ is the net longwave flux $F_{LW}$ derived from an externally specified downwelling flux $F_{LW\downarrow}$ minus the upwelling flux $F_{LW\uparrow}$ given by the Stefan-Boltzmann (SB) equation linearized around 0°C: $F_{LW\uparrow} = A + BT_G$, with $A$ and $B$ the value and derivative of the equation at 0°C (320 W m$^{-2}$ and 4.6 W m$^{-2}$ K$^{-1}$, respectively). $W$ is defined by $W = F_{LW\downarrow} - A$ so that the net flux is $F_{LW} = F_{LW\downarrow} - (A + BT_G) = W - BT_G$.

The heat conduction equation is complemented by an annual mean surface energy budget in which the heat loss during ice growth is balanced by heat gain over the melt season:

$$\tau_G(W - BT_G) + \tau_M S = 0. \quad (2)$$

$S$, the average melt season net surface energy flux, is the sum of $F_{SW}$ and $F_{LW} = F_{LW\downarrow} - A$, since melting is assumed to occur at 0°C (here $F_{LW\downarrow}$ refers to the downwelling longwave flux averaged over the melt season, which is, of course, different from $F_{LW\downarrow}$ used in the definition of $W$, which is averaged over the ice growth season). Also, the heat capacity is ignored so that the ice warms up to 0°C at the start of the sunlit portion of the year.

The implications of (1) and (2) can be understood by plotting contours of constant $S$ and $W$ in the $(h, T_G)$ plane, as shown in Figure 2. For fixed $W$, $F_{cond} \rightarrow 0$ as $T_G \rightarrow W/B$ since the net longwave flux from the ice is zero when $T_G = W/B$. With a nonzero surface temperature $F_{cond} \rightarrow 0$ requires $h \rightarrow \infty$ since a small net surface flux must be matched by a small conductive flux, which requires large $h$. (1) and (2) can be combined to obtain $S = -kT_G/h \cdot (\tau_G/\tau_M)$, so that contours of constant $S$ are straight lines emanating from the origin. The intersection of $S$ and $W$ contours at point $A$ represents a valid climate state in which energy balance and flux continuity are both satisfied.

The dashed curves in Figure 2 show the sensitivity of $h$ and $T_G$ to 20 W m$^{-2}$ changes in $W$ and $S$. It is clear that $h$ is considerably more sensitive to a decrease in $S$ (compare $h$ values at $A$ and $B$) than to an equivalent decrease in $W$ (compare $h$ at $A$ and $C$). The larger sensitivity to $S$ can be understood as a consequence of the fact that $T = 0$ during ice melt, so that the ice cannot compensate for changes in the imposed energy flux by adjusting its summer temperature. Instead, a reduction in $S$ must be compensated by a reduction in the net winter surface flux $W - BT_G$, which must become smaller as $S$ is reduced. As stated above, a small net flux means a small conductive flux which requires large $h$. Holding $W$ fixed while reducing net flux by $\delta S$ means moving to the right along the $W$ curve in Figure 2, on which $h$ increases without bound as the net flux approaches zero.

On the other hand, if $S$ is held fixed while $W$ is reduced, the net winter flux (the left hand side of (1)) must remain

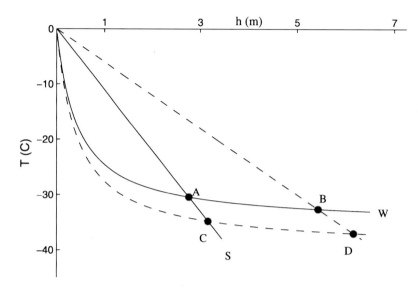

**Figure 2.** Solid curves are isolines of $W$ and $S$ in the $(h, T)$ plane derived from equations (1) and (2) for $W = -162$ W m$^{-2}$ and $S = 44$ W m$^{-2}$ assuming that the ice melt season (MJJA) is half as long as the ice growth season (SONDJFMA). The intersection of the isolines at $A$ represents a valid climate state satisfying energy balance and surface flux continuity. Dashed curve through $C$ is the isoline for $W = -182$ W m$^{-2}$. Dashed line through $B$ is the isoline for $S = 24$ W m$^{-2}$.

the same to satisfy (2). Unlike the $S$ reduction case, wintertime temperature can adjust to the reduction in $W$, and the increase in $h$ is just enough to keep the same conductive flux despite the lower temperature. The required increase in $h$ is proportional to the decrease in $T$, $\delta h / h = \delta T / T$, so no singularity is encountered when $h$ increases to balance a decrease in $W$.

In reality, $F_{LW\downarrow}$ is strongly dependent on $T_G$, as longwave radiation emitted by the surface warms the overlying atmosphere and increases its longwave emission. This dependence is documented by *Key et al.* [1996] and used by *Lindsay* [1998] (hereinafter referred to as L98) to infer $F_{LW\downarrow}$ from $T_G$. The implied dependence of $W$ on $T_G$ means that a reduction in $S$ leads to a lower value of $W$, which leads to a larger increase in $h$ for a decrease in $S$ than would otherwise occur. The dependence of $W$ on $T_G$ is incorporated into the version of (1) and (2) presented by T92 through a simple atmosphere in radiative equilibrium with the underlying surface. In this atmosphere, the surface downwelling longwave flux is $F_{LW\downarrow} = F_E + \left(\frac{N}{N+2}\right) F_{LW\uparrow}$, an externally specified portion $F_E$ plus a contribution due to the upwelling surface flux in which $N$ is the longwave optical depth of the atmosphere. Using this formula the change in $W$ for an imposed change in $S$ is $\delta W = (N / 2)(\tau_M / \tau_G)\delta S$. (This result is obtained by setting $W = F_E + \frac{N}{N+2}(A + BT_G) - A$, from which the sensitivity of $W$ to $T_G$ can be determined. The sensitivity of $T_G$ to $S$ can then

be found using (2) with the above formula for $W$, resulting in a sensitivity of $W$ to $S$. Alternatively, (1) and (2) can be recast in terms of $F_E$ and $n = (N + 2)/N$ to express the wintertime net flux as $F_{LW} = F_E - (A + BT_G)/n \equiv \tilde{W} - BT_G/n$. $\tilde{W} - BT_G/n$ can then be substituted for $W - BT_G$ in (1) and (2), with $\tilde{W} = -78$ W m$^{-2}$ (the value used by T92) when $W = -162$ W m$^{-2}$ as in Figure 2.) Using $N = 3$ as given by T92 (for the winter season) and $\tau_G = 2\tau_M$, since melting typically occurs from May to August, we have $\delta W = 0.75\delta S$. Taking this feedback into account, the $h$ value for a 20 W m$^{-2}$ decrease in $S$ lies between points $B$ and $D$ on the dashed line in Figure 2, an increase which does not qualitatively change the results of the analysis.

## 3. EXPECTED THICKNESS SPREAD DUE TO LONGWAVE FLUX ERRORS

The sensitivity of $h$ to errors in surface energy fluxes can be examined by solving (1) and (2) for $h$ assuming errors of order $\delta F$ in the annual mean surface energy flux:

$$h = -\frac{k}{B} \cdot \left[ 1 + \frac{\tau_G}{\tau_M} \cdot \frac{W(1 + \delta F / W)}{S(1 + \delta F / S)} \right]. \quad (3)$$

In a multimodel ensemble of climate simulations, model errors in energy fluxes contribute to the range $\delta F$ of model-to-model variation in surface fluxes, which, in turn, should produce an intermodel spread of $h$ values approximated by the above formula (for simplicity we consider the impact of annual mean flux errors, although flux errors and thickness sensitivity are expected to have strong seasonality). In an unbiased ensemble in which the ensemble mean values of $W$ and $S$ are close to their real-world counterparts, (3) expresses the expected relationship between flux errors and thickness errors.

EUW use (3) to estimate the ensemble spread in $h$ which should occur as a result of cloud-induced longwave flux errors, which they estimate to be about $\delta F = \pm 20$ W m$^{-2}$. Assuming $\tau_M = \tau_G$ and ensemble mean $S = 22$ W m$^{-2}$ and $W = -162$ W m$^{-2}$ (values taken from T92), this range implies an $h$ range from 1 to 39 m, a full order of magnitude larger than the range of 1 to 4 m which they find in the spread of mean Arctic sea ice thickness over the 20C3M coupled model ensemble. To account for this order of magnitude discrepancy, they propose that modelers must be adjusting $S$ by tuning sea ice albedo. Following T92, they use $F_{SW\downarrow} = 100$ W m$^{-2}$ so that with $F_{SW} = (1-\alpha)F_{SW\downarrow}$ an albedo decrease of 0.1 yields a 10 W m$^{-2}$ increase in $S$. Thus, a mean thickness of 39 m could be corrected to 6 m through a decrease of 0.1 in surface albedo. If modelers apply these albedo adjustments to avoid excessive thickness, the spread in $h$ is compressed through a nonphysical relationship between cloud-induced longwave flux and sea ice albedo. Such spread compression is clearly quite misleading if models are evaluated according to their ability to produce a reasonable mean Arctic sea ice thickness in the present-day climate.

Fortunately, the order-of-magnitude discrepancy can be resolved by considering the different lengths of the ice growth and ice melt seasons. Melting in the 20C3M ensemble, much like in reality, occurs only from May to August so that $\tau_G / \tau_M \approx 2$. To maintain the values of $T_G = -30.8°C$ and $h = 2.8$ m assumed by EUW (following T92) with a growth season that is twice the length of the melt season would require an average melt season surface flux $S = -kT_G / h \cdot (\tau_G/\tau_M)$ of 44 W m$^{-2}$. With $S = 44$ W m$^{-2}$, $W = -162$ W m$^{-2}$, and $\delta F = \pm 20$ W m$^{-2}$, (3) yields a thickness range of 1 to 6 m, in qualitative agreement with the actual ensemble spread. Thus, there is no need to assume a nonphysical relationship between longwave cloud forcing and sea ice albedo to reconcile the ranges of $h$ and $\delta F$, provided that the ensemble mean $S$ value is close to 44 W m$^{-2}$.

For the growth season ($W$) and melt season ($S$) flux values assumed here, much of the ice thickness spread in $h$ implied by (3) comes from the variation of the flux error ($\delta F$) in the denominator, the expected spread in $h$ approaching infinity when $\delta F$ approaches $-S$. This flux error causes a singularity because it reduces the summertime energy gain to zero,

which implies zero wintertime conductive heat loss in (1) resulting in infinite thickness. Thus, the ensemble mean net summer flux $S$ is the critical factor in determining the range of $h$ values corresponding to the ensemble range of flux errors, at least in this simple diagnostic analysis.

## 4. 20C3M SURFACE ENERGY FLUXES AND FLUX-DERIVED THICKNESS

### 4.1. Ensemble Spread of Flux-Derived Thickness

With this insight, we consider the multimodel 20C3M ensemble, applying a modified version of the simple model to relate the spread in $h$ to the ensemble spread in surface fluxes. We calculate a diagnostic thickness $h_D$ for each model using area-averaged fluxes and temperatures for grid points north of 70°N in which 1980–1999 climatological September sea ice fraction for that model is at least 85%. The diagnostic thickness equation is derived from (1) and (2) but includes a complete surface energy budget:

$$h_D = -\frac{kT_G(\tau_G/\tau_M)}{\hat{S}}, \qquad (4)$$

where

$$\hat{S} \equiv (F_{SW} + F_{LW} + F_{SL})_M + R(\tau_G + \tau_M)/\tau_M. \qquad (5)$$

$F_{SL}$ is the sum of latent and sensible heat flux from the surface and $R$ is the annual mean energy budget residual (calculated as minus the annual mean sum of surface fluxes), including ocean heat flux convergence, energy transport by ice motion, and any error in the model budgets (variables given in Table 1). The presumption in (1) is that the net energy flux from the ice is negative (out of the ice), while (2) is an annual mean budget in which $S$ is a mean over the months when (1) does not apply, i.e., when the surface flux is positive. Thus $(\ )_M$ is the average for May, June, July, and August (MJJA), the months of positive surface flux for all but two models (models 1 and 2 in Figure 3a, discussed further in section 4.2; similar results were obtained using a JJA average). If desired, an equation analogous to (3) relating thickness to the ratio of summer and winter energy fluxes can be formed by combining (5) with an energy budget equation like (2) but including all surface fluxes and the budget residual $R$ in (5). However, this equation is unnecessarily complicated, since the spread in (1) is determined primarily by the spread in $\hat{S}$, as shown below.

Figure 3a compares $h_D$ with the coupled model simulated 1980–1999 mean ice thickness over the same grid points. Model output is obtained from the Program for Climate

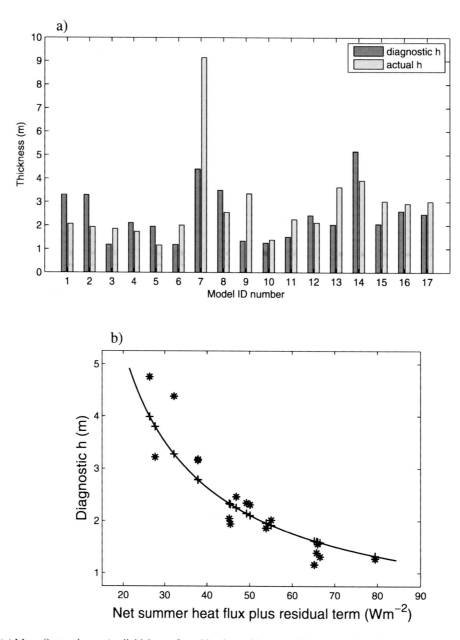

**Figure 3.** (a) Mean September to April thickness for grid points with perennial Arctic sea ice from model output (light shading) and from diagnostic calculation (dark shading), described in section 4.1. Model output is obtained from the Program for Climate Model Diagnosis and Intercomparison, and data are from run 1 of each model. (b) Scatter of $\hat{S}$ (denominator in (4)) and diagnostic thickness $h_D$ from 20C3M model output (asterisks), $h_D$ calculated using the ensemble mean value of the numerator in (4)(plus signs), and $h_D$ as a continuous function of $\hat{S}$ using the ensemble mean value of the numerator in (4) (solid line).

Model Diagnosis and Intercomparison, and data are from run 1 of each model (models are listed in Table 2). For most models, the diagnostic calculation produces a thickness estimate which is loosely consistent with the actual model output, the most obvious exception being model 7. There are, of course, a variety of factors neglected in (4) which could account for discrepancies between $h$ and $h_D$, like the insulating effect of snow cover, the linearization of the SB equation, and the use of the ratio seasonal-mean $h$ and $T$ values instead of the seasonal mean of $(T/h)$ in (1).

**Table 2.** Models Enumerated as in Figures 3a and 5

| | Model | Center | Country |
|---|---|---|---|
| 1, | cccma_cgcm3_1 | Canadian Centre for Climate Modeling and Analysis | Canada |
| 2, | cccma_cgcm3_1_t63 | Canadian Centre for Climate Modeling and Analysis | Canada |
| 3, | cnrm_cm3 | Centre National de Recherches Météorologiques | France |
| 4, | gfdl_cm2_0 | Geophysical Fluid Dynamics Laboratory | United States |
| 5, | gfdl_cm2_1 | Geophysical Fluid Dynamics Laboratory | United States |
| 6, | giss_aom | Goddard Institute for Space Studies | United States |
| 7, | iap_fgoals1_0_g | Institute of Atmospheric Physics | China |
| 8, | inmcm3_0 | Institute for Numerical Mathematics | Russia |
| 9, | ipsl_cm4 | Institute Pierre Simon Laplace | France |
| 10, | miroc3_2_hires | Center for Climate System Research | Japan |
| 11, | miroc3_2_medres | Center for Climate System Research | Japan |
| 12, | miub_echo_g | Meteorological Institute of the University of Bonn | Germany |
| 13, | mpi_echam5 | Max Plank Institute for Meteorology | Germany |
| 14, | mri_cgcm2_3_2a | Meteorology Research Institute | Japan |
| 15, | ncar_ccsm3_0 | National Center for Atmospheric Research | United States |
| 16, | ukmo_hadcm3 | Hadley Center for Climate Prediction and Research | United Kingdom |
| 17, | ukmo_hadgem1 | Hadley Center for Climate Prediction and Research | United Kingdom |

To relate the ensemble spread of $h_D$ to the spread of $\hat{S}$, we scatter $h_D$ against $\hat{S}$ in Figure 3b (asterisks). As expected, there is a strong inverse relationship between $\hat{S}$ and $h_D$. The plus signs show the values of $h_D$ obtained using the ensemble mean value of $-kT_G$ ($\tau_G/\tau_M$) in (4) so that $h_D$ becomes a unique function of $\hat{S}$. According to Figure 3b, the ensemble spread of $h$ expected from the flux variations is between 1 and 5 m (asterisks), of which a spread of 1 to 4 m in thickness can be explained by $\hat{S}$ variations alone (plus signs),

without considering variations in growth season temperature and energy fluxes.

### 4.2. Realism of 20C3M Surface Energy Fluxes

As noted in section 3, the ensemble mean value of $\hat{S}$ is the critical factor in relating the spread in $h_D$ to the spread in surface fluxes. If the ensemble mean $\hat{S}$ is close to its real-world counterpart, then we have some confidence that the ensemble spread in $h_D$ correctly represents the uncertainty in $h_D$ which results from errors in $\hat{S}$.

To assess the accuracy of $\hat{S}$ as simulated by the climate models, Table 3 compares the values of MJJA mean surface fluxes from L98's analysis of Russian ice station data (21 stations over 45 annual cycles), *Persson et al.*'s [2002] analysis of data from the 1997–1998 Surface Heat Budget of the Arctic (SHEBA) field campaign, and the mean of the 17 models in Figure 3. While discrepancies exist in the individual fluxes, the ensemble mean $S$ agrees quite closely in all three data sets, 36 (SHEBA), 35 (L98), and 37 (20C3M ensemble mean) W m$^{-2}$. It should be noted that the L98 albedo is high (0.74) because the value does not include ponds, thin ice, or leads. Their high albedo value may lead to their higher value of $F_{SW\downarrow}$ (234 compared to 214 W m$^{-2}$ for Persson et al.), since albedo is used in the formula from which they calculate $F_{SW\downarrow}$. To account for melt ponds and leads in the formula for $F_{SW\downarrow}$, they reduce the albedo value in the formula by 0.05 for July and August, which may be an insufficient reduction. We note, however, that the net shortwave $F_{SW}$ agrees closely between L98 (59 W m$^{-2}$) and Persson et al. (60 W m$^{-2}$), so

**Table 3.** Comparison of Surface Energy Flux Quantities for MJJA Between Observations From SHEBA [*Persson et al.*, 2002], *Lindsay* [1998], and the 20C3M Ensemble Mean, Plus Maximum and Minimum Quantities From 20C3M[a]

| | $F_{SW\downarrow}$ | $F_{LW\downarrow}$ | $F_{SW}$ | $F_{LW}$ | $F_{RAD}$ | $S$ | Cloud (%) | $\alpha$ (%) |
|---|---|---|---|---|---|---|---|---|
| SHEBA | 214 | 282 | 60 | −20 | 40 | 36 | – | 61 |
| Lindsay | 234 | 284 | 59 | −16 | 43 | 35 | 85 | 74 |
| 20C3M | 202 | 273 | 76 | −28 | 47 | 37 | 83 | 61 |
| Max | 248 | 289 | 106 | −17 | 72 | 61 | 93 | 72 |
| Min | 176 | 250 | 50 | −38 | 12 | 13 | 66 | 46 |

[a] $F_{RAD}$ is the net surface radiation and $S = (F_{RAD} + F_{SL})_M$. Maximum (Max) and minimum (Min) values for $F_{SW}$ and $F_{LW}$ are not constrained to sum to the Max and Min $F_{RAD}$ values, since the extreme values for $F_{SW}$ and $F_{LW}$ will not, in general, come from the same simulation. Albedo from 20C3M is the ratio of monthly mean $F_{SW\uparrow}$ and $F_{SW\downarrow}$. Albedo from SHEBA is from the albedo line measurements. Albedo from Lindsay does not include leads or melt ponds. Fluxes are positive down, i.e., positive when the surface gains energy from above. Values in $F_{LW}$ and $F_{SW}$ columns are net fluxes.

the combination of high albedo and high downwelling short-wave radiation apparently yields a net shortwave estimate in agreement with the SHEBA data.

A more detailed comparison of the 20C3M surface energy fluxes is presented in Figures 4a–4e, which shows the annual cycle of surface radiative fluxes and turbulent (latent and sensible) fluxes from L98 and the 20C3M ensemble. All fluxes have relatively large ensemble spread, and down-welling longwave and shortwave fluxes are underestimated in the ensemble mean compared to L98. L98 computes downwelling shortwave radiation using a formula which accounts for the dependence of $F_{SW\downarrow}$ on surface albedo, as higher albedo leads to increased multiple scattering and hence increased downwelling shortwave radiation [*Shine*, 1984; see also DHH]. As noted above, the L98 $F_{SW\downarrow}$ values may be overestimates because of the treatment of melt ponds, and are high compared to Persson et al.'s SHEBA-derived values.

For net radiation ($F_{SW} + F_{LW}$, Figure 4c), the ensemble mean cold season (SONDJFMA) flux is too low by 6 to 9 W m$^{-2}$. The ensemble mean overestimates the net flux in June by 13 W m$^{-2}$, but there is relatively close agreement between the ensemble mean and L98 in the remainder of the warm season. June is also the month of largest ensemble spread (96 W m$^{-2}$), possibly because of intermodel differences in melt onset date. For surface turbulent flux, the ensemble mean is in reasonable agreement with L98, capturing the minima in May-June and in August. The ensemble mean is below the L98 value in most months, with a maximum underestimate of 6 W m$^{-2}$ in January. Radiative and turbulent fluxes are combined into a net surface flux in Figure 4e, which closely resembles the net radiative fluxes. The ensemble spread in turbulent fluxes is of the same order as the annual cycle. Since the ensemble mean underestimates both radiative and turbulent fluxes in the cold season, the maximum underestimate in Figure 4e is greater than that in Figure 4c at 13 W m$^{-2}$ in January. The June overestimate is 11 W m$^{-2}$, accompanied by the largest spread, 81 W m$^{-2}$.

Figures 4f–4h compare nonflux quantities of relevance to the surface energy budget. The ensemble spread in surface albedo (Figure 4f) is quite high, with melt season values ranging from 0.56 to 0.81 in May, 0.45 to 0.8 in June, and 0.4 to 0.65 in July and August. The largest spread is in June, also the month of maximum spread in net radiation. L98's albedo values are considerably higher, presumably because his values do not include leads and melt ponds. Cloud fraction (Figure 4g) also shows large ensemble spread, particularly in winter, with values ranging from 34 to 95% in January. Arctic cloud biases are a well-known shortcoming of climate models, as noted by *Vavrus* [2004], *Walsh et al.* [2002], and *Beesley and Moritz* [1999] [see also *Gorodets-*

*kaya and Tremblay*, this volume]. The spread in cloud fraction is smallest in the melt season, with values between 65 and 95% in JJA and good agreement between the ensemble mean and L98. From the simple model perspective, the fact that the best agreement occurs over the melt season is beneficial, since the impact of cloud biases on $\hat{S}$ and the spread of $h_D$ is thereby minimized. However, cloud radiative forcing is more dependent on cloud microphysics than on cloud fraction [e.g., *Gorodetskaya et al.*, 2006; see also *Gorodetskaya and Tremblay*, this volume], so the consequences of cloud biases for the radiation budget cannot be assessed from cloud fraction alone.

Finally, the surface air temperature from L98 and the 20C3M ensemble are compared in panel *h*. The comparison is motivated by the close association between surface temperature and surface flux, and earlier findings of cold biases in climate models [e.g., *Randall et al.*, 1998]. The ensemble mean is close to L98, with differences between ±2°C and no clear preference for colder temperature. From these comparisons we conclude that while monthly flux values for individual models can disagree dramatically with L98, the ensemble mean net fluxes are in reasonable agreement with observations.

The budget residual $R$, which includes ocean heat flux convergence and latent heat transported in sea ice motion, plotted in Figure 5, has values that are generally between 0 and 10 W m$^{-2}$. Observations are not available to verify $R$, but the contribution of ocean heat flux could, in principle, be calculated from the 20C3M ocean model output. For CCSM3, the heat flux due to ice motion was found to be about 5 W m$^{-2}$ in a long present-day control run (run b30.009, for which thickness tendency due to dynamics is available), and DHH obtained a value of 4.4 W m$^{-2}$ in a version of CCSM3 with a slab ocean but full ice dynamics. T92 gives a range of reasonable ocean heat flux convergence values of 2 to 10 W m$^{-2}$. Notable features in Figure 5 are the extreme value for model 9 (about 15 W m$^{-2}$) and two models with negative $R$ values, presumably indicating that the ocean is exporting heat from the Arctic.

Following *Maykut and Untersteiner* [1971], EUW and T92 assume that shortwave and longwave radiation dominate the surface energy budget, and changes in $S$ are compensated exclusively by changes in $W$ as shown in Figure 2. The extent to which the surface energy budget and its intermodel variation in the 20C3M ensemble is dominated by radiative fluxes is examined in Figure 6a, in which the annual mean net shortwave and longwave fluxes are scattered against each other. With the exception of models 1 and 2, the assumption holds to a reasonable degree, and the correlation between shortwave and longwave fluxes excluding these two models is 0.75 (note also the large spread in radiative fluxes, from 20 W m$^{-2}$ to over 40 W m$^{-2}$). For the outlier models (models 1 and 2, which are actually the same

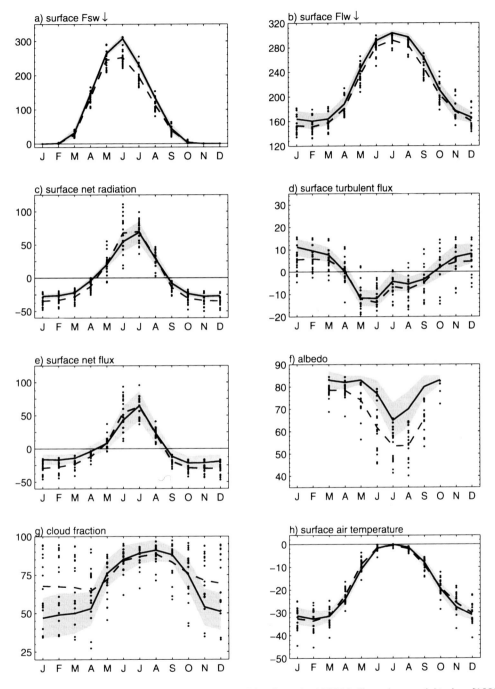

**Figure 4.** Monthly surface energy fluxes and related quantities from the 20C3M climatology and *Lindsay* [1998]. The solid line is the L98 value, with shading for ±1σ (spatial and temporal standard deviation of the drifting stations). The dashed line is the 20C3M ensemble mean, and dots are values for individual models. The 20C3M albedo is calculated at each grid point and then spatially averaged, so albedo values are missing in months when the lowest grid point value of insolation is zero. All fluxes are positive down, i.e., positive when the surface gains energy from above.

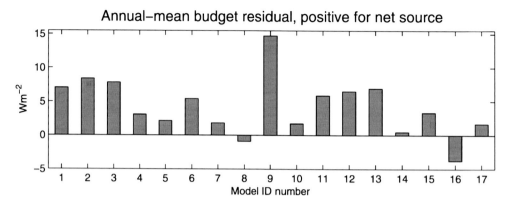

**Figure 5.** The 20C3M values for the energy budget residual in (5), which includes ocean heat flux convergence and latent heat transfer due to sea ice motion.

model at different resolutions), outgoing longwave radiation (36 and 37 W m$^{-2}$, respectively) exceeds absorbed shortwave radiation (19 and 21 W m$^{-2}$) by 15.5 and 18 W m$^{-2}$. Part of the compensation comes from the budget residual (Figure 5), for which models 1 and 2 rank fourth (7.0 W m$^{-2}$) and second (8.4 W m$^{-2}$), respectively.

The remainder of the budget deficit for models 1 and 2 must be compensated by the sensible and latent heat fluxes, of which the annual mean sensible heat flux dominates, with values of 10.4 W m$^{-2}$ and 11.5 W m$^{-2}$ for the two models, compared to about –1.8 W m$^{-2}$ for annual mean latent heat flux for both. Figure 6b shows the sensible heat flux for the 17 models, and it is apparent that models 1 and 2 (thick lines) have the highest values of sensible heat flux in all months. It is difficult to infer physical relationships between the high sensible heat gain and low insolation from available model outputs. One factor determining sensible heat flux is the temperature difference between air at the surface and the surface itself, plotted in Figure 6c. Models 1 and 2 are the only models for which surface air temperature exceeds surface temperature throughout the melt season. It is possible that the low values of melt season insolation in these models are responsible for the fact that the surface remains colder than the air at the surface, so that the shortwave flux deficiency induces a compensating enhancement in sensible heat flux.

### 4.3. Correlations in the Ensemble Spread of S

The spread of S values among the coupled climate models (Table 3) is quite large (13 to 61 W m$^{-2}$), and our results suggest that a characterization of this spread may be helpful in understanding the spread of $h_D$ and ultimately h. Table 4 shows the correlations between the spread of S and the spread

of several flux-related MJJA quantities. Table 4 shows that the spread of S is most strongly correlated with the spread in the net shortwave flux ($F_{SW}$, $r = 0.95$) which explains 90% of its variance. The high correlation with albedo ($\alpha$, $r = -0.91$) and much lower correlation with the incoming shortwave flux ($r = 0.43$) suggests that albedo differences account for a large part of the dependence of S on $F_{SW}$. Longwave fluxes, which include the effects of longwave cloud forcing, are less important for the spread in the net MJJA surface flux.

The quantities specified externally in our diagnostic calculation are not, of course, truly external, and most of the variables in Table 4 have substantial correlations with the MJJA surface temperature, $T_M$. The high correlation with $F_{LW\downarrow}$ is consistent with the idea that upwelling longwave flux warms the overlying atmosphere which reradiates to the surface (e.g., L98).

S and $\alpha$ are also closely correlated with $T_M$ ($r = 0.72$ and –0.77, respectively), as colder ice surfaces are expected to have more snow and fewer melt ponds and hence higher albedo. As the surface warms, albedo becomes smaller and S increases, leading to the standard sea ice-albedo feedback [e.g., *Ebert and Curry*, 1993]. In contrast to these results, the radiation-albedo-thickness correlations inferred by EUW's albedo tuning would be positive: to maintain a given thickness under increased downwelling longwave radiation, the albedo must increase (see their Figure 3). Along the same lines, one might expect albedo tuning to lead to a positive correlation between albedo and downwelling shortwave flux, since larger (smaller) values of albedo could be used to counteract the effects of higher (lower) levels of insolation. But the correlation of $F_{SW\downarrow}$ and $\alpha$ found here is negative and very small ($r = -0.09$). Instead, the correlations in Table 4 are consistent with the idea that the sea ice-albedo feedback

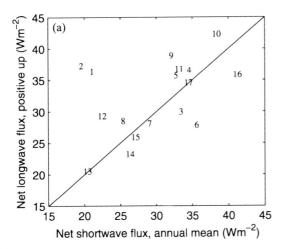

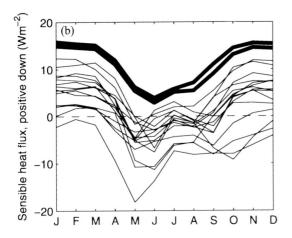

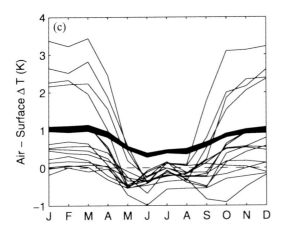

**Table 4.** Correlations Across the 20C3M Ensemble Between Surface Fluxes (Radiative, Sensible, and Latent), Albedo, and Net Surface Radiation for the MJJA and JJA Seasons[a]

| | $F_{SW\downarrow}$ | $F_{LW\downarrow}$ | $F_{SW}$ | $F_{LW}$ | $\alpha$ (%) | $T_M$ |
|---|---|---|---|---|---|---|
| | | | *MJJA* | | | |
| $S$ | 35 | 43 | 95 | 8 | −91 | 72 |
| $\alpha$ | −9 | −48 | −90 | −7 | 100 | −77 |
| $F_{RAD}$ | 30 | 56 | 93 | 27 | −90 | 75 |
| $\hat{S}$ | 14 | 30 | 59 | 19 | −60 | 43 |
| $T_M$ | −24 | 82 | 61 | 43 | −77 | 100 |
| | | | *JJA* | | | |
| $S$ | 48 | −26 | 94 | −44 | −85 | 74 |
| $\alpha$ | −9 | 19 | −87 | 42 | 100 | −64 |
| $F_{RAD}$ | 51 | −17 | 96 | −28 | −85 | 67 |
| $\hat{S}$ | 28 | −4 | 52 | 2 | −45 | 32 |
| $T_M$ | 35 | −21 | 68 | −33 | −64 | 100 |

[a]Correlations are multiplied by 100. Values in $F_{LW}$ and $F_{SW}$ columns are net fluxes.

is a prominent contributor to the ensemble spread in $S$, $\hat{S}$, and hence $h_D$, while longwave cloud effects play a smaller role. Similar results are obtained if the correlations are taken for the JJA months, excluding May from the melt season, as shown in Table 4 (bottom).

## 5. DISCUSSION AND CONCLUSIONS

The conclusion of the diagnostic analysis is that the ensemble spread in Arctic sea ice thickness is roughly consistent with the spread in the surface energy budget, with a range of 20 to 80 W m⁻² in $\hat{S}$ (the melt season surface energy fluxes plus a budget residual term) corresponding to a thickness range of 1 to 6 m. The ensemble mean value of $\hat{S}$ is one key factor in relating the ranges, and this value conforms to observations to the extent that such comparisons can be made. Another key factor is the difference in the lengths of the ice growth and melt seasons. EUW's finding of inconsistency between the thickness and flux ranges stems from their assumption of equal season lengths, rather than a short (MJJA) melt season and a long growth season.

**Figure 6.** (a) Scatter of annual mean net longwave (positive upward) and net shortwave fluxes (positive downward) for the 17 20C3M models, labeled by the model numbers in Figures 3a and 5, with a solid line to denote exact balance between longwave and shortwave fluxes. (b) Monthly sensible heat flux for the models, with thick lines for models 1 and 2. (c) Difference of surface air temperature and surface temperature for all models, with thick lines for models 1 and 2.

The consistency between thickness and flux ranges is not definitive, given the highly simplified form of the diagnostic equation. One important feedback mechanism not discussed in section 2 is the latent heat transport associated with ice motion. The effective heat flux due to ice transport is given approximately by $\nabla \cdot L\rho h\vec{V}$, where $L$ is the latent heat of fusion, $\rho$ is ice density, and $\vec{V}$ is ice velocity. Assuming that the ice density and thickness are spatially uniform and ice motion is independent of thickness, the heating rate becomes $(\rho L\nabla \cdot \vec{V})h$, and the heat energy added per year is $\gamma h$, where $\gamma = (\tau_M + \tau_G)(\rho L\nabla \cdot \vec{V})$ [Bitz and Roe, 2004]. The term $\gamma h$ can be added to the left-hand side of (2), which can then be combined with (1) to produce a modified equation for the variation of $T$ and $h$ along an isoline of $S$:

$$kT + S\frac{\tau_M}{\tau_G}h + \gamma h^2 = 0. \qquad (6)$$

Instead of the straight lines in Figure 2, isolines of $S$ are downturned parabolas, and $h\rightarrow\infty$ only if $T\rightarrow-\infty$ so the extreme ice thickness discussed in section 2 is avoided. Physically, extreme thickness is avoided because rapid ice growth due to low insolation leads to thicker ice which, for the same ice velocity, means an increase in ice transport.

The effect of sea ice motion cannot be easily determined from the 20C3M simulation archive, although it is included implicitly in (4) through the budget residual $R$. The effect of ice motion is to prevent low $\hat{S}$ values, since low $S$ leads to thicker ice and more ice export, thus higher $R$ compensates lower $S$. Despite this compensation, low values of $S$ do occur in the ensemble. Furthermore, Figure 6 shows that the dominant balance in most models is between longwave and shortwave fluxes, leaving a smaller role for motion. For the two models in which longwave loss dramatically exceeds shortwave gain, sensible heat flux makes a stronger contribution to the budget deficit than the residual term. The role of ice export was examined by Hibler and Hutchings [2002] (discussed by Bitz and Roe [2004]). They found that the assumption of thickness-independent motion is reasonable for $h<2$ m but fails for thicker ice, as strong thick ice resists motion induced by wind stress.

Discussions of uncertainty in sea ice simulations such as those of Randall et al. [2007], DeWeaver [2007], and Kattsov et al. [2005] typically address ensemble spread in sea ice by discussing uncertainties in the numerous processes which affect the ice, like cloud radiative properties, surface albedo, and the stability of the upper ocean. The analysis presented here is intended to complement such discussions by considering the inherent sensitivity of sea ice to surface flux errors. In the simple diagnostic framework, the sensitivity can be understood as a consequence of two factors: First, there

is a need to balance the upwelling surface energy flux during ice growth with a conductive flux through the ice that depends on ice thickness. This dependence means that the small upwelling flux associated with cold surface temperature must be accompanied by large ice thickness. Second, since the melt season temperature is held to the freezing point, the system cannot regulate its energy gain in summer, and the melt season gain must be compensated by growth season loss. Thus, low summer insolation must be balanced by low longwave radiation in winter, leading to very thick ice. While the equations used to formalize this assessment are severely simplified, the basic conceptual insight may still be helpful in understanding the large spread of sea ice in the 20C3M simulations.

The approximately inverse relationship between $\hat{S}$ and $h_D$ in Figure 3 is analogous to the derivation of uncertainty in global climate sensitivity derived by Roe and Baker [2007] [see also Bitz, this volume]. In their analysis, the inverse relationship is between climate sensitivity (the change in global temperature for a unit change in radiative forcing) and the factor $1-f$, where $f$ is the feedback factor representing the net effect of climate feedbacks. They show that this inverse relationship produces a long-tailed distribution in climate sensitivity for a normally distributed range of $f$ values. The same derivation applied to the $S$-$h$ curve in Figure 3 would show that a large model ensemble with normally distributed $S$ values would have positively skewed $h$ values, so one might always expect to see a few models with extreme thickness. Randall et al. [2007] note that 20C3M sea ice simulations do not show dramatic improvement over the preceding generation of climate models despite considerable model development, and they ascribe the lack of improvement to persistent problems in the atmosphere and ocean component models. But our sensitivity analysis suggests a role for intrinsic sea ice sensitivity, since a relatively large reduction in flux uncertainty is required to reduce thickness uncertainty. Hopefully, improvements in physical understanding and model construction will narrow the range of $S$ and $\hat{S}$ values. However, the close association between $S$ and surface temperature poses a challenge to such model improvement, since the implied sea ice-albedo feedback will amplify errors in the surface energy budget caused by inadequacies in parameterization.

The potential implications of our analysis for anthropogenic climate change simulations can be understood by regarding $\delta F$ in (3) as the longwave flux increase due to anthropogenic greenhouse gas emissions. Following the logic of section 4.1, the primary effect of a greenhouse gas–induced flux increase is to shift the $\hat{S}$ value to the right in Figure 3b. The resulting decrease in $h$ is nearly proportional to $h^2$ (if $h\propto 1/\hat{S}$ then $\delta h\propto -h^2\delta\hat{S}$), so that thick ice thins faster, as shown by *Bitz and Roe* [2004]. Figure 1 of *Bitz* [this volume] shows that the thinning of sea ice due to global warming is indeed faster for models which produce thicker sea ice in their 20th century climate simulations. A faster thinning rate for initially thicker ice should be good news for climate change simulations, since the differential thinning reduces the future consequences of present-day thickness errors. Nevertheless, Bitz's analysis shows that much of the spread in future thickness projections can be understood as a consequence of thickness spread in present-day simulations.

The large spread of Arctic sea ice thickness in the 20C3M ensemble is clearly undesirable, particularly when model projections are used for policy decisions. However, uncertain projections can still provide useful guidance, so long as their uncertainties are properly understood and acknowledged. But uncertainties in climate modeling can never be fully understood, and there is a danger that the uncertainties will be underestimated or overlooked. As *Oreskes* [2000, p. 81] puts it, "Modeling may lead to greater rigor in the evaluation of earth processes, but it may also propagate the illusion that things are better known than they really are." For the 20C3M model ensemble, uncertainty is commonly assessed through the models' ability to simulate present-day observations (e.g., sea ice thickness and concentration), with the spread of the ensemble serving as an error bar. There is, of course, no absolute guarantee that fidelity to observations means fidelity to the underlying physics, particularly given the large number of parameterizations and parameter settings which must be chosen by the modeler. Likewise, there is no guarantee that the ensemble spread is a good measure of the uncertainty. Thus, diagnostics which relate simulation quality for gross features like sea ice thickness to underlying physical processes can play a valuable role in making sure that the simulations really are as good (albeit marginally good) as they look.

*Acknowledgment.* This research was supported by the Office of Science (BER), U.S. Department of Energy, grant DE-FG02-03ER63604 to E. DeWeaver. We thank Cecilia Bitz, Steve Vavrus, Bruce Briegleb, Dave Bailey, and two anonymous reviewers for sharing their insights. We also thank the Program for Climate Model Diagnosis and Intercomparison (PCMDI) for collecting and archiving the model data. The 20C3M Data Archive at Lawrence Livermore National Laboratory is supported by the Office of Science, U.S. Department of Energy.

## REFERENCES

Amstrup, S. C., B. G. Marcot, and D. C. Douglas (2008), A Bayesian network modeling approach to forecasting the 21st century worldwide status of polar bears, this volume.

Beesley, J. A., and R. E. Moritz (1999), Toward an explanation of the annual cycle of cloudiness over the Arctic Ocean, *J. Clim.*, *12*, 395–415.

Bitz, C. M. (2008), Some aspects of uncertainty in predicting sea ice retreat, this volume.

Bitz, C. M., and G. H. Roe (2004), A mechanism for the high rate of sea ice thinning in the Arctic Ocean, *J. Clim.*, *17*, 3623–3632.

DeWeaver, E. (2007), Uncertainty in climate model projections of Arctic Sea ice decline: An evaluation relevant to polar bears, administrative report, 47 pp., Alaska Sci. Cent., U.S. Geol. Surv., Anchorage. (Available at www.usgs.gov/newsroom/special/polar_bears/)

DeWeaver, E. T., E. C. Hunke, and M. M. Holland (2008), Comment on "On the reliability of simulated Arctic sea ice in global climate models" by I. Eisenman, N. Untersteiner, and J. S. Wettlaufer, *Geophys. Res. Lett.*, *35*, L04501, doi:10.1029/2007GL031325.

Ebert, E. E., and J. A. Curry (1993), An intermediate one-dimensional thermodynamic sea ice model for investigating ice-atmosphere interactions, *J. Geophys. Res.*, *98*, 10,085–10,109, doi:10.1029/93JC00656.

Eisenman I., N. Untersteiner, and J. S. Wettlaufer (2007), On the reliability of simulated Arctic sea ice in global climate models, *Geophys. Res. Lett.*, *34*, L10501, doi:10.1029/2007GL029914.

Gorodetskaya, I., and L.-B. Tremblay (2008), Arctic cloud properties and radiative forcing from observations and their role in sea ice decline predicted by the NCAR CCSM3 model during the 21st century, this volume.

Gorodetskaya, I., B. Tremblay, B. Liepert, and M. Cane (2006), Can Arctic sea ice summer melt be accelerated by changes in spring cloud properties?, *Eos Trans. AGU*, *87*(52), Fall Meet. Suppl., Abstract C32A-05.

Hibler, W. D., and J. K. Hutchings (2002), Multiple equilibrium Arctic ice cover states induced by ice mechanics, paper presented at Ice in the Environment: Proceedings of the 16th IAHR Intternational Symposium on Ice, Int. Assoc. of Hydraul. Eng. and Res., Dunedin, New Zealand.

Kattsov, V. M., et al. (2005), Future climate change: Modeling and scenarios for the Arctic, in *Arctic Climate Impact Assessment*, chap. 4, pp. 99–150, Cambridge Univ. Press, New York.

Key, J. R., R. A. Silcox, and R. S. Stone (1996), Evaluation of surface radiative flux parameterizations for use in sea ice models, *J. Geophys. Res.*, *101*, 3839–3849.

Lindsay, R. W. (1998), Temporal variability of the energy balance of thick Arctic pack ice, *J. Clim.*, *11*, 313–333.

Maykut, G. A., and N. Untersteiner (1971), Some results from a time-dependent thermodynamic model of sea ice, *J. Geophys. Res.*, *76*, 1550–1575.

Oreskes, N. (2000), Why believe a computer? Models, measures and meaning in thet natural world, in *The Earth Around Us: Maintaining a Livable Planet*, edited by J. Schneiderman, pp. 70–82, W. H. Freeman, New York.

Persson, P. O. G., C. W. Fairall, E. L. Andreas, P. S. Guest, and D. K. Perovich (2002), Measurements near the Atmospheric Surface Flux Group tower at SHEBA: Near-surface conditions and surface energy budget, *J. Geophys. Res.*, *107*(C10), 8045, doi:10.1029/2000JC000705.

Randall, D., J. Curry, D. Battisti, G. Flato, R. Grumbine, S. Hakkinen, D. Martinson, R. Preller, J. Walsh, and J. Weatherly (1998), Status and outlook for large-scale modeling of atmosphere-ice-ocean interactions in the Arctic, *Bull. Am. Meteorol. Soc.*, *79*, 197–219.

Randall, D. A., et al. (2007), Climate models and their evaluation, in *Climate Change 2007, The Physical Science Basis: Contribution of Working Group I to the Fourth Assessment Report of the Intergovernmental Panel on Climate Change*, edited by S. Solomon et al., pp. 589–662, Cambridge Univ. Press, New York.

Roe, G. H., and M. B. Baker (2007), Why is climate so unpredictable?, *Science*, *318*, 629–632.

Shine, K. P. (1984), Parameterization of shortwave flux over high albedo surfaces as a function of cloud thickness and surface albedo, *Q. J. R. Meteorol. Soc.*, *110*, 747–764.

Thorndike, A. S. (1992), A toy model linking atmospheric thermal radiation and sea ice growth, *J. Geophys. Res.*, *97*, 9401–9410.

Vavrus, S. J. (2004), The impact of cloud feedbacks on Arctic climate under greenhouse forcing, *J. Clim.*, *17*, 603–615.

Walsh, J. E., V. M. Kattsov, W. L. Chapman, V. Govorkova, and T. Pavlova (2002), Comparison of Arctic climate simulations by uncoupled and coupled models, *J. Clim.*, *15*, 1429–1446.

E. T. DeWeaver, Center for Climatic Research, University of Wisconsin-Madison, 1225 West Dayton Street, Madison, WI 53706, USA. (deweaver@aos.wisc.edu)

M. M. Holland, National Center for Atmospheric Research, Boulder, CO 80307-3000, USA.

E. C. Hunke, Los Alamos National Laboratory, Los Alamos, NM 87545, USA.

# The Atmospheric Response to Realistic Reduced Summer Arctic Sea Ice Anomalies

Uma S. Bhatt,[1] Michael A. Alexander,[2] Clara Deser,[3] John E. Walsh,[4] Jack S. Miller,[5]
Michael S. Timlin,[6] James Scott,[2] and Robert A. Tomas[3]

The impact of reduced Arctic summer sea ice on the atmosphere is investigated by forcing an atmospheric general circulation model, the Community Climate Model (CCM 3.6), with observed sea ice conditions during 1995, a low-ice year. The 51 experiments, which spanned April to October of 1995, were initiated with different states from a control simulation. The 55-year control was integrated using a repeating climatological seasonal cycle of sea ice. The response was obtained from the mean difference between the experiment and control simulations. The strongest response was found during the month of August where the Arctic displays a weak local thermal response, with warmer surface air temperatures and lower sea level pressure (SLP). However, there is a significant remote response over the North Pacific characterized by an equivalent barotropic (anomalies are collocated with height and increase in magnitude) structure, with anomalous high SLP collocated with a ridge in the upper troposphere. The ice anomalies force an increase (decrease) in precipitation north of (along) the North Pacific storm track. A linear baroclinic model forced with the transient eddy vorticity fluxes, transient eddy heat fluxes, and diabatic heating separately demonstrated that transient eddy vorticity fluxes are key to maintaining the anomalous high over the North Pacific. The model's sensitivity to separately imposed ice anomalies in the Kara, Laptev–East Siberian, or Beaufort seas includes SLP, geopotential height, and precipitation changes that are similar to but weaker than the response to the full sea ice anomaly.

[1]Geophysical Institute, Department of Atmospheric Sciences, University of Alaska Fairbanks, Fairbanks, Alaska, USA.

[2]NOAA Earth System Research Laboratory, Boulder, Colorado, USA.

[3]National Center for Atmospheric Research, Boulder, Colorado, USA.

[4]International Arctic Research Center, University of Alaska Fairbanks, Fairbanks, Alaska, USA.

[5]ARSC, University of Alaska Fairbanks, Fairbanks, Alaska, USA.

[6]Department of Atmospheric Sciences, University of Illinois at Urbana-Champaign, Urbana, Illinois, USA.

Arctic Sea Ice Decline: Observations, Projections, Mechanisms, and Implications
Geophysical Monograph Series 180
Copyright 2008 by the American Geophysical Union.
10.1029/180GM08

## 1. INTRODUCTION

Summer sea ice in the Arctic decreased at a rate of 4–6% per decade [*Deser et al.*, 2000] through the 1990s, and the melt rate has accelerated to 10% per decade [*Stroeve et al.*, 2007; National Snow and Ice Data Center, press release, 1 October 2007, available at nsidc.org/news/press/2007_seaiceminimum/20071001_pressrelease.html) in the 2000s. In the 1990s the melting of Arctic ice was consistent with the positive phase of the North Atlantic Oscillation (NAO), which is characterized by enhanced storminess and warm moist air penetration into the Arctic. The NAO has approached more neutral values since 2000, yet the ice melt has accelerated. The observed influx of plugs of warm Atlantic layer water into the Arctic provides one likely mechanism for the continued ice melt [*Polyakov et al.*, 2005], and recent work shows that heat from the Atlantic layer can penetrate through the halocline into the upper ocean [*Walsh et al.*, 2007]. *Maslanik et al.* [2007] demonstrate that even though the NAO is in the negative phase, the net local atmospheric circulation in the Arctic is consistent with continued ice reduction. Another contributor that could hasten ice retreat in summer is enhanced moisture in the Arctic leading to increased downward longwave fluxes [*Francis and Hunter*, 2006]. Circulation trends since 1979 were found to be weak by *Deser and Teng* [this volume], consistent with the view that multiple mechanisms have led to recent ice declines. The summer sea ice is expected to continue its decline based on the recently documented decreases in winter ice [*Comiso*, 2008] and the warm Atlantic water headed for the Arctic that is being tracked by various ocean observing programs [*Polyakov et al.*, 2007]. In a warmer climate large decreases in summer sea ice may become more common, and while the ice anomalies initially result from both atmospheric and oceanic forcing, we hypothesize that they can, in turn, markedly alter the air-sea exchanges of heat and moisture to subsequently influence the large-scale climate.

The summer warming and sea ice reductions are correlated with cold season circulation anomalies [*Wallace et al.*, 1996; *Rigor et al.*, 2002; *Maslanik et al.*, 2007], which lead to changes in low-level horizontal temperature advection. For example, reduced summer sea ice in the Barents-Kara-Laptev seas is associated with anomalously low pressure centered in the Arctic the preceding spring (April–June) [*Deser et al.*, 2000]. The general tendency toward lower pressure in the Arctic [*Walsh et al.*, 1996] from 1960 to 2000 is consistent with enhanced penetration of storms into the Arctic. There has been an increase of warm season cyclone count and intensity in the Arctic (north of 60°N) since late 1950s [*Serreze et al.*, 1997; *Zhang et al.*, 2004]. *Maslanik et al.* [1996] find an increase of cyclone activity over the central Arctic Ocean, which advects warm southerly winds into the Laptev and East Siberian seas as well as transports ice away from the coast. Ice that is particularly thin as a result of wintertime circulation patterns can be easily broken down and transported because of summer storms, further reducing ice area/concentration.

When high-albedo ice is replaced by low-albedo ocean, there is significantly more net solar flux at the surface, increasing the heat stored in the upper layer of the ocean. This heat stored during the summer can then slow the freezeup the following winter as well as melt ice at the ice/ocean interface. There is still a reasonably strong correlation (+0.6) between the time series of EOF1 of sea ice concentration during the summer and that of the following winter [*Deser et al.*, 2000]. By August the sea surface temperature (SST)

can warm in the marginal seas by several degrees [*Steele et al.*, 2008] when ice extent is low. Fluxes of sensible and latent heat into the atmosphere increase with a warmer ocean, which, we hypothesize, could exert some influence back on the atmosphere.

Climate models are ideal tools for understanding the influence of sea ice on the atmosphere because in the observations, climate anomalies are dominated by the atmospheric forcing of the ice. *Singarayer et al.* [2006] ran the Hadley Centre Atmospheric Model (HadAM3) with climatological SSTs and observed sea ice concentrations from 1978 to 2000 to investigate the impact of sea ice on the atmospheric circulation. The model surface air temperature (SAT) response to ice forcing most closely matches the observed SAT variability over the 1993–1995 period [see *Singarayer et al.*, 2006, Figure 4a]. This suggests that sea ice forcing played a more important role than SST (note that this simulation used climatological SSTs) in shaping the SAT anomalies. The observed sea ice anomalies display large interannual variability in the mid-1990s and reached a low for the decade in 1995. *Singarayer et al.* [2006] argue that the ice anomalies were likely large enough that sea ice forcing dominated the atmospheric response. The SAT response during summer strengthened and become statistically significant when observed above-normal SSTs replace climatological values. *Sewall* [2005] investigated the response to reduced Arctic sea ice in the Community Climate System Model, version 3 (CCSM3) plus a suite of coupled Intergovernmental Panel on Climate Change Fourth Assessment simulations and found a robust pattern of reduced wintertime precipitation for the western United States by ~30%.

*Magnusdottir et al.* [2004] and *Deser et al.* [2004] investigated the response to sea ice and SST anomalies during winter in the North Atlantic using CCM3. The ice anomaly pattern corresponds to an enhanced observed trend with ice reductions (increases) east (west) of Greenland. *Magnusdottir et al.* [2004] found a significant model circulation response to sea ice that resembled the negative phase of the North Atlantic Oscillation, which is opposite of the atmospheric pattern that forced the observed sea ice trend, suggesting that sea ice has a negative feedback on the atmosphere. There is growing evidence that a model's internal variability influences its forced response. To investigate this further, *Deser et al.* [2004] decomposed the atmospheric response to sea ice into the part that projects on the leading mode of model variability and the residual from this projection. The leading mode has an equivalent barotropic vertical structure and resembles the NAO, while the residual is baroclinic. A subsequent study by *Deser et al.* [2007] examines the transient response to wintertime sea ice anomalies in the North Atlantic. They analyzed the general circulation model (GCM) output using a linear baroclinic model (LBM) to show that the initial local response is baroclinic and forced by the diabatic heating anomalies associated with surface heat fluxes resulting from reduced sea ice area. The equilibrium response is large scale in extent, barotropic, and primarily maintained by the transient eddy vorticity fluxes. *Peng and Whittaker* [1999] elucidated this eddy-driven mechanism to describe the atmospheric response to midlatitude SSTs in an idealized GCM, which can be applied to surface changes resulting from decreased sea ice. These studies show that the atmosphere responds to surface boundary conditions in ways that can influence the storm track.

*Alexander et al.* [2004] forced CCM3 with realistic sea ice conditions, characterized by negative (positive) ice extent anomalies east (west) of Greenland, from 1982 to 1983 that had a similar pattern but with a smaller ice area than the anomalies from *Magnusdottir et al.* [2004] and *Deser et al.* [2004]. The pattern of response is similar in the three studies, with positive (negative) height anomalies in the Arctic (midlatitudes). A comparison of ice area to the strength of 500-hPa response reveals a linearly increasing relationship [see *Alexander et al.*, 2004, Figure 9].

*Alexander et al.* [2004] also examined the response to ice anomalies in the North Pacific and found that the atmospheric response suggested a positive feedback of the ice on the atmosphere. The different atmospheric responses to ice in the North Atlantic and North Pacific may arise from the position of the storm track relative to the ice edge. In the

North Atlantic the ice edge is in the vicinity of the storm track, whereas in the North Pacific the ice edge is well north of the storm track. A thorough discussion of additional studies of the response to winter sea ice is presented by *Alexander et al.* [2004].

Numerous GCM simulations have investigated the impact of winter sea ice on the atmosphere but few have examined the atmospheric response to sea surface temperature or sea ice during the summer months. Several studies find the response during summer to be much weaker than winter and focus their analysis on winter [*Parkinson et al.*, 2001; *Singarayer et al.*, 2006]. *Raymo et al.* [1990] reduced the ice to paleoclimatic conditions throughout the year that reached an ice-free Arctic during the month of September. During June, July, and August (JJA) they found a 3°K warming over Greenland and an overall warming over the polar region. They found no significant differences in sea level pressure, evaporation/precipitation ratios, or cloudiness in the North Atlantic.

This study employs CCM3 to investigate the atmospheric response to reduced realistic summer sea ice in the Arctic from the summer of 1995, which had the lowest June–September ice area (based on both extent and concentration) with the exception of the summer of 2007 (Figure 1). Note that the sea ice minimum in September has been close to or well below the 1995 levels since 2002 [*Stroeve et al.*, 2008]. In addition to using realistic sea ice extents and concentrations in the Arctic, the other unique features of our study

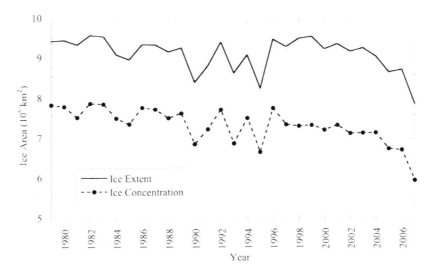

**Figure 1.** Observed Arctic-wide ice cover (multiplied by $10^6$ km$^2$) based on ice extent (solid line) and concentration (dashed line) during summer (June–September) over the period 1979–2007 in the HadISST1 1° × 1° data set. Ice is defined to extend over a grid square when the ice concentration is 15% or greater. The summer of 1995 had the overall minimum June–September ice extent with the exception of 2007, which was significantly lower.

include the summer focus and the use of a large number (51) of ensemble members for each set of experiments to enhance the signal-to-noise ratio. We chose to employ CCM3 for this study to facilitate a comparison with winter sea ice forcing studies that used the same model [*Alexander et al.*, 2004; *Deser et al.*, 2004; *Magnusdottir et al.*, 2004]. In addition, a suite of further experiments is conducted to investigate the sensitivity of the model to the location of the ice anomaly and a LBM is used to diagnose the forcing to assist in the interpretation of the results.

Some key questions that we address in this study are the following:

- Does the Northern Hemisphere atmosphere respond to realistic summertime Arctic sea ice anomalies? Is there a remote response as well as a local response? How does the response during summer differ from winter?
- Is the atmospheric response sensitive to the placement (latitude/longitude) of the summer Arctic sea ice anomalies?
- How does the response to sea ice extent compare with that to concentration?
- Does the atmospheric response have any implications for feedback mechanisms?

The model experiments are described in section 2, and the results are discussed in section 3. The summary and a discussion of mechanisms are presented in section 4.

## 2. MODEL EXPERIMENTS

### 2.1. Boundary Conditions and Experiment Design

Boundary conditions for the simulations are from the Hadley Centre sea ice concentration and sea surface temperature data set (HadISST version 1.1 [*Rayner et al.*, 2003]), and climatologies are based on the 1979–1999 period. Observed monthly mean values were interpolated to the model grid using bilinear interpolation over the open ocean and by averaging nearby grid values in coastal regions. Arctic sea ice area varies while thickness is specified to be 2.5 m. It is not expected that specifying ice thickness will significantly influence the results since the summer atmosphere in a regional climate model was shown to be insensitive to changes in sea ice thickness [*Rinke et al.*, 2006]. The conductive heat flux through sea ice is small during summer regardless of ice thickness because the ocean and near-surface air temperatures are similar. Global SSTs and sea ice in the Southern Hemisphere (specified to be 1 m) evolve according to the mean seasonal cycle in order to isolate the influence of Arctic sea ice. In regions where the ice extent was lower than

the mean extent, the exposed ocean was set to the climatological SST; when the ice area expanded above normal, SSTs were blended from −1.8°C (the temperature at which there is 100% ice cover) at the ice edge with climatological values from two grid boxes (2.8° latitude × 2.8° longitude) seaward from the ice edge. This method was employed to smooth the temperature gradient between ice and ocean. In the extent experiments, the monthly Arctic sea ice values were specified to cover 100% of the grid square if the observed monthly averaged concentration exceeded 15%; otherwise, the grid square was set to be ice free. Monthly mean ice and SST values were linearly interpolated in time from mid monthly values to obtain smoothly varying daily extents and concentrations. As a result, the transition from no ice to complete ice cover in a grid square is not instantaneous in the extent simulations; instead, the amount of ice linearly evolves between 0% and 100% within the 30-day period when ice forms or melts. While this provides for a smooth transition of the ice edge in space and time, and is probably more realistic than an instantaneous transition, it also introduces fractional ice cover into the extent experiments.

### 2.2. Experiment Design

We focus this study on the summer of 1995, which contained minimum sea ice extent over the entire Arctic from June to September during recent years, with the exception of 2007 (Figure 1). Ice area during 2005 and 2006 was just slightly above that of 1995. Typically, Arctic summer sea ice extent reaches a minimum in mid September of approximately $5 \times 10^6$ km$^2$, though recent minima (e.g., 2005 through 2007) have been consistently lower. Three model experiments have been performed in which Arctic sea ice varies according to the following observations:

- Ice extent varies over April to October of 1995 (Sum95e).
- Ice concentration varies over April to October of 1995 (Sum95c).

The experiments are designated, in parentheses above, by the season, year, and ice configuration. We also performed an extended (55 years) control simulation in which ice extent repeats the same seasonal cycle each year based on the average of the 1979–1999 period (Cntle),

The Sum95e and Sum95c experiments consist of an ensemble of 51 CCM3 simulations that extend from April to October. Each member of the ensemble is initialized from a different 1 April from years 5–55 of the control extent experiment (Cntle). A discussion quantifying the relationship between signal-to-noise ratio and ensemble members

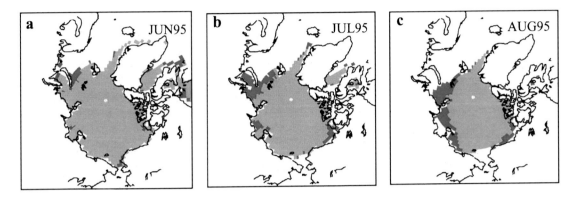

**Plate 1.** Evolution of sea ice extent during the boreal summer months of June–August of 1995. Blue (red) squares indicate enhanced (reduced) ice when compared to the monthly mean ice extent (represented by gray plus red areas).

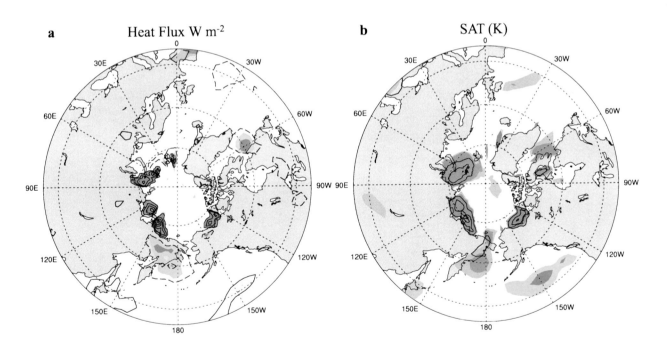

**Plate 2.** (a) Net surface heat flux anomalies (sensible plus latent plus longwave) felt by the atmosphere and (b) surface air temperature (SAT) anomalies in August of Sum95e. Dark (light) red or blue indicates statistical significance at the 99% (95%) or greater level based on a pooled variance $t$ test. Confidence interval (CI) is 5 W m$^{-2}$ and 0.5 K for heat flux and SAT, respectively. This is a polar stereographic view from 40° to 90°N.

**a**      SLP hPa      **b**   500-hPa Height (m)      **c**   200-hPa Height (m)

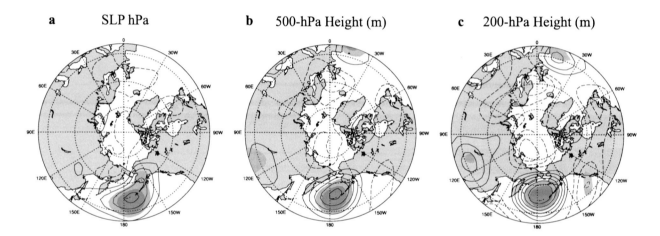

**Plate 3.** Sum95e (a) SLP anomaly response to reduced sea ice during August and geopotential height anomalies at (b) 500 hPa and (c) 200 hPa. Dark (light) shading indicates statistical significance at the 99% (95%) or greater level based on a pooled variance *t* test. The CI is 0.5 hPa for SLP and 5 m for geopotential height. This is a polar stereographic view from 40° to 90°N.

is given by *Alexander et al.* [2004]. The modeling results are generally presented as monthly anomalies constructed by averaging over the 51 ensembles and subtracting the corresponding long-term monthly mean over the last 51 years of the control simulation.

The discussion in this paper focuses on the model response during August of 1995. Ice anomalies evolve during the simulation based on observed April to October ice conditions. In 1995, ice was below normal in the Kara-Barents seas in July and throughout the Eurasian Arctic and in the Chukchi and Beaufort seas during August (Plate 1) and September. The atmospheric response in June and July was generally weak and will not be discussed. This may be a consequence of overall smaller sea ice anomalies during these months.

## 2.3. Atmospheric General Circulation Model

The CCM (version 3.6) is the atmospheric GCM used in this study; it has 18 vertical levels and a horizontal spectral resolution of T42, which is approximately 2.8° latitude by 2.8° longitude. *Kiehl et al.* [1998] describe the model physics, while *Hack et al.* [1998] and *Hurrell et al.* [1998] evaluate the model's climate with a global perspective while *Briegleb and Bromwich* [1998a, 1998b] evaluate the polar climate. Model evaluations relevant for this study will be briefly outlined.

*Hurrell et al.* [1998] find that while the subtropical summer time SLPs are higher than observed, CCM3 captures the key interseasonal shifts of the subtropical highs. The newer Community Atmosphere Model (CAM3) displays similar SLP features as CCM3 [*Hurrell et al.*, 2006]. SLP over the Arctic is higher than observed, and none of the Atmospheric Model Intercomparison Project models investigated by *Bitz et al.* [2002] capture the observed closed low over the central Arctic during summer. An investigation by *DeWeaver and Bitz* [2006] shows that JJA Arctic SLP in the Community Climate System Model, version 3 (CCSM3) is too high, a feature that is particularly prominent at T42 resolution. They find that in the model there is subsidence due to a thermally direct mean meridional circulation while reanalysis data indicate rising motion with an indirect Ferrel cell in the Arctic. Consistent with these studies, the Cntle simulation SLP is 5–7 hPa too high compared to the National Centers for Environmental Prediction/National Center for Atmospheric Research (NCAR) reanalysis over the Arctic during August (not shown).

*Briegleb and Bromwich* [1998b] find that CCM3 summer time tropospheric temperatures in the Arctic are cooler than observed by 2°–4°K, while precipitation minus evaporation (P − E) compares favorably with observations. The Arctic July total cloud amount in CCM3 is similar to observations [*Briegleb and Bromwich*, 1998a, Figure 10b], but the cloud water path is too high resulting in clouds that reflect (emit) excessively in the shortwave (longwave) range [also see *Gorodetskaya and Tremblay*, this volume].

While the model has some deficiencies over the Arctic, e.g., it is colder and wetter than observed (which also occurs in most other atmospheric general circulation models [*Randall et al.*, 1998]), many aspects of the Earth's climate are well simulated. This is a well-documented model that has been used in numerous studies of the impact of sea ice on the atmosphere [e.g., *Deser et al.*, 2004; *Magnusdottir et al.*, 2004; *Alexander et al.*, 2004].

## 2.4. Linear Baroclinic Model

To understand the mechanism for the large-scale response over the North Pacific to reduced Arctic sea ice in August, we forced a LBM with daily mean diabatic heating and transient eddy heat and vorticity fluxes, similar to *Deser et al.* [2007], from Cntle and Sum95e. The LBM [see *Peng et al.*, 2003] is based on the primitive equations configured with T21 horizontal resolution and 10 equally spaced pressure levels from 950 to 50 hPa. The model is linearized about the CCM3 basic state obtained from the long-term August mean in Cntle. The LBM includes dissipation in the form of Rayleigh friction in the momentum equation and Newtonian cooling in the thermodynamic equation, as well as biharmonic thermal diffusion. The Rayleigh and Newtonian damping time scales are 1 day at 950 hPa transitioning linearly to 7 days above 700 hPa. The LBM is integrated for 31 days.

The pattern of the CCM3 response to sea ice forcing is diagnosed by comparing the LBM responses to anomalous diabatic heating and transient eddy fluxes from Cntle and Sum95e. The transient eddies are based on 14-day high-pass-filtered data, constructed by subtracting the 11-day running means from the raw daily data (the half-power point of this filter is 14 days).

## 3. RESULTS AND DISCUSSION

### 3.1. Local Arctic Response

The model atmosphere displays a local thermal response to reduced western Arctic sea ice extent. The net heat flux anomalies resulting from the reduced sea ice are 10–25 W m$^{-2}$ from the ocean to the atmosphere (Plate 2a). The sensible heat flux is the dominant form of heating contributing about 4–8 W m$^{-2}$, followed by latent heat flux at 2–6 W m$^{-2}$ and then longwave at 2–4 W m$^{-2}$. Increased upward (downward) directed longwave radiation of 2 W m$^{-2}$ is associated with a decrease (enhanced) in low-level clouds of 2%. It is not surprising that daily model turbulent heat fluxes differ by an order of magnitude between summer and winter. Ignoring wind speed and drag coefficients, sensible heat flux is proportional to the temperature difference between the ocean and near-surface air, which is on the order of 1°K in summer and 10°K in winter. The winds are generally stronger in winter, which increases the fluxes, but the polar atmosphere is also relatively stable which damps turbulent fluxes. The sensible heat flux differences between winter and summer can be explained by the seasonal difference in vertical temperature gradients. Parallel arguments can be made for latent heat fluxes. Plots of model net surface solar flux, albedo, surface temperature, cloud cover, and specific humidity are described in the text but will not be shown. Anomalies of net surface solar heat flux are directed into the ocean and are on the order of 15–30 W m$^{-2}$ where high-albedo ice is replaced by a lower-albedo ocean. However, the shortwave anomalies do not impact our simulation since the ocean temperature and ice are fixed and are not included in the net heat flux calculation. In nature, the enhanced solar flux into the surface would melt more ice or act to warm the ocean in the shallow ice-free seas. However, we specify the observed evolution of sea ice and argue that any ice melt from increased net solar radiation into the ocean is represented by the observed sea ice conditions.

The surface temperature anomalies associated with the reduced ice area are between 0.5° and 1.5°C where climatological ocean sea surface temperatures replace sea ice. The surface air temperature (Plate 2b) warms throughout the Arctic with strongest warming present over the Kara-Barents and East Siberian seas and over eastern Siberia with anomalies between 0.5° and 1.5°C. This low-level warming is associated with small decreases in sea level pressure and geopotential heights that are not statistically significant (Plate 3). The air temperature warming is relatively shallow over the Arctic with no significant anomalies at or above 925 hPa. There is a significant increase in convective precipitation (1 mm d$^{-1}$), convective clouds (1–2%), and middle-level clouds (2–4%) in the Laptev Sea where sea ice is reduced. In the Kara Sea over the reduced ice extent, there is a significant decrease in total cloud cover (2–4%), which results from less low, medium, and high clouds. There are no significant changes in large-scale precipitation over the Arctic. The CCM3 positive SLP bias in the control simulation discussed earlier could play a role in the small convective response over the Arctic.

### 3.2. Midlatitude Response

The model response to reduced western Arctic sea ice in the North Pacific is characterized by changes in the large-scale circulation and displays significant anomalies in east-

ern Siberia (65°N, 165°E) and over the ocean storm track region (55°–60°N). In far eastern Siberia, negative sensible and longwave heat flux anomalies total 5 W m$^{-2}$ (Plate 2a) while downward solar heat flux is reduced by 5 W m$^{-2}$, resulting in a net heat flux change of near zero. Surface temperature and surface air temperatures are warmer by up to 1.0°C and 0.5°C (Plate 2b), respectively. Increased convective and large-scale precipitation is collocated with increases in specific humidity (up to 2 g kg$^{-1}$). Total cloudiness in eastern Siberia increases by up to 4% with more clouds at all levels. The SLP and geopotential height anomalies are weakly negative and not significant over eastern Siberia. Since the net surface heat flux anomalies are weak, the warmer moister atmosphere results from southerly advection associated with the circulation changes over the North Pacific.

The SLP response is characterized by a significant anomalous high over the North Pacific with a central maximum of 2 hPa (Plate 3a). At 500 and 200 hPa the anomalous high in the North Pacific reaches 20 and 30 m (Plates 3b and 3c), respectively, displaying an equivalent barotropic structure. This pattern is characteristic of the equilibrium response to a midlatitude heating source attributed to transient eddy feedbacks that results from the interaction of the forced anomalous flow and the storm tracks [Kushnir and Lau, 1992; Ting and Peng, 1995; Peng and Whittaker, 1999]. The response to reduced sea ice does not project on the dominant modes of model variability, and this is discussed further in section 4. Additionally, these SLP and 500-hPa patterns compare favorably with observed circulation anomalies associated with reduced Eurasian sea ice, and this is briefly addressed at the end of section 4.

The model displays a response in the North Pacific storm track region with significant total precipitation anomalies (Plate 4). Anomalies of large-scale precipitation are about twice as large as those of convective precipitation (not shown). The magnitude of the total precipitation response reaches values of 25% of the mean climatological precipitation (Plate 4, contours). The mean and anomalous precipitation patterns suggest a weakening of the main North Pacific storm track and a slight enhancement on the poleward side. In other words, the storm track shifted northward and weakened, which is consistent with a weakened meridional temperature gradient [see Hartmann, 1994, section 9.5]. The mechanisms associated with the strengthening of the subtropical high and the storm track changes are not well understood, and a further discussion is included in section 4. Note that the precipitation maximum on the south coast of Alaska is associated with orography, and the reduced onshore winds associated with the SLP response is likely responsible for the coastal precipitation anomalies.

Storm track variability as indicated by 2- to 8-day band-pass-filtered variance statistics for 500-hPa height variance,

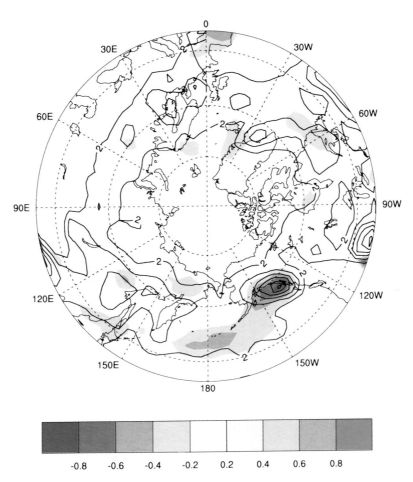

**Plate 4.** Sum95e total precipitation anomalies (shaded) are overlaid with contours of mean total precipitation from the control simulation (Cntle) during August. Anomaly magnitudes greater than 0.2 mm d$^{-1}$ are statistically significant at the 95% or greater level based on a pooled variance $t$ test. CI is 0.2 mm d$^{-1}$ for total precipitation anomalies, where red (blue) shading represents positive (negative) anomalies and values between −0.2 and +0.2 are white. CIs for mean precipitation are 2, 3, 4, 5, 6, and 7 mm d$^{-1}$.

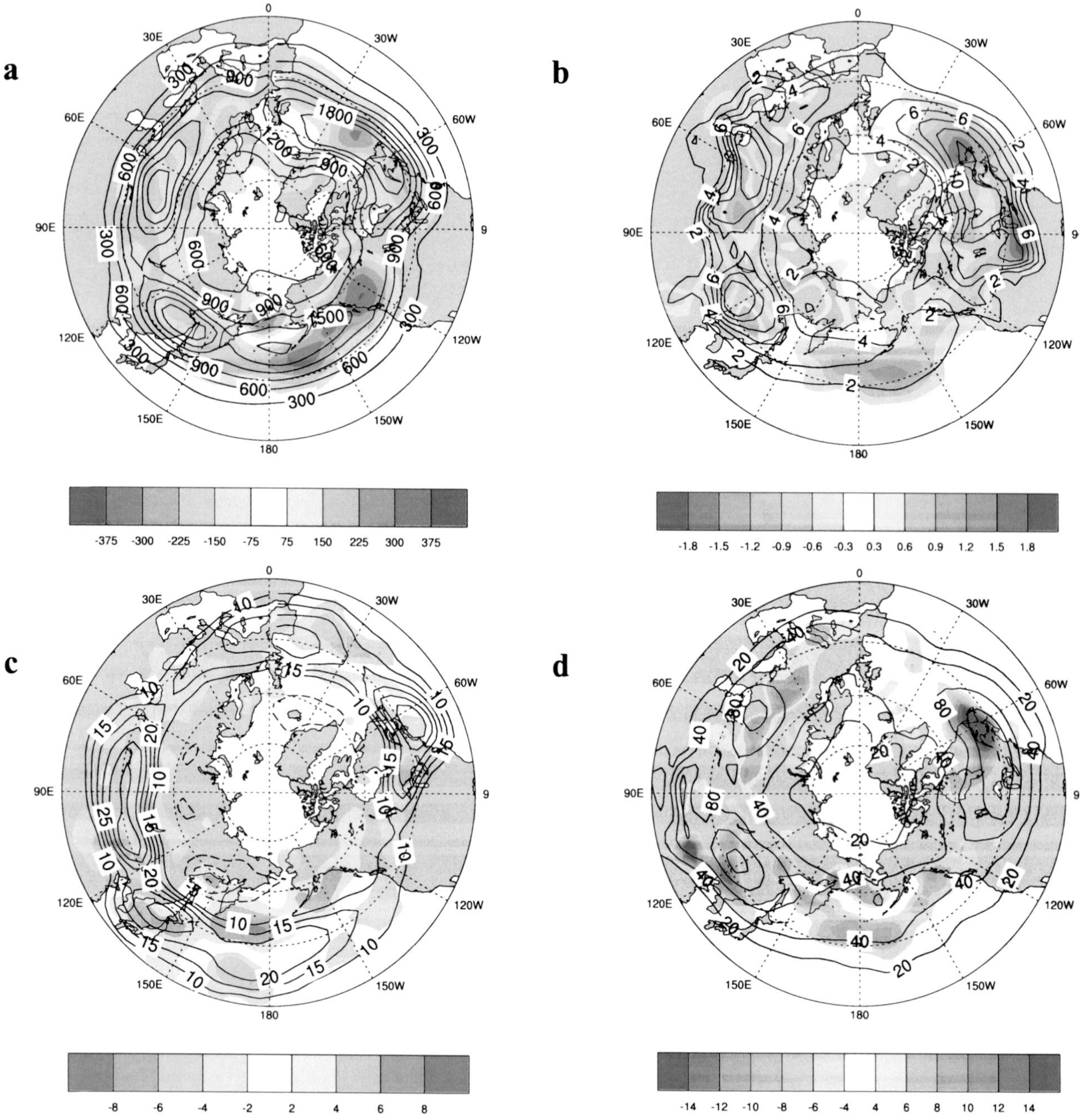

**Plate 5.** August storm track variability as indicated by 2- to 8-day band-passed (a) 500-hPa geopotential height variance, (b) 850-hPa v'T', (c) 200-hPa u'v', and (d) 500-hPa omega variance. The Cntle mean for each quantity is shown by contours, and the Sum95e-Cntle anomalies are shown by shaded values. The units are $m^2$, $m\ s^{-1}\ C$, $m^2\ s^{-2}$, and $10^{-4}\ Pa^2\ s^{-2}$ for Plates 5a, 5b, 5c, and 5d, respectively. In the Pacific sector, statistically significant anomalies have approximate magnitudes greater than $225\ m^2$, $0.9\ m\ s^{-1}\ C$, $5\ m^2\ s^{-2}$, and $6\ (10)^4\ Pa^2\ s^{-2}$ for Plates 5a, 5b, 5c, and 5d, respectively.

850-hPa transient heat flux, 200-hPa transient momentum flux, and 500-hPa omega variance are shown in Plate 5. All of the mean model storm track measures in the control simulation (Cntle) display maxima in the eastern Pacific, eastern North Atlantic, and central Eurasia (Plate 5, contours). These maxima are qualitatively similar to observed maxima [see *Zhang et al.*, 2004, Figure 2b]. Decreased ice (Sum95e) leads to a general weakening of the storm tracks throughout the hemisphere (Plate 5, blue shading); however, the primary significant response is in the Pacific sector. There is a small region of enhanced storm track activity over the east Siberia–Bering Sea region (Plate 5, red shading). Note that an increase in 500-hPa height variance signifies both the passage of more highs as well as lows. The band-passed 500-hPa omega vertical velocity variance anomalies (Plate 5d) are consistent with the height variances. The 2- to 8-day band-passed 850-hPa v'T' or transient eddy heat fluxes are reduced over the mean storm track in the North Pacific and enhanced to the north in eastern Siberia–Bering Sea. In addition, the transient eddy heat fluxes at 850 hPa display significant reductions in storm track activity over North America into the North Atlantic. The 2- to 8-day band-passed 200-hPa u'v' transient eddy momentum fluxes are characterized by increased (decreased) poleward momentum flux to the north

(south) of the mean storm track in the North Pacific, which is consistent with the anomalous high in geopotential height response (Plate 5c). Referring to the geopotential tendency equation, the convergence of vorticity (or momentum) fluxes north of the mean storm track is consistent with the positive equivalent barotropic height anomalies [*Lau and Nath*, 1991]. The significant precipitation anomalies (Plate 4) are consistent with the weakening and northward displacement of the North Pacific storm track (Plate 5).

### 3.3. Diagnosis of Forcing

One possible mechanism for the remote response over the North Pacific involves a Rossby wave train (albeit weak in this case) that is initially excited by diabatic heating anomalies in the Arctic. This wave train propagates into the North Pacific, where through interactions with the storm tracks, an anomalous high is generated over the center of the basin. This mechanism resembles the large-scale eddy feedback described by *Peng et al.* [2003] with the exception that the boundary forcing was close to the storm track in their study.

Diabatic heating anomalies are constructed to investigate the forcing of the atmosphere by reduced sea ice extent. The Cntle mean vertically integrated diabatic heating is shown by contours in Plate 6a and displays cooling of 50–100 W m⁻² over the Arctic. The vertically integrated diabatic heating displays positive anomalies where Arctic sea ice is reduced of 15–25 W m⁻², which is about 10–20% of mean. There is a decrease in the region of the North Pacific storm track (Plate 6a) A vertical cross section through the largest diabatic heating anomalies indicates that in the Arctic the positive heating anomalies are located below 800 hPa, and the negative anomalies in the North Pacific penetrate up to 400 hPa (Plate 6b).

The linear baroclinic model described in section 2 was forced with the transient eddy vorticity fluxes, transient eddy heat fluxes, and mean diabatic heating separately to diagnose the key forcing behind the atmospheric response patterns. The LBM response (Plates 7c–7h) is compared to the full GCM anomalies (Plates 7a and 7b) at 500 and 950 hPa. This diagnostic model analysis reveals that the transient eddy vorticity fluxes are responsible for maintaining the anomalous high in the North Pacific, whereas transient eddy heat fluxes and diabatic heating yield a negligible response. The LBM response to the total transient eddy and diabatic heating is nearly indistinguishable from the response to the transient eddy vorticity fluxes. The primary role of transient eddy vorticity fluxes has been noted in previous studies [*Peng and Whitaker*, 1999; *Deser et al.*, 2007]. The LBM analysis does not reveal how the reduced Arctic ice anomalies induced the eddy momentum fluxes over the North Pacific, perhaps in-

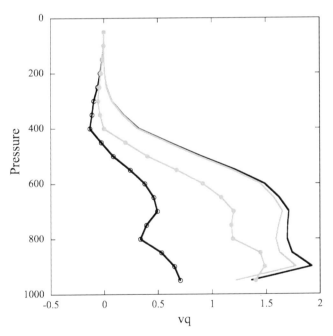

**Figure 2.** August meridional moisture transport in units of g kg⁻¹ m s⁻¹. Vertical profiles are shown at 70°N for Cntle (black lines) and Sum95e (grey). Zonal averages over all longitudes (0°–360°) are shown by the lines with no circles. Averages over the Pacific sector (160°–200°E) are displayed by the lines with circles.

dicating that the diabatic heating anomalies over the Arctic are too shallow and weak to drive the large-scale response. In other words, we argue for an indirect mechanism via the storm track for the ice anomalies to impact the North Pacific rather than impacting the flow directly.

### 3.4. Partial Ice Reduction Sensitivity Experiments

To investigate the sensitivity of the atmospheric response to the placement of the ice anomalies, three experiments were conducted where CCM3 was forced with partial sea ice anomalies from Sum95e ice conditions. The ice was removed in the Kara (Sum95ke) (Plate 8a), Laptev–East Siberian (Sum95le) (Plate 8b), and Beaufort (Sum95be) (Plate 8c) seas. The integration and processing procedure for the partial ice experiments was similar to one used for the full anomaly case (Sum95e) to construct a 51 ensemble member response. The largest positive net surface heat flux anomalies (not shown) into the atmosphere are located directly over grid boxes where ice was removed and are identical to the analogous anomalies from the Sum95e (Plate 2a) simulation. Sum95ke and Sum95le display weak negative heat flux anomalies over eastern Siberia, and Sum95be has significant negative anomalies around 5 W m$^{-2}$.

Surface temperature and SAT responses to the partial ice anomalies are characterized by warming in the vicinity of the reduced ice anomaly, and the magnitudes are nearly identical to those from the full ice experiment. Warm SAT anomalies in far eastern Siberia–Bering region are significant in Sum95le and Sum95be, with the Beaufort ice forcing the largest response in eastern Siberia. The eastern Siberia positive SAT anomalies are consistent with positive advection associated with the anomalous low (high) over the Siberia (North Pacific).

The atmospheric SLP and geopotential height responses to the partial ice experiments resemble that of Sum95e. A weak high over the Kara Sea, a weak low over east Siberia stretching into the Chukchi Sea, and the anomalous high in the North Pacific are all common features of the SLP and geopotential height response patterns to partial ice anomalies (Plates 8d–8i). The individual responses shown in Plate 8 are weaker than the response in Sum95e; however, the sum of these three partial ice experiment response patterns in Plate 8 for SLP and 500-hPa geopotential height is nearly twice as strong as the response to the total ice anomaly (Sum95e). Ice reductions in the Laptev–East Siberian and Beaufort seas produce a statistically significant response in the North Pacific. The anomalous low (SLP and 500-hPa height) over east Siberia is stronger in the Beaufort partial ice experiment than in Sum95e. This suggests that the model atmosphere is sensitive to ice reductions in all three

of these regions and the induced climate anomalies are fairly similar.

The precipitation response (not shown) patterns to the partial ice anomalies are sensitive to the location of the ice anomalies. The positive precipitation anomalies over eastern Siberia are weakly evident in Sum95le and are significant in Sum95be. The negative precipitation anomalies in the mean model storm track zone are overall largest for Sum95be, largest over south coastal Alaska for Sum95ke, and significant for a limited area over the ocean for Sum95le. The sum of the precipitation anomalies for the three partial ice experiments is slightly larger than the precipitation anomalies for Sum95e.

### 3.5. Ice Concentration Experiments

The August 1995 experiment was repeated using ice concentration anomalies (Sum95c) (Plate 9a). The net surface heat fluxes (not shown) and SAT were similar to Sum95e. One feature different from the Sum95e ensemble average is an area of significant negative surface air temperature anomaly between 120° and 150°E in eastern Siberia (Plate 9b). The Sum95c SLP response has a weaker anomalous high in the North Pacific and a stronger anomalous low in eastern Siberia compared to the extent experiment. The atmospheric response at 500 hPa is similar though it looks more like a wave train in the Pacific (Plate 9c). During summer the contrast between using ice extent and concentration is small, whereas the differences are larger in winter [*Alexander et al.*, 2004]. The area of open water is slightly larger during summer than winter (compare the difference between the two time series in Figure 1 in 1995 with the difference for 1996 in Figure 1 of *Alexander et al.* [2004]), but the larger air-sea temperature contrast in winter strongly influences the turbulent heat fluxes. Anomalies for Sum95c are constructed by taking the difference between the concentration experiment (Sum95c) and an extent control (Cntle). Large ice anomaly differences exist between the concentration control (Cntlc) and Cntle, complicating the interpretation of the differences when a concentration control is used, and thus the Cntlc experiments are not used as a baseline here.

### 4. CONCLUSIONS

This study employs an atmospheric global climate model (CCM 3.6) to examine the atmospheric response to observed variations in sea ice during the summer of 1995, which had the lowest ice extent during June–September in the Arctic over the last ~30 years with the exception of 2007. (The September ice minimum has been near or well below the 1995 levels since 2002 [*Stroeve et al.*, 2008]). Sea ice was

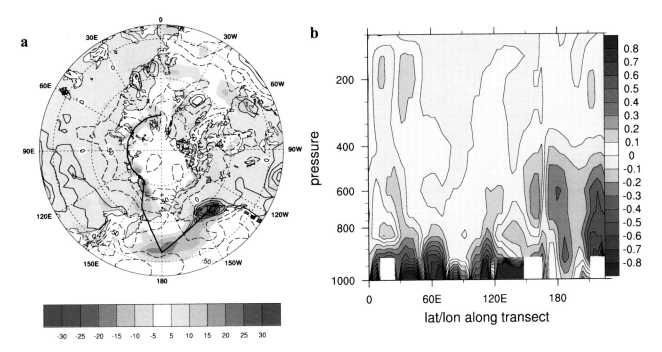

**Plate 6.** August (a) mean (Cntle, contours) and anomalous (Sum95e-Cntle, shaded) vertically integrated diabatic heating rate. The path of the transect is shown by a thick black line in Plate 6a. (b) Transect through the Arctic into the North Pacific showing total diabatic heating rate anomalies. The units are W m$^{-2}$ in Plate 6a for both shaded and contoured fields. CI is 0.1 K d$^{-1}$ in Plate 6b. The total diabatic heating rate is the sum of convective adjustment, solar heating, longwave heating, vertical diffusion, and horizontal diffusion.

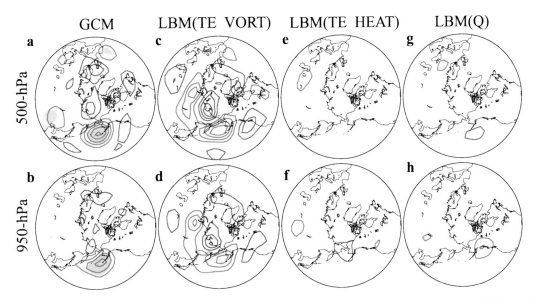

**Plate 7.** LBM results in August for Sum95e. Geopotential height GCM response to reduced sea ice at (a) 500 hPa and (b) 950 hPa. Individual LBM response to (c and d) transient eddy vorticity, (e and f) transient eddy heat fluxes, and (g and h) diabatic heating. CI is 5 m where red (blue) signifies positive (negative) height anomalies. The LBM response has been multiplied by 2.5 to match the magnitude of the GCM response.

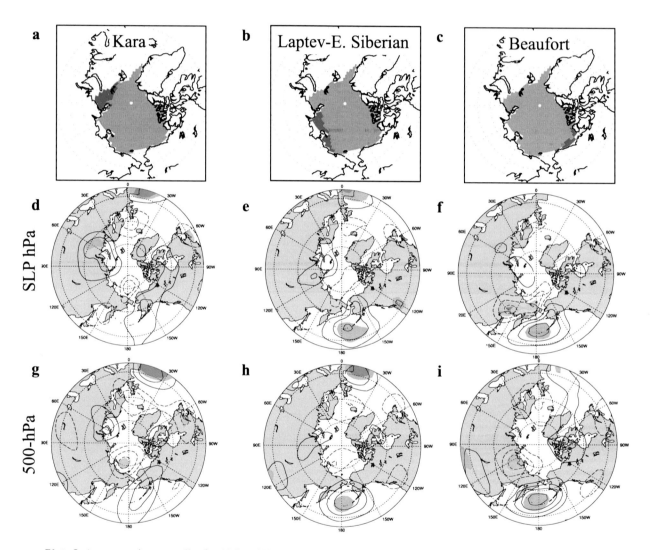

**Plate 8.** August sea ice anomalies for (a) Sum95ke, (b) Sum95le, and (c) Sum95be where ice is reduced only in the Kara Sea, Laptev–East Siberian seas, and the Beaufort Sea, respectively. (d–f) SLP anomalies and (g–i) 500-hPa geopotential height anomalies in response to reduced sea ice in individual seas. Dark (light) shading indicates statistical significance at the 99% (95%) or greater level based on a pooled variance *t* test. CI is 0.5 hPa for SLP and 5 m for geopotential height.

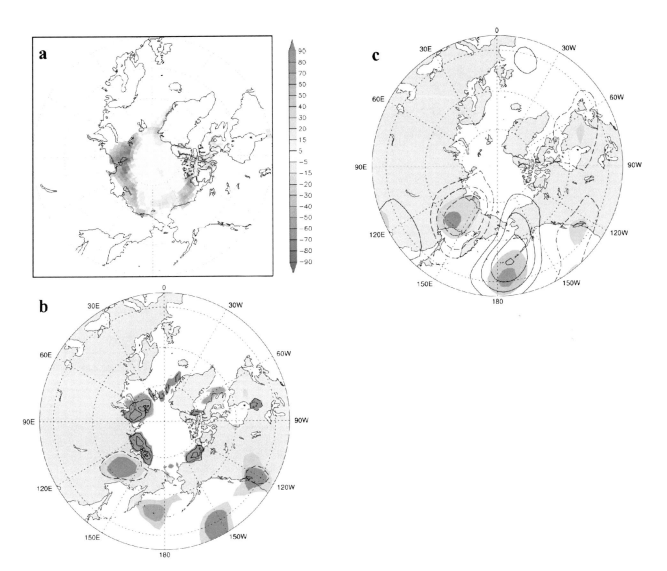

**Plate 9.** Sum95c August ice concentration anomalies in (a) percent of area, (b) surface air temperature with a CI of 0.5 K, and (c) 500-hPa height with a CI of 5 m.

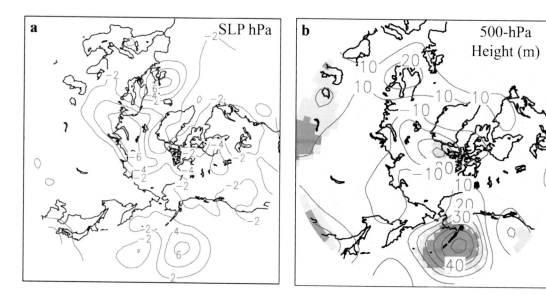

**Plate 10.** (a) Observed SLP anomalies in August 1995 in hPa. The mean climatology is based on the years 1968–1996, and the CI is 2 hPa. (b) Observed 500-hPa geopotential height composites based on reduced ice in the Kara Sea region. The 500-hPa height is in meters. Shading indicates statistical significance at the 95% or greater level based on a pooled variance *t* statistic. The years used in this composite are 1979, 1984, 1985, 1994, 1995, 1997, and 2000.

prescribed as ice extent (ocean grid box is either completely covered or totally ice free) or ice concentration (partial grid box covered in ice allowed) based on monthly observations. Fifty-one ensemble members were integrated from April to October 1995 using climatological sea surface temperatures. The control simulation was integrated with global climatological sea ice extent and SSTs. The strongest response was found during the month of August when the ice area is nearing its minimum for the year.

The Arctic displays a local thermal response with increased surface heat fluxes (sensible plus latent plus longwave) into the atmosphere, warmer SATs, and a weak decrease in SLP. The atmospheric response is also characterized by an anomalous high in sea level pressure in the North Pacific, which is part of a northward expansion of the summertime subtropical high. The atmospheric response with height is equivalent barotropic, and the anomalous high increases in amplitude with height and is significant at 200 hPa. There is a significant decrease (increase) of precipitation along the eastern (northwestern) part of the mean North Pacific storm track, consistent with the 500-hPa geopotential height variances and 850-hPa transient eddy heat fluxes that indicate enhanced storminess north of the mean storm track and a decrease over the mean storm track in the North Pacific.

Additional climate experiments were conducted to determine the model sensitivity to the location of sea ice anomalies. When ice reduction is limited to only the Kara Sea, the Laptev–East Siberian seas, or the Beaufort Sea the atmospheric response patterns for SLP, geopotential height, and precipitation are similar but weaker than when the sea ice is reduced for all the seas, suggesting that the model is sensitive to sea ice anomalies in all three regions. The area of the significant response increases from the Kara to the Beaufort, which is closest to the North Pacific. These results are analogous to a GCM study by *Geisler et al.* [1985] where the model Pacific North American response pattern (magnitude) is insensitive (sensitive) to the longitude of the tropical Pacific SST anomaly. The August 1995 experiment was repeated using ice concentration anomalies. The atmospheric response is similar though it resembles a wave train in the Pacific, similar to what *Alexander et al.* [2004] found for the response during winter to sea ice concentration extremes during winter of 1995–1996.

There has been increased interest recently in understanding mechanisms that force and maintain the summertime subtropical highs. In a zonal average, the subtropical highs are strongest in winter when subsidence associated with the Hadley circulation is most vigorous [*Grotjahn and Osman*, 2007]. However, the North Pacific (NP) high is strongest during boreal summer [see *Grotjahn*, 2004, Figure 1] and forms to the west of a region with strong thermal contrast

between the cool ocean water and the warm North American landmass. *Miyasaka and Nakamura* [2005] employed a nonlinear spectral primitive equation model driven by zonally asymmetric diabatic heating and demonstrated that the strong surface thermal contrast can explain ~70% of the strength of the subtropical high, consistent with ideas first proposed by *Hoskins* [1996]. *Grotjahn* [2004] proposed that extratropical storms could provide forcing through transient eddies to maintain the subtropical high. *Grotjahn and Osman* [2007, Figure 2] present a conceptual picture of how ageostrophic motions arising from developing storms converge at the jet level, leading to sinking motion on the east side of the subtropical high and low-level divergence and southward motion that strengthens the subtropical high. They demonstrate that the variability of the NP high is dominated by midlatitude forcing during summer. Some of the features found in a warm season SLP composite analysis of observations by *Grotjahn and Osman* [2007] are qualitatively similar to circulation anomalies forced by reduced sea ice in CCM3. They find that SLP is weaker in parts of the Arctic Ocean when the North Pacific high is stronger and a stronger North Pacific high is associated with positive SLP anomalies on the northern flank of the high.

The LBM analysis suggests that the far field response is not forced directly by the Arctic ice but could rather be a consequence of the local Arctic response, which acts to reduce the flow between the Arctic and the lower latitudes. There may be some parallel with modeling studies of the response to Antarctic sea ice extremes. *Hudson and Hewitson* [2001] have examined the response to realistic monthly varying sea ice and SST anomalies in the Antarctic. They found that where the sea ice has been reduced and ocean exposed, the SAT increases and there is a strengthening and a southward extension of the subtropical high-pressure belt. *Raphael* [2003] found complementary results using the NCAR CCSM.

The dominant mode of variability determined from Cntle empirical orthogonal function (EOF1) of SLP in August resembles the Arctic Oscillation. The model response to reduced summer ice does not correspond to the dominant mode for SLP or 500-hPa heights. Given results from previous studies we hypothesize that the reason that this occurs is because the ice anomaly is located far from the storm tracks. *Honda et al.* [1999] and *Alexander et al.* [2004] found that wintertime North Pacific ice edge anomalies, located well north of the average storm track, do not project on the dominant modes of the GCM. In contrast, *Deser et al.* [2004], *Magnusdottir et al.* [2004], and *Alexander et al.* [2004] found that the GCM response to ice edge anomalies in the North Atlantic during winter strongly project on the dominant modes of variability. The storm track is located nearly above the ice edge. *Glowienka-Hense and Hense* [1992] forced a GCM with a polynya in the Kara Sea. Their ice anomaly was in the pack ice far from the ice edge and their response was weak local heating with a general weakening of the Atlantic storm track, very similar to our Sum95e response. They argue that open water in the ice pack yields a different response than at the ice edge.

Perhaps, a parallel can be drawn from the better understood topic of the atmospheric response to midlatitude SSTs, where it has been shown that the atmospheric response is highly sensitive to the location of the SST forcing with respect to the climatological flow [see *Kushnir et al.*, 2002, and references therein]. A conundrum in our results is that the partial ice anomaly experiments (Sum95ke, Sum95le, and Sum95be) all yield very similar patterns to each other and the full ice anomaly. This finding would be consistent with the response projecting on a key mode of natural variability. So, having examined the first four EOF patterns, it is unclear at this point whether a less dominant mode of variability is being excited by the sea ice anomalies.

The atmospheric circulation response to extreme sea ice anomalies is explored in the context of how they may feed back onto the sea ice. A strong negative feedback was suggested in the winter sea ice forcing GCM studies [*Alexander et al.*, 2004; *Deser et al.*, 2004] where the atmospheric response was of the opposite sign to the circulation that initially forced the sea ice anomalies. The exchanges of latent heat between the Arctic north of 70°N and the midlatitudes are largest during August as shown in a study of the observed energy budget of the Arctic [see *Serreze et al.*, 2007, Figure 6]. Increased moisture in the Arctic has been shown to enhance downward longwave fluxes and possibly impact the sea ice [*Francis and Hunter*, 2006]. Figure 2 presents the ensemble averaged vertical profiles of meridional moisture transport in Sum95e (grey line) and Cntle (dark line) at 70°N averaged over all longitudes (plain lines) and in the Pacific sector (lines with dots) for 160°–200°E. The global average moisture transports into the Arctic cap do not differ much between Cntle and Sum95e. However, in the Pacific sector the poleward moisture transport is enhanced notably in the lower 500 hPa. This increase of moisture would trap more longwave radiation and would work to delay ice formation, suggesting a positive feedback.

Observed atmospheric circulations present during reduced Arctic sea ice summers resemble the model response found in our study. During August 1995 the observed SLP field displays a negative anomaly over the Arctic and an anomalous high over the North Pacific (Plate 10a), which compares favorably with the model response to reduced sea ice. Plate 10b presents a 7-year composite of August 500-hPa anomalies based on summers with anomalously low sea ice in the Kara Sea. The anomalous high in the North Pacific is strikingly similar to the model response at 500 hPa (Plate 3b). This pair of panels was chosen to illustrate that the model response compares well with observations during 1995 as well as in a more robust measure based on composites. The similarity between the observations and the model results suggests that realistic Arctic sea ice decreases may force circulation changes in the North Pacific and warrants further examination.

*Acknowledgments.* This work benefited from discussions with H. Nakamura, W. Robinson, S. Peng, R. Grotjahn, N. Mölders, and I. Polyakov. S. Bourne is thanked for a careful reading of the manuscript. We deeply appreciate the through critiques received from an anonymous reviewer and E. DeWeaver that improved this paper. This research was supported by a grant from the NOAA's Arctic Research Office issued through the International Arctic Research Center (IARC), the Frontier Research System for Global change through IARC, and by the Geophysical Institute. Support was also provided by the National Science Foundation through grant ARC-0327664. We also thank Steve Worley at NCAR for providing the HadISST data set for N. Raynor of the Hadley Centre. This work was supported in part by a grant of HPC resources from the Arctic Region Supercomputing Center at the University of Alaska Fairbanks as part of the Department of Defense High Performance Computing Modernization Program. We thank G. Robinson, C. Swingley, and W. Chapman for their assistance with various computer issues. Plots have been prepared using the open source software packages NCL (www.ncl.ucar.edu) and GrADS (www.iges.org/grads/).

## REFERENCES

Alexander, M. A., U. S. Bhatt, J. Walsh, M. Timlin, and J. Miller (2004), The atmospheric response to realistic Arctic sea ice anomalies in an AGCM during winter, *J. Clim.*, *17*, 890–905.

Bitz, C. M., J. C. Fyfe, and G. M. Flato (2002), Sea ice response to wind forcing from AMIP models, *J. Clim.*, *15*, 522–536.

Briegleb, B. P., and D. H. Bromwich (1998a), Polar radiation budgets of the NCAR CCM3, *J. Clim.*, *11*, 1246–1269.

Briegleb, B. P., and D. H. Bromwich (1998b), Polar climate simulation of the NCAR CCM3, *J. Clim.*, *11*, 1270–1286.

Comiso, J. C., C. L. Parkinson, R. Gersten, and L. Stock (2008), Accelerated decline in the Arctic sea ice cover, *Geophys. Res. Lett.*, *35*, L01703, doi:10.1029/2007GL031972.

Deser, C., and H. Teng (2008), Recent trends in Arctic sea ice and the evolving role of atmospheric circulation forcing, 1979–2007, this volume.

Deser, C., J. E. Walsh, and M. Timlin (2000), Arctic sea ice variability in the context of recent atmospheric circulation trends, *J. Clim.*, *13*, 617–633.

Deser, C., G. Magnusdottir, R. Saravanan, and A. Phillips (2004), The effects of North Atlantic SST and sea ice anomalies on the winter circulation in CCM3. Part II: Direct and indirect components of the response, *J. Clim.*, *17*, 877–889.

Deser, C., R. A. Thomas, and S. Peng (2007), The transient atmospheric circulation response to North Atlantic SST and sea ice anomalies, *J. Clim.*, *20*, 4751–4767.

DeWeaver, E., and C. M. Bitz (2006), Atmospheric circulation and its effect on Arctic sea ice in CCSM3 simulations at medium and high resolution, *J. Clim.*, *19*, 2415–2436.

Francis, J. A., and E. Hunter (2006), New insight into the disappearing Arctic sea ice, *Eos Trans. AGU*, *87*(46), 509.

Geisler, J. E., M. L. Blackmon, G. T. Bates, and S. Muñoz (1985), Sensitivity of January climate response to the magnitude and position of equatorial Pacific sea surface temperature anomalies, *J. Atmos. Sci.*, *42*, 1037–1049.

Glowienka-Hense, R., and A. Hense (1992), The effect of an Arctic polynya on the Northern Hemisphere mean circulation and eddy regime: A numerical experiment, *Clim. Dyn.*, *7*(3), 155–163.

Gorodetskaya, I. V., and L.-B. Tremblay (2008), Arctic cloud properties and radiative forcing from observations and their role in sea ice decline predicted by the NCAR CCSM3 model during the 21st century, this volume.

Grotjahn, R. (2004), Remote weather associated with South Pacific subtropical sea-level high properties, *Int. J. Climatol.*, *24*, 823–839.

Grotjahn, R., and M. Osman (2007), Remote weather associated with North Pacific subtropical sea-level high properties, *Int. J. Climatol.*, *27*, 587–602.

Hack, J. J., J. T. Kiehl, and J. W. Hurrell (1998), The hydrologic and thermodynamic characteristics of the NCAR CCM3, *J. Clim.*, *11*, 1151–1178.

Hartmann, D. (1994), *Global Physical Climatology*, 411 pp., Academic, London.

Honda, M., K. Yamazaki, H. Nakamura, and K. Takeuchi (1999), Dynamic and thermodynamic characteristics of atmospheric response to anomalous sea-ice extent in the Sea of Okhotsk, *J. Clim.*, *12*, 3347–3358.

Hoskins, B. (1996), On the existence and strength of the summer subtropical anticyclones, *Bull. Am. Meteorol. Soc.*, *77*, 1287–1292.

Hudson, D. A., and B. C. Hewitson (2001), The atmospheric response to a reduction in summer Antarctic sea-ice extent, *Clim. Res.*, *16*, 79–99.

Hurrell, J. W., J. J. Hack, B. A. Boville, D. L. Williamson, and J. T. Kiehl (1998), The dynamical simulation of the NCAR Community Climate Model version 3, *J. Clim.*, *11*, 1207–1236.

Hurrell, J. W., J. J. Hack, A. S. Phillips, J. Caron, and J. Yin (2006), The dynamical simulation of the Community Atmosphere Model version 3 (CAM3), *J. Clim.*, *19*, 2162–2183.

Kiehl, J. T., J. J. Hack, G. B. Bonan, B. A. Boville, and P. J. Rasch (1998), The National Center for Atmospheric Research Community Climate Model: CCM3, *J. Clim.*, *11*, 1131–1149.

Kushnir, Y., and N.-C. Lau (1992), The general circulation model response to a North Pacific SST anomaly: Dependence on time scale and pattern polarity, *J. Clim.*, *5*, 271–283.

Kushnir, Y., W. A. Robinson, I. Bladé, N. M. J. Hall, S. Peng, and R. Sutton (2002), Atmospheric GCM response to extratropical SST anomalies: Synthesis and evaluation, *J. Clim.*, *15*, 2233–2256.

Lau, N.-C., and M.-J. Nath (1991), Variability of the baroclinic and barotropic transient eddy forcing associated with monthly changes in the midlatitude storm tracks, *J. Atmos. Sci.*, *48*, 2589–2613.

Magnusdottir, G., C. Deser, and R. Saravanan (2004), The effects of North Atlantic SST and sea ice anomalies on the winter circulation in CCM3. Part I: Main features and storm track characteristics of the response, *J. Clim.*, *17*, 857–876.

Maslanik, J., S. Drobot, C. Fowler, W. Emery, and R. Barry (2007), On the Arctic climate paradox and the continuing role of atmospheric circulation in affecting sea ice conditions, *Geophys. Res. Lett.*, *34*, L03711, doi:10.1029/2006GL028269.

Maslanik, J. A., M. C. Serreze, and R. G. Barry (1996), Recent decreases in Arctic summer ice cover and linkages to atmospheric circulation anomalies, *Geophys. Res. Lett.*, *23*, 1677–1680.

Miyasaka, T., and H. Nakamura (2005), Structure and formation mechanisms of the Northern Hemisphere summertime subtropical highs, *J. Clim.*, *18*, 5046–5065.

Parkinson, C., D. Rind, R. J. Healy, and D. G. Martinson (2001), The impact of sea ice concentration accuracies on climate model simulations with the GISS GCM, *J. Clim.*, *14*, 2606–2623.

Peng, S., and J. S. Whitaker (1999), Mechanisms determining the atmospheric response to midlatitude SST anomalies, *J. Clim.*, *12*, 1393–1408.

Peng, S., W. A. Robinson, and S. Li (2003), Mechanisms for the NAO responses to the North Atlantic SST tripole, *J. Clim.*, *16*, 1987–2004.

Polyakov, I., et al. (2007), Observational program tracks Arctic Ocean transition to a warmer state, *Eos Trans. AGU*, *88*(40), 398.

Polyakov, I. V., et al. (2005), One more step toward a warmer Arctic, *Geophys. Res. Lett.*, *32*, L17605, doi:10.1029/2005GL023740.

Randall, D., J. Curry, D. Battisti, G. Flato, R. Grumbine, S. Hakkinen, D. Martinson, R. Preller, J. Walsh, and J. Weatherly (1998), Status of and outlook for large-scale modeling of atmosphere-ice-ocean interactions in the Arctic, *Bull. Am. Meteorol. Soc.*, *79*, 197–219.

Raphael, M. N. (2003), Impact of observed sea-ice concentration on the Southern Hemisphere extratropical atmospheric circulation in summer, *J. Geophys. Res.*, *108*(D22), 4687, doi:10.1029/2002JD003308.

Raymo, M. E., D. Rind, and W. F. Ruddiman (1990), Climatic effects of reduced Arctic sea ice limits in the GISS II general circulation model, *Paleoceanography*, *5*, 367–382.

Rayner, N. A., D. E. Parker, E. B. Horton, C. K. Folland, L. V. Alexander, D. P. Rowell, E. C. Kent, and A. Kaplan (2003), Global analyses of sea surface temperature, sea ice, and night marine air temperature since the late nineteenth century, *J. Geophys. Res.*, *108*(D14), 4407, doi:10.1029/2002JD002670.

Rigor, I. G., J. M. Wallace, and R. L. Colony (2002), Response of sea ice to the Arctic oscillation, *J. Clim.*, *15*(18), 2648–2668.

Rinke, A., W. Maslowski, K. Dethloff, and J. Clement (2006), Influence of sea ice on the atmosphere: A study with an Arctic atmospheric regional climate model, *J. Geophys. Res.*, *111*, D16103, doi:10.1029/2005JD006957.

Serreze, M. C., F. Carse, and R. Barry (1997), Icelandic low cyclone activity: Climatological features, linkages with the NAO, and relationships with recent changes in the Northern Hemisphere circulation, *J. Clim.*, *10*, 453–464.

Serreze, M. C., A. P. Barrett, A. G. Slater, M. Steele, J. Zhang, and K. E. Trenberth (2007), The large-scale energy budget of the Arctic, *J. Geophys. Res.*, *112*, D11122, doi:10.1029/2006JD008230.

Sewall, J. O. (2005), Precipitation shifts over western North America as a result of declining Arctic sea ice cover: The coupled system response, *Earth Interact.*, *9*, paper 26, doi:10.1175/EI171.1.

Singarayer, J. S., J. L. Bamber, and P. J. Valdes (2006), Twenty-first-century climate impacts from a declining Arctic sea ice cover, *J. Clim.*, *19*, 1109–1125.

Steele, M., W. Ermold, and J. Zhang (2008), Arctic Ocean surface warming trends over the past 100 years, *Geophys. Res. Lett.*, *35*, L02614, doi:10.1029/2007GL031651.

Stroeve J., M. M. Holland, W. Meier, T. Scambos, and M. Serreze (2007) Arctic sea ice decline: Faster than forecast, *Geophys. Res. Lett.*, *34*, L09501, doi:10.1029/2007GL029703.

Stroeve J., M. Serreze, S. Drobot, S. Gearheard, M. Holland, J. Maslanik, W. Meier, and T. Scambos (2008), Arctic sea ice extent plummets in 2007, *Eos Trans. AGU*, *89*(2), 13.

Ting, M., and S. Peng (1995), Dynamics of the early and middle winter atmospheric responses to northwest Atlantic SST anomalies, *J. Clim.*, *8*, 2239–2254.

Walsh, D., I. Polyakov, L. Timokhov, and E. Carmack (2007), Thermohaline structure and variability in the eastern Nansen Basin as seen from historical data, *J. Mar. Res.*, *65*, 685–714.

Walsh, J. E., W. L. Chapman, and T. L. Shy (1996), Recent decrease of sea level pressure in the central Arctic, *J. Clim.*, *9*, 480–486.

Wallace, J. M., Y. Zhang, and L. Bajuk (1996), Interpretation of interdecadal trends in Northern Hemisphere surface air temperature, *J. Clim.*, *9*, 249–259.

Zhang X., J. E. Walsh, J. Zhang, U. S. Bhatt, and M. Ikeda (2004), Climatology and interannual variability of Arctic cyclone activity: 1948–2002, *J. Clim.*, *17*, 2300–2317.

M. A. Alexander and J. Scott, NOAA Earth System Research Laboratory, Boulder, CO 80305, USA.

U. S. Bhatt, Geophysical Institute, Department of Atmospheric Sciences, University of Alaska Fairbanks, P.O. Box 75-7320, Fairbanks, AK 99775-7320, USA. (bhatt@gi.alaska.edu)

C. Deser and R. A. Tomas, National Center for Atmospheric Research, Boulder, CO 80307, USA.

J. S. Miller, ARSC, University of Alaska Fairbanks, Fairbanks, AK 99775, USA.

M. S. Timlin, Department of Atmospheric Sciences, University of Illinois at Urbana-Champaign, Urbana, IL 61801, USA.

J. E. Walsh, International Arctic Research Center, University of Alaska Fairbanks, Fairbanks, AK 99775, USA.

# Sea Ice–Albedo Feedback and Nonlinear Arctic Climate Change

Michael Winton

*Geophysical Fluid Dynamics Laboratory, NOAA, Princeton, New Jersey, USA*

The potential for sea ice–albedo feedback to give rise to nonlinear climate change in the Arctic Ocean region, defined as a nonlinear relationship between polar and global temperature change or, equivalently, a time-varying polar amplification, is explored in the Intergovernmental Panel on Climate Change climate models. Five models supplying Special Report on Emissions Scenario A1B ensembles for the 21st century are examined, and very linear relationships are found between polar and global temperatures (indicating linear polar region climate change) and between polar temperature and albedo (the potential source of nonlinearity). Two of the climate models have Arctic Ocean simulations that become annually sea ice–free under the stronger $CO_2$ increase to quadrupling forcing. Both of these runs show increases in polar amplification at polar temperatures above $-5°C$, and one exhibits heat budget changes that are consistent with the small ice cap instability of simple energy balance models. Both models show linear warming up to a polar temperature of $-5°C$, well above the disappearance of their September ice covers at about $-9°C$. Below $-5°C$, effective annual surface albedo decreases smoothly as reductions move, progressively, to earlier parts of the sunlit period. Atmospheric heat transport exerts a strong cooling effect during the transition to annually ice-free conditions, counteracting the albedo change. Specialized experiments with atmosphere-only and coupled models show that the main damping mechanism for sea ice region surface temperature is reduced upward heat flux through the adjacent ice-free oceans resulting in reduced atmospheric heat transport into the region.

## 1. INTRODUCTION

The speculation that Arctic climate has nonlinear behaviors associated with sea ice albedo feedback has deep roots in climatology [*Brooks*, 1949; *Donn and Ewing*, 1968]. Energy balance models (EBMs) were used to study ice albedo effects starting in the late 1960s, and one of the first uses of atmospheric global climate models (GCMs) was to explore

Arctic Sea Ice Decline: Observations, Projections, Mechanisms, and Implications
Geophysical Monograph Series 180
This paper is not subject to U.S. copyright. Published in 2008 by the American Geophysical Union.
10.1029/180GM09

the climatic impact of the Arctic sea ice cover (see review by *Royer et al.* [1990]). Although climate models show that global temperature change is mainly linear in climate forcing over a broad range [*Manabe and Stouffer* 1994; *Hansen et al.*, 2005], the nonlinear relationship between ice albedo and temperature may introduce local nonlinearity. Simple diffusive energy balance models, that represent this relationship with a step function, produce an abrupt disappearance of polar ice as the global climate gradually warms [*North*, 1984]. The phenomenon is known as the small ice cap instability (SICI) as it disallows polar ice caps smaller than a certain critical size related to heat diffusion and radiative damping parameters. *Thorndike* [1992] coupled an atmospheric energy balance model to a simple analytical model of sea ice in an ocean mixed layer, thereby simulating rather than parameterizing the albedo temperature relationship, and found that seasonally ice-free states were unstable. Under increased forcing, Thorndike's "toy" model transitions directly from annually ice-covered to annually ice-free states inducing a large and abrupt increase in surface temperature.

The Arctic sea ice cover has been in decline since the 1950s [*Vinnikov et al.*, 1999]. This decline is more pronounced in the summer, and recent years have produced striking record minima [*Stroeve et al.*, 2005]. Some researchers have noted that nonlinear behaviors such as thresholds and tipping points may be associated with this decline [*Lindsay and Zhang*, 2005; *Serreze and Francis*, 2006]. The goal of this paper is to assess the potential for nonlinearity of Arctic climate change in the Intergovernmental Panel on Climate Change (IPCC) fourth assessment report (AR4) climate models. In section 2 we demonstrate potential nonlinearities in a simple model and develop a strategy for assessment. In section 3 we examine 21st century simulations for signs of nonlinearity. Section 4 continues this search by examining two strongly forced experiments as they become annually ice free. Section 5 shows that the nonlinear behavior of one of these experiments is similar to the EBM SICI. Section 6 explores the stabilizing effect of ocean surface fluxes and atmospheric heat transport on the sea ice with special GCM experiments designed to illuminate the climate response to sea ice region changes. Section 7 summarizes and discusses the results.

## 2. ELEMENTARY ARCTIC CLIMATE DYNAMICS

The potential for a nonlinear relationship between ice albedo and temperature to generate nonlinear climate change can be demonstrated with a very simple energy balance model. Consider the energy balance at the top of an isolated polar atmosphere:

$$A + BT = S[1 - \alpha(T)]. \qquad (1)$$

The model represents a balance between absorbed shortwave radiation, insolation (S) times a planetary coalbedo $(1 - \alpha)$, and parameterized outgoing longwave radiation with a linear dependence on surface temperature, T. The model is isolated in the sense that the atmospheric heat transport convergence is held fixed, bundled with the longwave intercept into A. The nonlinearity of the model comes from the nonlinear dependence of $\alpha$ on T. At very low mean temperatures, where snow never melts, albedo is insensitive to temperature. The same is true at high mean temperatures where there is no ice. Between these flat sections, there is a drop from snow to seawater albedos. One might expect, on the basis of the liquid/ice transition occurring at a fixed tem-

perature, that this drop would resemble a cliff. However, the seasonal cycle and other variability allow sampling of various ice-cover states at any given long-term mean temperature, smoothing the relationship. For simplicity, let us take this smoothed section to be linear and call it the ramp. The slope of the ramp depends on the drop in albedo between its endpoints and the temperature range over which the drop is experienced. The drop in planetary albedo, the albedo above the atmosphere, will be less than the jump in surface albedo because only part of the insolation reaches and interacts with the surface. There may also be changes in atmospheric properties with temperature that impact the planetary albedo drop. *Gorodetskaya et al.* [2006] have used satellite sea ice cover and shortwave data to estimate the albedo drop for Northern Hemisphere sea ice regions. They obtain a 0.22 planetary albedo change for a 100% change in sea ice cover. This is roughly half the surface albedo difference between a typical sea ice cover and seawater.

The balance expressed in (1) is depicted schematically in Plate 1. The steepness of the albedo ramp, the drop divided by the ramp temperature range, impacts the character of the nonlinearity. In particular, if the ramp is so steep that, as warming occurs, the extra shortwave absorption exceeds the extra loss of energy from outgoing longwave radiation (OLR), the total feedback will be positive, and there will be unstable transitions between ice-covered and ice-free states. This is an example of the slope stability theorem of energy balance models (see *Crowley and North* [1991, pp. 18–19] for an elementary discussion). We can form an expression for the critical ramp temperature range, $\Delta T_C$, between stable and unstable solutions:

$$\Delta T_C = S \Delta \alpha / B. \qquad (2)$$

A larger range is needed for stabilization when the insolation and the albedo drop are large and when the longwave damping is small.

Plate 1 shows schematic examples of subcritical and supercritical shortwave absorption profiles. When the albedo ramp steepness is supercritical, the total feedback is positive in the ramp temperature range so it will contain only unstable equilibria. As a result, the ramp range becomes a forbidden zone, inaccessible with any forcing. As forcing is slowly varied, these temperatures are skipped leading to a discontinuity in polar temperature. Since the polar temperature contributes to the global mean temperature, it would also have a (much smaller) temperature discontinuity when the ramp steepness is supercritical.

If we insert the insolation at the North Pole (173 W m$^{-2}$), the *Gorodetskaya et al.* [2006] planetary albedo drop, and a satellite-estimated OLR damping value (1.5 W m$^{-2}$ K$^{-1}$

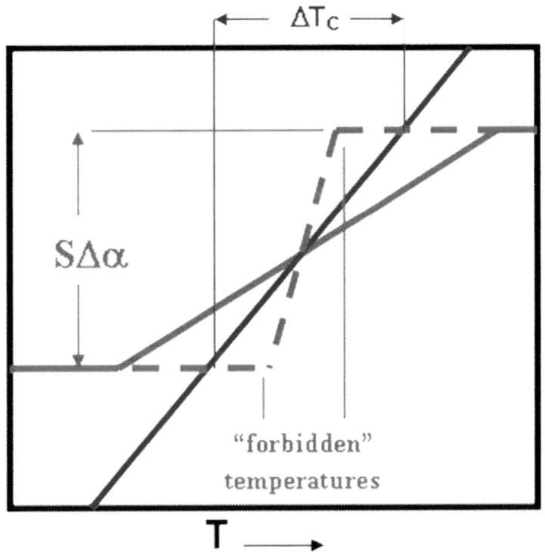

**Plate 1.** Schematic of the top-of-atmosphere absorbed shortwave radiation (red) and OLR (blue) as a function of surface temperature. The solid red absorbed shortwave profile is stable because the surface albedo transition occurs over a range of temperatures greater than $\Delta T_c$ and so is less steeply sloped than the OLR curve. The dashed profile is unstable since the transition occurs over a smaller range of temperatures, leading to a region where increased shortwave absorption exceeds OLR cooling as temperature increases (a net positive feedback). Within this temperature range is a "forbidden" zone where no equilibrium is possible under changes in forcing. (Forcing changes can be thought of as moving the blue OLR curve left or right without changing its slope.)

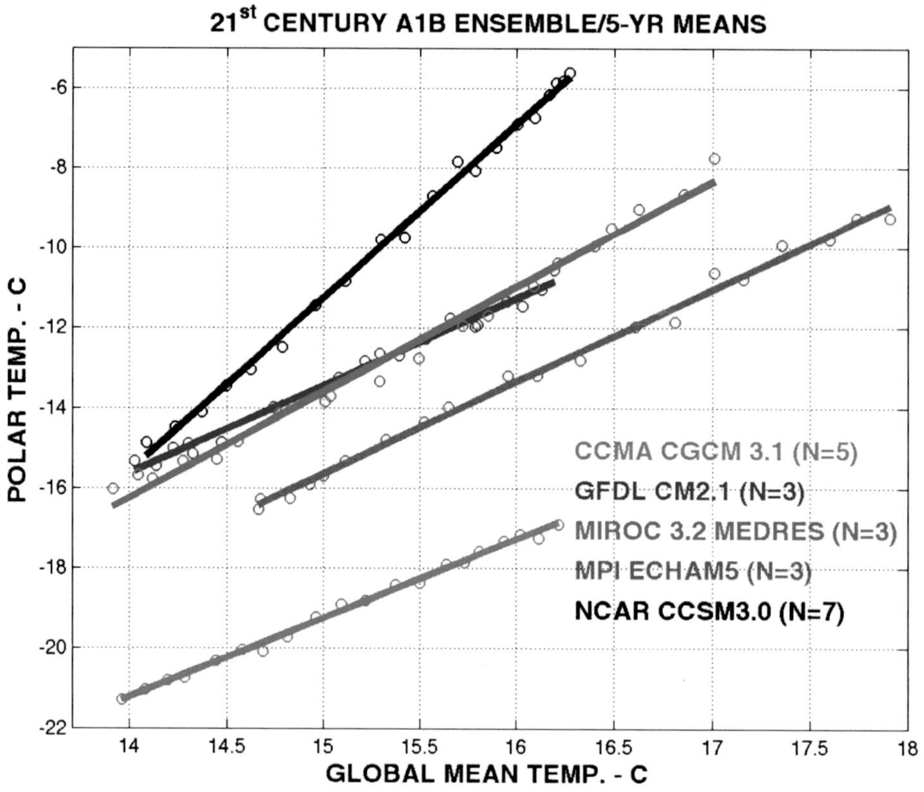

**Plate 2.** Polar versus global temperature for GCM ensembles forced with SRES A1B scenario. Each point represents an average over a 5-year period and over all ensemble members.

[*Marani*, 1999]) into (2), we get a critical transition temperature range of about 25°C. This value indicates instability of the transition to a sea ice–free climate because the annual mean temperature over the perennial sea ice today is about −18°C, and the perennial ice-free zone just beyond the maximum ice edge is at about 0°C. Since ice in this temperature range is producing the satellite-observed planetary albedo drop, the average slope of the albedo/temperature curve on the way to ice-free conditions exceeds the critical slope, and so the critical slope would have to be exceeded at some point, producing instability. However, this conclusion depends upon the assumption that heating from atmospheric transport remains fixed.

But it is unlikely that atmospheric heat transport would not respond to changes in shortwave absorption. The region north of 70°N receives more energy from the atmospheric transport than it absorbs from the Sun, and together they make up nearly all of the OLR; the surface flux is small [*Sereze and Barry*, 2005]. Since atmospheric heat transport is a big player in Arctic climate, it would not likely stand on the sidelines letting OLR completely balance a large change in absorbed shortwave radiation. Indeed, it is probable that the Arctic heat convergence is as high as it is because it is countering the smallness of Arctic shortwave absorption, which is about 1/3 of the global mean [*Serreze and Barry*, 2005].

Horizontal temperature diffusion is a simple method of representing heat transport in an EBM. However, adding diffusive heat transport to the EBM does not eliminate unstable transitions in all cases. Instead, these diffusive transport models can exhibit an unstable loss of a finite patch of polar ice as forcing is increased. The instability is called the SICI and in some ways is a companion to the large ice cap instability whereby the globe becomes ice covered after the ice reaches a critical maximum extent. The ice edge lies in a temperature boundary zone having a length scale determined by the diffusivity and longwave damping parameters [*North*, 1984]. Both instabilities occur when this zone impinges on a boundary, either the equator or the pole. The instability can be removed by reducing the albedo ramp slope, but the main point here is that the instability can occur in spite of down-gradient (warm to cold) transport. At least in some configurations, the instability is also robust to the inclusion of a seasonal insolation cycle [*Lin and North*, 1990].

Furthermore, we expect that the transport changes in response to $CO_2$ increase will have a significant up-gradient component. In the atmosphere this comes about because warmer air allows for an increase in the latent heat transport. *Held and Soden* [2006] show that increased latent transport drives an increase in heat transport to the polar regions, in spite of enhanced warming there, in both equilibrium and transient $CO_2$ increase experiments. Additionally, *Holland*

*and Bitz* [2003] have shown that the ocean also transports more heat into the Arctic, even as the heat transport is being reduced at lower latitudes in association with the weakened meridional overturning circulation. Thus, it is not clear that heat transport can be relied upon to stabilize Arctic climate by exerting a cooling influence on the region as it warms at a larger-than-global rate.

When evaluating the linearity of polar climate, it will be useful to note the well-established fact that the global temperature response to forcing is linear. This was clearly shown for the Goddard Institute for Space Studies model by *Hansen et al.* [2005], who calculated forcing efficacy, the ratio of global temperature change to forcing magnitudes for various forcing types and magnitudes. The efficacy was constant over a large range of magnitudes including the Last Glacial Maximum and the anthropogenic future.

We can make use of the global linearity as follows: since global temperature change is linear in forcing, if polar temperature change is linearly related to it, then polar temperature change must also be linear. The ratio of polar to global temperature change is called the polar amplification. It is typically larger than one for a number of reasons including the ice-albedo feedback. If the polar amplification is also constant, then polar climate change is linear. If there is a nonlinear relationship between polar and global temperature, a nonconstant polar amplification, then the polar change must be nonlinear. Unstable behavior is a subcategory of nonlinear behavior. If the relationship between polar and global temperature is nonlinear and shows a temperature discontinuity, we have evidence of an unstable polar climate change.

## 3. ARCTIC LINEARITY IN 21ST CENTURY EXPERIMENTS

Now we turn to the GCMs to see whether the projected 21st century polar climate change exhibits nonlinearity. Since Arctic climate is quite variable, it will be useful to do some averaging to bring out the forced signal. First, to form $\Delta T_p$, the change in polar surface air temperature, we average over the "half-cap" polar region north of 80°N between 90°E and 270°E in the coldest part of the Arctic Ocean. In the remainder of this paper this region is referred to as the polar region. Next we take 5-year averages, and we also average over the separate runs of the individual models made available in the AR4 experiment archive; thus each point represents an average over 15 to 35 years, depending upon the model's ensemble size. Plate 2 shows the results for five models that supplied multiple runs to the archive for the Special Report on Emissions Scenario (SRES) A1B experiment. In spite of differences in global warming, polar warming, and polar amplification, all of the models show a very linear relationship between polar and global temperature change. From this close relationship, it is clear that simulated 21st century polar climate change is very linear. While linear, the polar temperature changes are quite large in some of the models, comparable to the magnitude of the larger warmings at Greenland during the Dansgaard-Oeschger cycles [*Alley*, 2000].

Plate 3 shows the relationship between polar region effective albedo and surface temperature. The effective albedo is the long-term ratio of surface-up to surface-down shortwave flux. Effective albedo can be shown to be the time-averaged albedo weighted with the surface downward shortwave flux. This weighting is especially important in the Arctic where the insolation has a very large seasonal cycle. In general, the albedo and temperature have close linear relationships. Four of the five models become ice free during September in the polar region over the course of the 21st century without disturbing this relationship. In terms of the EBM discussion of the last section, the simulated Arctic climate changes are linear because the albedo/temperature relationship stays entirely within a subcritical linear ramp region during the 21st century.

The National Center for Atmospheric Research (NCAR) Community Climate System Model, version 3 (CCSM3), which spans the largest range of temperatures and albedos, shows some gentle downward arcing in its albedo/temperature relationship (Plate 3). Since this arcing is not apparent in the polar amplification plot, other feedbacks must be compensating for its slightly nonlinear impact. An examination of a similar plot for the planetary albedo (not shown) does not show this arcing behavior, so the compensation may occur between atmospheric and surface shortwave terms.

*Holland et al.* [2006] have noted that there are abrupt declines in September Arctic sea ice cover in the individual ensemble members of the NCAR CCSM3 SRES A1B experiments. Part of the steepness of the ice cover decline must be related to an acceleration of global warming in the early 21st century under SRES A1B forcing. Holland et al. report that the annual mean ice cover in the NCAR CCSM3 is linearly related to global mean temperature, a result earlier found in the UK Met Office HadCM3 model by *Gregory et al.* [2002], but that September ice cover is not so related. Therefore another factor must be involved in these sharp declines. They note that ice cover responds more sensitively to melting when it is thin. Part of the acceleration of the ice cover decline and its increase in variability are likely due to this increased sensitivity. It is to be expected that a binary variable such as ice cover will show some degree of nonlinearity when confined spatial and temporal averaging is done. The abrupt September ice cover declines are perhaps best characterized as a nonlinear response to linear climate dynamics.

## 4. ARCTIC NONLINEARITY IN ANNUALLY SEA ICE–FREE EXPERIMENTS

From Plate 3 we note that, even with the complete loss of September ice in most of the models, effective albedo has a long way yet to fall to approach open water values of about 0.1. Furthermore, following their linear trends, the models would achieve this albedo at temperatures well above freezing, between 11°C and 29°C. The curves would therefore likely experience considerable steepening under further warming, potentially inducing nonlinear climate changes.

We can only be sure of establishing the presence or absence of nonlinear behaviors associated with ice-albedo feedback in experiments that warm to the point of complete ice removal. Beyond this point, there can be no further reductions in polar ocean surface albedo. The presence or absence of sea ice is easily determined by examining air temperatures in the coldest month and annual effective surface albedos (the ratio of annual surface-up to annual surface-down shortwave fluxes). If the coldest month temperature is at freezing and the effective albedo is near an open ocean value (about 0.1), then we can be assured that there is little sea ice in the particular region in either summer or winter. Seventy-nine runs of four standard experiments (1% per year $CO_2$ increase to doubling, 1% per year $CO_2$ increase to quadrupling, SRES A1B, and SRES A2) were examined for annually ice-free conditions in their polar regions (80°N–90°N, 90°E–270°E) based on these criteria. Of these, only two, had years with February polar region temperatures at freezing temperature and annual surface albedos below 0.15. Thus, it is quite uncommon for a model's Arctic Ocean to become sea ice–free year-round in these climate change experiments. By contrast, it is common in these runs for the Arctic sea ice to disappear in September; about half of the runs had Septembers with surface albedos less than 0.15. Unlike *Thorndike*'s [1992] "toy" model, the seasonally ice-free state is apparently quite stable in GCMs.

The two runs which lose their Arctic sea ice year-round are the 1% per year $CO_2$ increase to quadrupling experiments of the Max Planck Institute (MPI) ECHAM5 and the NCAR CCSM3. Eleven other models supplying data for this experiment did not lose all Arctic sea ice. Of the four forcing scenarios, the quadrupling experiment attains the highest forcing level, over 7 W m$^{-2}$. Both models are run for nearly 300 years, well past the time of quadrupling at year 140. The atmospheric $CO_2$ is held constant after quadrupling, but temperatures are generally still rising in the models as the ocean heat uptake declines [*Stouffer*, 2004].

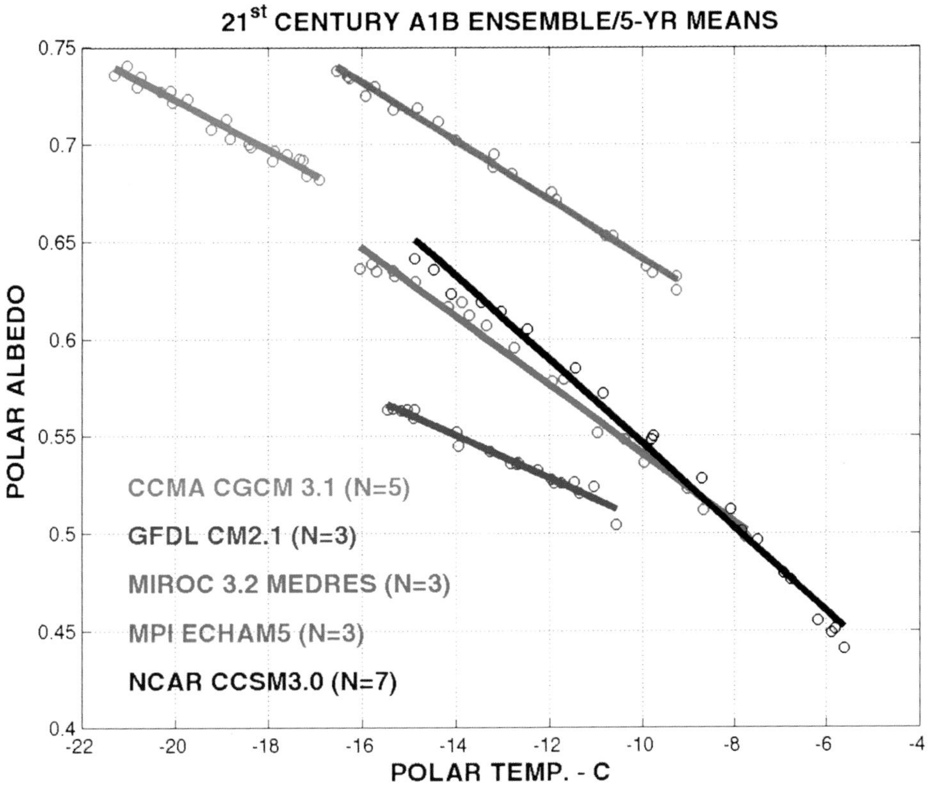

**Plate 3.** Polar effective albedo (annual surface-up to surface-down ratio) versus polar temperature for GCM ensembles forced with SRES A1B scenario. Each point represents an average over a 5-year period and over all ensemble members.

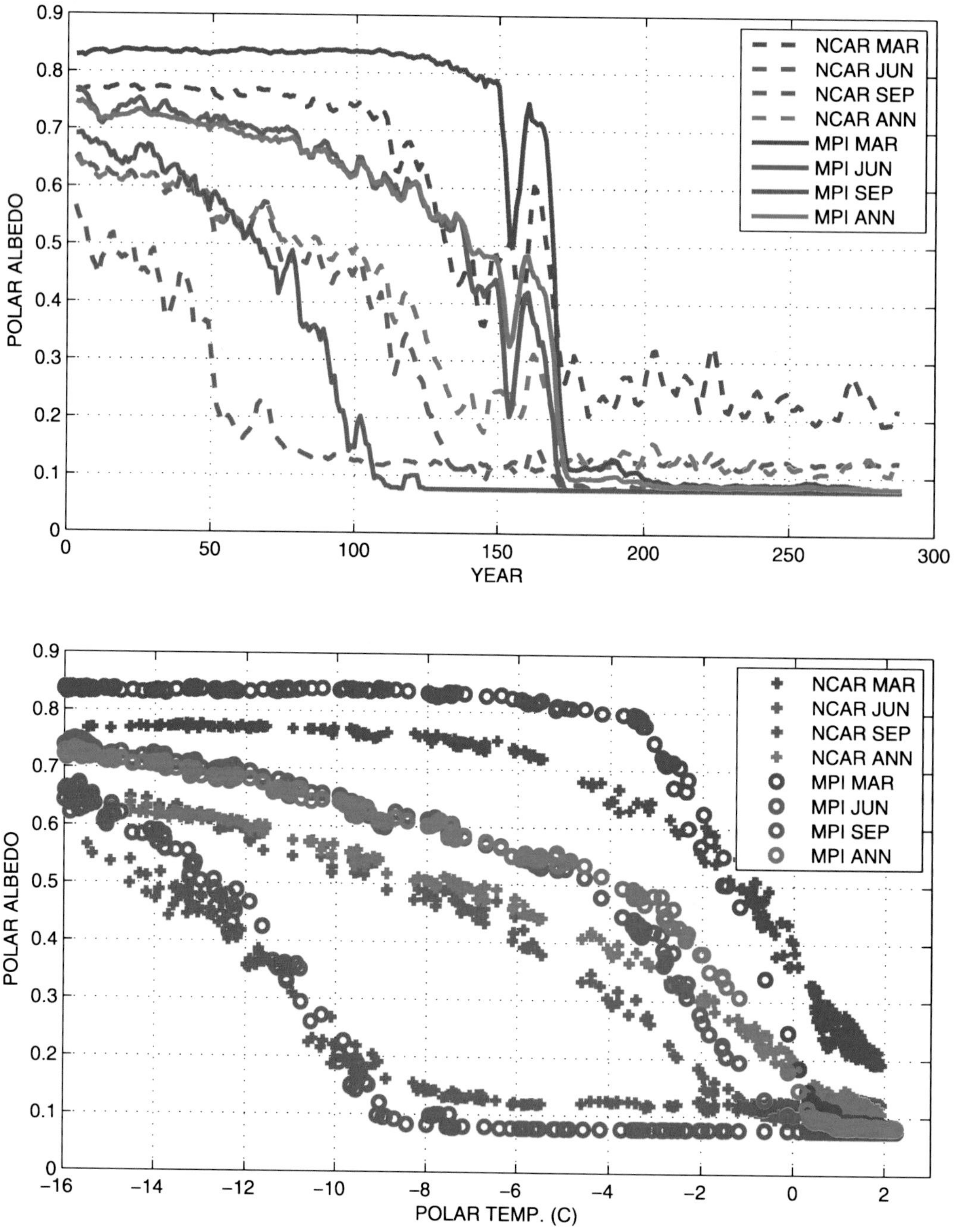

**Plate 4.** Polar region albedo as a function of (top) time and (bottom) annual mean polar region surface temperature for the MPI ECHAM5 (circles) and NCAR CCSM3 (plusses) models. All data have been boxcar filtered over a 5-year period.

Plate 4 (top) shows surface albedo for the polar region as a function of time for the two model experiments. The seasonal ice state is indicated by the albedos for three months: March (blue), June (green), and September (red). The annual effective albedo (light blue) characterizes the time mean reflective capacity of the ice pack. The NCAR model loses its September sea ice near year 50; the MPI model loses it later, at about year 100. Both models have a progression of albedo reductions moving to earlier months in the sunlit season over the course of the integrations. The March sea ice is lost abruptly in the MPI model in the $CO_2$ stabilized period.

The March decline is more gradual in the NCAR model. The variability of March albedos after the decline indicates occasional reappearance of ice in the NCAR model but not in the MPI model.

Plate 4 (bottom) shows the albedos as a function of polar region surface air temperature. The albedo changes are more similar when viewed as a function of temperature rather than as a function of time, but differences remain. The MPI model albedo declines are more abrupt in temperature as well as in time. Both models become seasonally ice free (the September albedo flattens) at an annual polar temperature of about

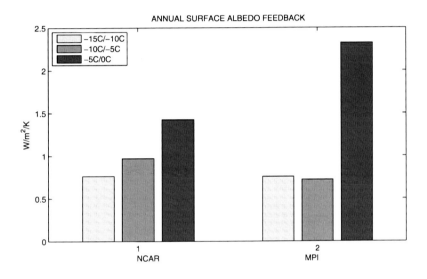

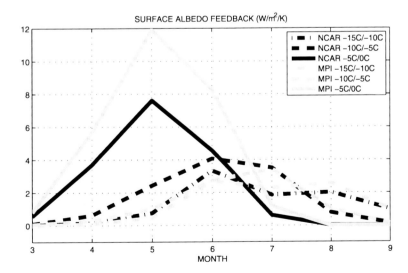

**Figure 1.** (top) Polar surface albedo feedback in three temperature eras. (bottom) Monthly contribution to polar surface albedo feedback for surface temperatures less than $-5°C$ (dashed) and between $-5$ and $0°C$ (solid) for the NCAR CCSM3 (black) and MPI ECHAM5 (gray) models. All data have been boxcar filtered over a 5-year period.

−9°C. It is noteworthy that this loss of ice does not alter the nature of the decline in effective annual albedo in either model. This behavior was also noted in the 21st century plot (Plate 3). Here we see that the extension to warmer temperatures does involve nonlinear effective albedo changes: steepened arcing in the CCSM3 and a kink-like turn in ECHAM5. The total fall in effective surface albedo is over 0.5 in both models, but the effective planetary albedo drop over the experiments is about 0.1 for both models (not shown), indicating a large role for atmospheric shortwave masking and shortwave property changes.

The rapidity of the transition to annually ice-free conditions in the ECHAM5 model and the failure of subsequent variability to produce significant ice are suggestive of an unstable transition to a new equilibrium. Since the lifetime of sea ice in the Arctic (about 10 years) is short compared to the timescale of $CO_2$ increase (70 years for $CO_2$ doubling), we can view the ice as passing through a series of quasi-equilibrated states as the warming progresses. Under this interpre-

tation the rapid transition to the annually ice-free state in the MPI model bears some resemblance to the SICI of simple energy balance models which occurs abruptly as a global forcing is gradually raised above a threshold value.

To explore further the connection between the transition and SICI, we look at the changes in surface albedo feedback (SAF) as the transition progresses. Using the fact that the model transitions are more similar in temperature than in time (Plate 4), we evaluate the surface albedo feedback in three (annual mean) temperature eras: −15°C to −10°C (perennial to seasonal ice transition), −10°C to −5°C (seasonal ice), and −5°C to 0°C (transition to ice free). A method of *Winton* [2005] is used to estimate the SAF. This method uses standard model output to fit a simple optical model and estimate the impact of surface albedo changes on shortwave absorption.

In both models, surface albedo feedback makes an increasing contribution to the decline in sea ice as air temperatures approach freezing (Figure 1, top). In the NCAR model the

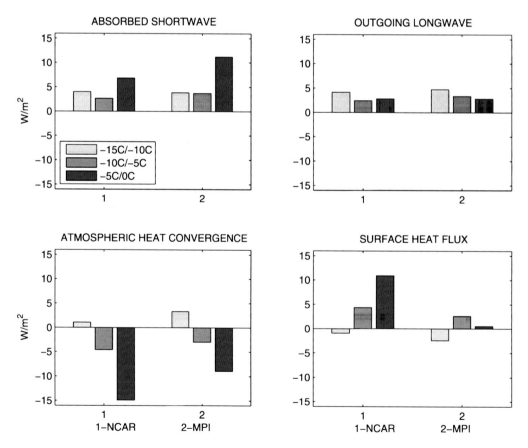

**Figure 2.** Polar atmosphere heat balance changes over three temperature eras: (top left) top-of-atmosphere absorbed shortwave and (top right) outgoing longwave radiation and (bottom left) atmospheric heating from sides and (bottom right) upward heat flux from the surface.

increase is gradual, consistent with the arcing decrease in effective annual surface albedo (Plate 4). In the MPI model a sharp increase occurs in the transition to ice-free temperature range, consistent with the kinked shape of the effective annual albedo decline for that model. In the MPI model, the SAF becomes very large (2.3 W m$^{-2}$ °C$^{-1}$) in the warmest temperature range.

Figure 1 (bottom) shows the monthly contributions to the SAF of the two models in the three temperature ranges. As the warming progresses, there is a shift to earlier months in the sunlit season. This shift allows the SAF to increase even as the ice-free season appears and grows. Aside from seasonal insolation variation, the early months of the sunlit season potentially contribute more to SAF than the later months for two reasons:

1. Surface albedos are initially larger so there is the potential for a larger albedo reduction as the ice is removed exposing the low-albedo seawater. Plate 4 shows that September albedos are 0.1 to 0.2 lower than those in March at the beginnings of the runs.

2. Atmospheric transmissivities are largest in the spring and decline through the summer to a minimum in September in both models. Ignoring multiple cloud-ground reflection, the SAF is the product of the downward atmospheric transmissivity and the surface albedo change, so these two factors compound each other.

The pattern of surface albedo decline in the CCSM3 model (not shown) shows a plume of reduced albedo penetrating into the half-cap region from the Kara Sea, indicating an oceanic influence in the decline. This interpretation is borne out by

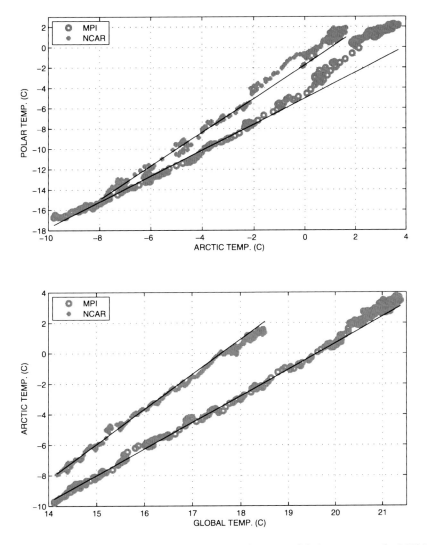

**Figure 3.** (top) Polar versus Arctic temperature and (bottom) Arctic versus global temperature for MPI ECHAM5 (circles) and NCAR CCSM3 (plusses). All data have been boxcar filtered over a 5-year period.

an examination of the polar region heat budgets for the two models shown in Figure 2. These budgets are constructed by regressing the fluxes on temperature in the three temperature ranges. The slopes are then multiplied by 5°C to give a representative flux change between the beginning and end of the specified temperature era. Figure 2 shows that the large increase of SAF in the MPI model at warmer temperatures also appears in the overall shortwave budget of the region and that there is a smaller increase in the shortwave budget of the NCAR model. The outgoing longwave radiation has a small damping effect on the warming of the region in both models. The surface budget changes are quite different: the MPI model has only small changes, while the NCAR model has a large increase in surface forcing as the ocean supplies increased heating. This ocean heating contributes more to the warming of the NCAR model in the warmest temperature era than the SAF. The atmospheric heat transport convergence shifts from a forcing for the warming in the coldest temperature era to damping the warming in the two warmer eras in both models. It is this change in atmospheric convergence of heat, rather than the OLR, that does the most to balance the forcing factors: shortwave flux in MPI and shortwave plus surface flux in NCAR. All of the surface flux changes are opposite to the atmosphere flux convergences in their impacts on the warming.

To gain a sense of the regional extent of nonlinear climate changes, we split the polar amplification into two factors: polar to Arctic (60°N–90°N) and Arctic to global amplifications. Figure 3 (top) shows the relationship between polar and Arctic temperatures and Figure 3 (bottom) shows Arctic and global temperatures for the two models over the course of the 1% per year to 4 times $CO_2$ runs. The warmest polar temperature attained in the two models is about the same, but the global temperature rise is considerably larger in the MPI model, while the Arctic/global and polar/Arctic amplifications are correspondingly smaller. The lines in Figure 3 are fits to the relationships for data with polar temperatures less than −5°C. A deviation from this fit at warmer temperatures might reflect the enhanced warming due to the dramatic changes in sea ice cover above this temperature in both mod-

els. The relationship is mainly linear in both models, but in the ECHAM5 model the polar temperature rises above the reference line starting at a polar temperature of −4°C until it is about 2°C larger and then begins to parallel the fitted line at a polar temperature of 0°C. Apparently, the large increase in surface albedo feedback in this range of temperatures (discussed above) plays a role in this extra warming of the polar region. After the ice is eliminated, the SAF drops to zero, and further warming falls below the −4°C to 0°C ratio. The behavior of the CCSM3 is somewhat different. At −5°C the polar temperature rises slightly above the fitted line but then parallels it as both regions warm further. In both cases, the transition to seasonally ice free at a polar temperature of −9°C does not disturb the linear relationship between warming in the two regions. The relationship between Arctic and global temperatures (Figure 3, bottom) is quite linear in both models indicating that the nonlinear changes in the Arctic Ocean do not have significant impacts on the broader region temperatures. Although the elimination of Arctic sea ice would doubtless have enormous consequences for the local environment, these models do not show it to be particularly important for the larger-scale climate changes.

## 5. EBM INTERPRETATION OF THE TRANSITION TO ANNUALLY ICE FREE

Now, to provide a mechanistic comparison to the GCM behaviors of the last section, we examine polar amplification in a simple one dimensional EBM as it experiences small ice cap instability. Following *North* [1984], the temperature equation for the EBM is

$$-D\frac{d}{dx}(1-x^2)\frac{dT}{dx} + A + BT = S(x)[1-\alpha(T)]. \quad (3)$$

Table 1 defines the notation and gives parameter values. The value for the longwave sensitivity parameter, B, comes from a regression of International Satellite Cloud Climatology Project outgoing longwave on surface temperature [*Marani*, 1999]. The value used for the albedo jump with temperature is the *Gorodetskaya et al.* [2006] value for the

**Table 1.** Notation and Parameter Values

| Notation | Description |
|---|---|
| x | sin(latitude) |
| T(x) | surface temperature (°C) |
| A, B | longwave parameters (B = 1.5 W m$^{-2}$ °C$^{-1}$; A is variable (W m$^{-2}$) |
| D | atmospheric diffusion (= 0.36B, W m$^{-2}$) |
| α(T) | albedo (= 0.3 if T > 0°C; = 0.52 if T < −$\Delta T_M$; = (0.3(T + $\Delta T_M$) − 0.52T)/$\Delta T_M$, otherwise) |
| $\Delta T_M$ | temperature range over which albedos transition linearly from ice covered to ice free |
| S(x) | annual shortwave distribution (= 340(1 − 0.482$P_2$(x)); $P_2$(x) = (3x$^2$ − 1)/2) |

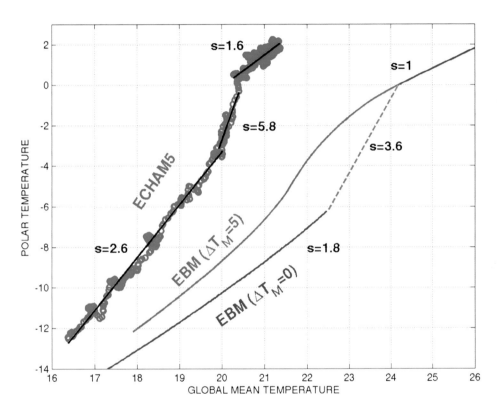

**Plate 5.** Polar versus global temperature for MPI ECHAM5 (green circles) and the EBM (step albedo, blue line; smoothed albedo, red line). ECHAM5 data have been boxcar filtered over a 5-year period.

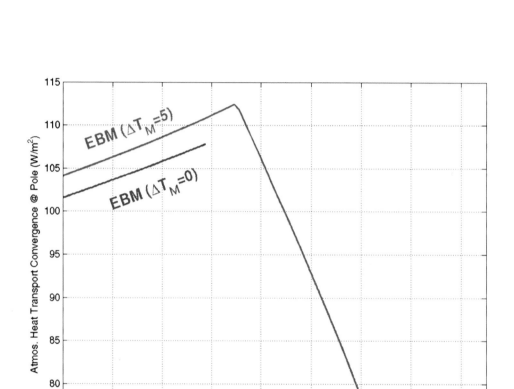

**Plate 6.** Polar (90°N) atmospheric heat transport convergence plotted against polar temperature for EBMs with step albedo (blue) and smoothed albedo (red).

radiative effectiveness of Northern Hemisphere sea ice: the impact on planetary albedo of the change from total to zero sea ice cover. This was determined using the Earth Radiation Budget Experiment shortwave measurements and HadISST1 sea ice concentration data. After setting these two parameters, the atmospheric diffusivity, D, is adjusted to give a reasonable planetary range of surface temperatures.

Most EBM studies explore climate sensitivity by varying the solar constant. Here, we are interested in exploring the relationship between temperature change at the pole (T($x$ = 1)) and the global mean temperature change ($\int_0^1 T(x)dx$) as climate warms. To this end, it is desirable to force in a manner that does not affect this relationship. So here we force climate in a meridionally uniform way by varying A, reducing A induces warming. Thus, we can think of A as a forcing for global mean temperature since

$$\int_0^1 T(x)dx = \left(\int_0^1 (1-\alpha)Sdx - A\right)\Big/B.$$

With this forcing, all polar amplification is due to ice-albedo feedback, the only spatially variable feedback in the system.

Initially, we configure the EBM with a step jump in albedo at $0°C$ ($\Delta T_M = 0°C$). North [1984] used $-10°C$ as the location of the step change. This lower value presumably represents the temperature needed to retain terrestrial snow through the summertime. Sea ice has a source (seawater freezing) that decreases with increased temperature but is positive while there are periods of below-freezing temperatures. This added source is a factor aiding the persistence of summer sea ice cover at higher annual temperatures than terrestrial snow.

Plate 5 shows the polar and global temperatures for the MPI ECHAM5 experiment discussed in the previous section (green) and the EBM with a step albedo jump (blue) and with the same jump smoothed over a transition zone of $5°C$ (red). The EBM changes have been forced by varying A in (3), while the GCM changes are forced by $CO_2$ increase, of course. The $CO_2$ forcing itself is generally somewhat reduced in the Arctic [Winton, 2006]. Nonetheless, the GCM line is the steepest at each polar temperature, so the polar amplification is always larger for the GCM than for the EBM. This is consistent with the findings of a number of studies that factors beside the surface albedo feedback contribute significantly to polar amplification of climate change [Alexeev, 2003; Holland and Bitz, 2003; Hall, 2004; Winton, 2006]. Further evidence of this can be seen in the MPI ECHAM5 curve where significant polar amplification remains even after the sea ice has been eliminated. The EBM does not represent these additional factors and so has smaller polar amplifications.

The EBM with a step albedo change has a discontinuity in polar and global temperatures where the small ice cap instability is encountered, and both warm abruptly with the removal of the reflective ice cap. The light blue dashed line spans this jump, and its slope defines a polar amplification across the instability. In the cooler part of the curve, to the left of this jump, the pole is always below freezing temperature so the local shortwave absorption does not change. The amplification of polar temperature change over global in this part of the curve, about 1.8, is due to the influence of increased absorption of shortwave energy at the ice edge, as the ice retreats poleward, conveyed to the pole by atmospheric transport. North [1984] shows that, as the instability is approached, the pole feels nearly as much warming impact from the ice retreat as the ice edge itself.

On the basis of the fact that the ice cap covers about 6% of the hemisphere before its elimination, we might expect the polar amplification across the jump to be about 16, since this increased absorption is the cause of both temperature jumps. The actual polar amplification is much less because atmospheric heating at the pole, which has been increasing to that point, collapses with the ice cap, countering its local impact to a large degree (Plate 6). After the ice cap collapse, there is no ice-albedo feedback, and polar and global temperatures rise in a one-to-one relationship. The sequence of changes in the polar energy budget encountered as the climate warms leads to a medium/high/none sequence of polar amplifications in the EBM.

The global and polar temperatures for the MPI ECHAM5 show a three-slope regime behavior similar to that of the EBM. However, the GCM does not show any discontinuity in these temperatures. This may be partly due to the GCM, unlike the EBM, not being fully equilibrated at each point in time and hence able to fill the "forbidden zone" with transient temperatures. However, taking note that the ECHAM5 sensitivity of polar albedo to temperature is steep but far from step-like (Plate 4, bottom), we explore the possibility that having the albedo changes occur over a finite range of temperatures stabilizes the transition while retaining enhanced polar sensitivity because of increased surface albedo feedback. EBM runs show that the multiple equilibria remain, with a reduced $\Delta T$ across the jump, for a ramp range of $\Delta T_M = 4°C$ but is eliminated when for $\Delta T_M = 5°C$. The plot for the polar amplification in the stabilized case (Plate 5, red line) shows that a continuous section of enhanced polar amplification fills the region occupied by the jump in the step albedo EBM. The enhanced sensitivity in this region is caused by the reduced overall (negative) feedback due to a positive, but subcritical, local ice-albedo feedback. Plate 6 shows that the diffusive term, operating as a negative feedback, provides less heating to the pole, opposing the enhanced shortwave absorption.

The change of atmospheric heat transport convergence with polar temperature in Figure 2 shows that, in the GCMs as well as the EBM, enhancement of polar warming by atmospheric transport at low temperatures gives way to a damping impact at higher temperatures. However, the GCM transition occurs at much lower temperature, perhaps partly because of its having a polar albedo response at lower temperatures. Other mechanisms, not present in the EBM, can significantly impact poleward transport in the GCMs, for example, the enhancement of latent heat transport with temperature [*Alexeev*, 2003; *Held and Soden*, 2006]. As the ice-free state is approached, the damping effect of the atmospheric heat transport change is much larger than longwave damping in the EBM as in the GCMs.

## 6. TETHERING EFFECT OF HEAT TRANSPORT

The previous section shows that atmospheric heat transport plays an important but complicated role in polar climate change: initially forcing the region to warm at a greater-than-global rate but eventually becoming a cooling influence at higher temperatures. *Held and Soden* [2006] show that the latent heat component of the transport scales up in a warming climate according, roughly, to the Clausius-Clapeyron relationship. This increase drives an increase of the total transport toward the North Pole in spite of polar amplification. However, it is possible that, even in the early warming, part of the transport is helping to maintain the very constant polar amplifications seen in Plate 2. To expose this moderating role, we perform two diagnostic experiments that force only the polar regions and examine the damping mechanisms.

The first is a modification of the Atmospheric Model Intercomparison Project (AMIP) experiment: an atmospheric model run with specified sea surface temperatures and sea ice cover. We perform a twin to this experiment where the sea ice boundary condition is replaced with seawater freezing temperature and albedo. The experiment is done with the atmospheric component of the Geophysical Fluid Dynamics Laboratory (GFDL) CM2.1 climate model. A similar experiment for the DJF season was performed earlier by *Royer et al.* [1990]. The impact on atmospheric temperatures and winds in the current experiment are in general agreement with those found by Royer et al. Plate 7 shows that there is an intense warming of the lower polar atmosphere, mainly confined below the 0°C potential temperature contour of the control, "ice-in," simulation. This is consistent with the regionally limited response of the GCMs to transition to ice-free conditions shown in Figure 3. Other features found by Royer et al., an equatorward shift of the jet, redistribution of sea level high pressure away from the central Arctic to adjacent land regions, and a reduction of cloud cover as the

Arctic Ocean becomes more convective, are also found in this experiment (not shown).

Our main interest in the experiment is to assess the stability of the Arctic ice and its causes. The net surface heat flux change in the ice-covered regions is the result of two competing changes: (1) increased shortwave absorption due to lowered albedo and (2) increased longwave and turbulent heat loss due to increased surface temperature. The net upward heat flux change of 24 W m$^{-2}$ (Figure 4) indicates that the second change is dominant, so the ice is stable and would grow back at an initial rate of 2.5 m a$^{-1}$. At the top of the atmosphere, the extra shortwave absorption is only partly balanced by increased OLR. Most of the damping influence comes from a reduction of heat transport into the ice-covered region by the atmosphere. This reduced heat transport, in turn, is mainly supported by a reduction in surface heat flux from the adjacent ice-free ocean, particularly in the North Atlantic (Plate 8).

The AMIP experiment fixes SSTs implicitly assuming that the ocean has an infinite heat capacity. This assumption may be reasonable here since the near-ice regions that are experiencing large heat flux changes are occupied by deep wintertime mixed layers with much larger heat capacity than the sea ice or the shallow atmospheric layer that interacts with it. Nonetheless, it is useful to relax this assumption by performing a similar experiment in a fully coupled climate model, a developmental version of GFDL CM2.1. In this 100-year experiment, we force the ice region by lowering the ice albedos. Figure 5 shows changes in climate model heat fluxes as in Figure 4. As expected, there is an increase in shortwave absorption at the top of the atmosphere, and as in the AMIP case, it is only partially offset by a local OLR change. Again, the main balancing effect is from a re-

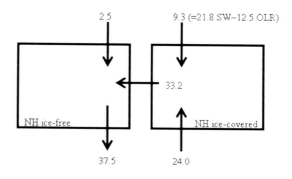

**Figure 4.** Difference in atmospheric heat fluxes over ice-covered and ice-free regions of the Northern Hemisphere between "ice-out" and "ice-in" AMIP runs. The "ice-covered" region is defined by the annual mean ice concentration of the "ice-in" experiment. All fluxes are in units of W m$^{-2}$ ice-covered region.

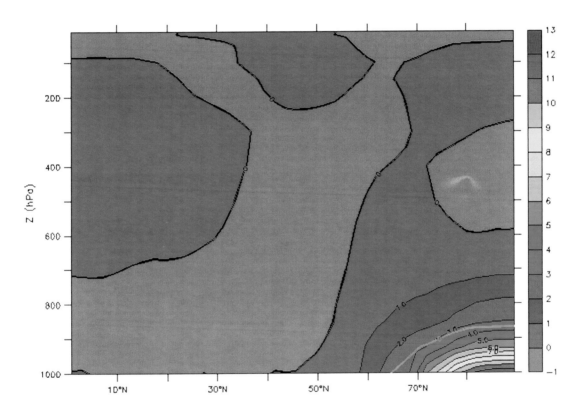

**Plate 7.** Change in zonal mean temperature due to sea ice removal in an AMIP experiment with the GFDL AM2.1 model. The freezing potential temperature contour of the "ice-in" experiment is shown for reference.

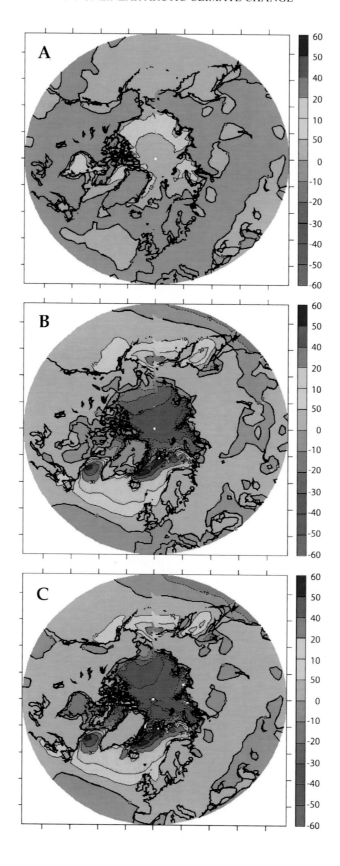

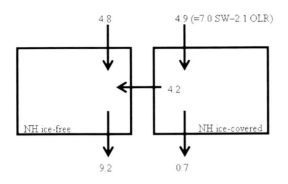

**Figure 5.** Difference in atmospheric heat fluxes over ice-covered and ice-free regions of the Northern Hemisphere between ice-albedo reduced and control coupled model runs. The "ice-covered" region is defined by the annual mean ice concentration of the control experiment. All fluxes are in units of W m$^{-2}$ ice-covered region.

duction of atmospheric heat transport convergence into the ice-covered region (defined from the control experiment). There is also some net downward flux at the surface in the ice-covered region which is mainly supported by reduced latent heating because of the reduced sea ice export in thinner ice. This change in sea ice transport has a salinifying influence near the sea ice edge. Again, the reduced heat transport into the Arctic is supported by a reduction in heat flux out of the adjacent ocean surface. But this response is now oversized compared with the changes in the sea ice region. The reason for this is that, in spite of the change in sea ice freshwater forcing, reduced ocean heat extraction caused by reducing sea ice albedo has induced a reduction of the meridional overturning circulation (MOC). This is shown in Plate 9 along with the change in deepwater ages averaged over the 100 years. The reduction in deepwater ventilation agrees with the result of a similar sea ice albedo reduction experiment performed by *Bitz et al.* [2006] with the NCAR CCSM3. The CCSM3 response was relatively larger in the Southern Ocean, perhaps because the CCSM3 has more Southern Ocean sea ice to feel the albedo reduction than GFDL's CM2.1.

## 7. SUMMARY AND DISCUSSION

The potential for sea ice–albedo feedback to give rise to nonlinear climate change in the Arctic Ocean, defined as a nonlinear relationship between polar and global temperature change or, equivalently, a time-varying polar amplification,

has been explored in the IPCC AR4 climate models. Five models supplying SRES A1B ensembles for the 21st century were examined, and very linear relationships were found between polar and global temperatures indicating linear Arctic climate change. The relationship between polar temperature and albedo is also linear in spite of the appearance of ice-free Septembers in four of the five models.

Two of the IPCC climate models have Arctic Ocean simulations that become annually sea ice–free under the stronger $CO_2$ increase to quadrupling forcing. Both runs show increases in polar amplification at polar temperatures above $-5°C$, and one exhibits heat budget changes that are consistent with the small ice cap instability of simple energy balance models. Both models show linear warming up to a polar temperature of $-5°C$, well above the disappearance of their September ice covers at about $-9°C$. Below $-5°C$, surface albedo decreases smoothly as reductions move, progressively, to earlier parts of the sunlit period. Atmospheric heat transport exerts a strong cooling influence during the transition to annually ice-free conditions.

Specialized experiments with atmosphere and coupled models show that perturbations to the sea ice region climate are opposed by changes in the heat flux through the adjacent ice-free oceans conveyed by altered atmospheric heat transport into the sea ice region. This, rather than OLR, is the main damping mechanism of sea ice region surface temperature. This strong damping along with the weakness of the surface albedo feedback during the emergence of an ice-free period late in the sunlit season are the main reasons for the linearity of Arctic climate change and the stability of seasonal ice covers found in the IPCC models.

Support for these mechanisms can be found in simple models. *Thorndike's* [1992] "toy" model shows their importance by demonstrating the consequences of their absence. In order to make the "toy" model analytically tractable, the ice experiences a constant insolation in the summer season. This disables the seasonal adjustment of albedo feedback just mentioned. Furthermore, the Thorndike model has only weak surface temperature damping to space through a gray body atmosphere. Atmospheric heat transport convergence is held fixed. The result is a model that has no stable seasonal cycle with ice-covered and ice-free periods: the model is either annually ice covered or annually ice free. *Eisenman* [2007] has enhanced the Thorndike model with a sinusoidal insolation and temperature sensitive atmospheric heat transport and finds stable seasonally ice-free seasonal cycles over a broad range of $CO_2$ forcings, from 2 to 15 times current levels.

---

**Plate 8.** (Opposite) Change in heat fluxes at the (top) top and (bottom) bottom of the atmosphere and (middle) the change in atmospheric heat transport convergence for the AMIP ice removal experiment.

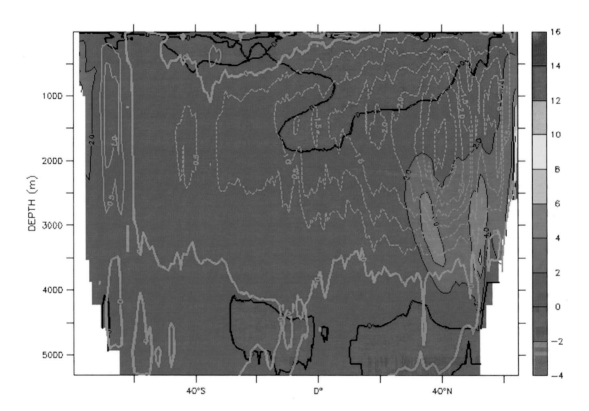

**Plate 9.** Change in coupled model overturning (contours) and 100-year mean age (shading) due to reducing sea ice albedo.

The stabilization of sea ice through opposing heating or cooling of the surrounding ocean was shown to have substantial impact on the overturning circulation and deep ventilation. This effect may provide a partial answer to an outstanding question raised by the Coupled Model Intercomparison Project group experiments [*Gregory et al.*, 2002]. These experiments separate MOC weakening under $CO_2$ increase into freshwater and thermally induced components by using two auxiliary experiments: control radiation with $CO_2$-induced increase in ocean freshwater forcing and the reverse. These experiments show that the two effects basically add linearly and that thermal forcing dominates the MOC response. The mechanisms for this thermal impact are yet to be explored, but ice-albedo feedback, and polar amplification more generally, may be one of them.

*Acknowledgments.* The author thanks Isaac Held, Ron Stouffer, Ian Eisenman, Eric DeWeaver, Gerald North, and an anonymous reviewer for helpful comments on the manuscript. The author also acknowledges the international modeling groups for providing their data for analysis, the Program for Climate Model Diagnosis and Intercomparison (PCMDI) for collecting and archiving the model data, the JSC/CLIVAR Working Group on Coupled Modeling (WGCM) and their Coupled Model Intercomparison Project (CMIP) and Climate Simulation Panel for organizing the model data analysis activity, and the IPCC WG1 TSU for technical support. The IPCC Data Archive at Lawrence Livermore National Laboratory is supported by the Office of Science, U.S. Department of Energy.

## REFERENCES

Alexeev, V. A. (2003), Sensitivity to $CO_2$ doubling of an atmospheric GCM coupled to an oceanic mixed layer: A linear analysis, *Clim. Dyn.*, *20*, 775–787.

Alley, R. B. (2000), Ice-core evidence of abrupt climate changes, *Proc. Natl. Acad. Sci. U. S. A.*, *97*, 1331–1334.

Bitz, C. M., P. R. Gent, R. A. Woodgate, M. M. Holland, and R. Lindsay (2006), The influence of sea ice on ocean heat uptake in response to increasing $CO_2$, *J. Clim.*, *19*, 2437–2450.

Brooks, C. E. P. (1949), *Climate Through the Ages*, 2nd ed., Dover, Mineola, N. Y.

Crowley, T. J., and G. R. North (1991), *Paleoclimatology*, Oxford Univ. Press, New York.

Donn, W. L., and M. Ewing (1968), The theory of an ice-free Arctic Ocean, *Meteorol. Monogr.*, *8*(30), 100–105.

Eisenman, I. (2007), Arctic catastrophes in an idealized sea ice model, in *2006 Program of Studies: Ice, Tech. Rep. 2007-2*, pp. 131–161, Geophys. Fluid Dyn. Program, Woods Hole Oceanogr. Inst., Woods Hole, Mass.

Gorodetskaya, I. V., M. A. Cane, L.-B. Tremblay, and A. Kaplan (2006), The effects of sea-ice and land-snow concentrations on planetary albedo from the Earth Radiation Budget Experiment, *Atmos. Ocean*, *44*, 195–205.

Gregory, J. M., P. S. Stott, D. J. Cresswell, N. A. Rayner, and C. Gordon (2002), Recent and future changes in Arctic sea ice simulated by the HadCM3 AOGCM, *Geophys. Res. Lett.*, *29*(24), 2175, doi:10.1029/2001GL014575.

Hall, A. (2004), The role of surface albedo feedback in climate, *J. Clim.*, *17*, 1550–1568.

Hansen, J., et al. (2005), Efficacy of climate forcings, *J. Geophys. Res.*, *110*, D18104, doi:10.1029/2005JD005776.

Held, I. M., and B. J. Soden (2006), Robust responses of the hydrological cycle to global warming, *J. Clim.*, *19*, 5686–5699.

Holland, M. M., and C. M. Bitz (2003), Polar amplification of climate change in coupled models, *Clim. Dyn.*, *21*, 221–232.

Holland, M. M., C. M. Bitz, and B. Tremblay (2006), Future abrupt reductions in the summer Arctic sea ice, *Geophys. Res. Lett.*, *33*, L23503, doi:10.1029/2006GL028024.

Lin, R. Q., and G. R. North (1990), A study of abrupt climate change in a simple nonlinear climate model, *Clim. Dyn.*, *4*, 253–261.

Lindsay, R. W., and J. Zhang (2005), The thinning of the Arctic sea ice, 1988–2003: Have we passed a tipping point?, *J. Clim.*, *18*, 4879–4894.

Manabe, S., and R. J. Stouffer (1994), Multiple-century response of a coupled ocean-atmosphere model to an increase of carbon dioxide, *J. Clim.*, *7*, 5–23.

Marani, M. (1999), Parameterizations of global thermal emissions for simple climate models, *Clim. Dyn.*, *15*, 145–152.

North, G. R. (1984), The small ice cap instability in diffusive climate models, *J. Atmos. Sci.*, *41*, 3390–3395.

Royer, J. F., S. Planton, and M. Deque (1990), A sensitivity experiment for the removal of Arctic sea ice with the French spectral general circulation model, *Clim. Dyn.*, *5*, 1–17.

Serreze, M. C., and R. G. Barry (2005), *The Arctic Climate System*, Cambridge Univ. Press, New York.

Serreze, M. C., and J. A. Francis (2006), The Arctic amplification debate, *Clim. Change*, *76*, 241–264.

Stouffer, R. J. (2004), Time scales of climate response, *J. Clim.*, *17*, 209–217.

Stroeve, J., M. C. Serreze, F. Fetterer, T. Arbetter, W. Meier, J. Maslanik, and K. Knowles (2005), Tracking the Arctic's shrinking ice cover: Another extreme September sea ice minimum in 2004, *Geophys. Res. Lett.*, *32*, L04501, doi:10.1029/2004GL021810.

Thorndike, A. S. (1992), A toy model linking atmospheric radiation and sea ice growth, *J. Geophys. Res.*, *97*, 9401–9410.

Vinnikov, K. Y., A. Robock, R. J. Stouffer, J. E. Walsh, C. L. Parkinson, D. J. Cavalieri, J. F. B. Mitchell, D. Garrett, and V. F. Zakharov (1999), Global warming and Northern Hemisphere sea ice extent, *Science*, *286*(5446), 1934–1937.

Winton, M. (2005), Simple optical models for diagnosing surface-atmosphere shortwave interactions, *J. Clim.*, *18*, 3796–3805.

Winton, M. (2006), Amplified Arctic climate change: What does surface albedo feedback have to do with it?, *Geophys. Res. Lett.*, *33*, L03701, doi:10.1029/2005GL025244.

M. Winton, Geophysical Fluid Dynamics Laboratory, NOAA, P.O. Box 308, Princeton University Forrestal Campus, Princeton, NJ 08542, USA. (Michael.Winton@noaa.gov)

# The Role of Natural Versus Forced Change in Future Rapid Summer Arctic Ice Loss

Marika M. Holland,[1] Cecilia M. Bitz,[2] L.-Bruno Tremblay,[3] and David A. Bailey[1]

Climate model simulations from the Community Climate System Model, version 3 (CCSM3) suggest that Arctic sea ice could undergo rapid September ice retreat in the 21st century. A previous study indicated that this results from a thinning of sea ice to more vulnerable conditions, a "kick" in the form of pulse-like increases in ocean heat transport and positive feedbacks that accelerate the retreat. Here we further examine the factors affecting these events, including the role of natural versus forced change and the possibility of threshold-like behavior in the simulated sea ice cover. We find little indication that a critical sea ice state is reached that then leads to rapid ice loss. Instead, our results suggest that the rapid ice loss events result from anthropogenic change reinforced by growing intrinsic variability. The natural variability in summer ice extent increases in the 21st century because of the thinning ice cover. As the ice thins, large regions can easily melt out, resulting in considerable ice extent variations. The important role of natural variability in the simulated rapid ice loss is such that we find little capability for predicting these events based on a knowledge of prior ice and ocean conditions. This is supported by results from sensitivity simulations initialized several years prior to an event, which exhibit little predictive skill.

## 1. INTRODUCTION

In recent years, the Arctic has undergone dramatic changes throughout the system that are consistent with increasing anthropogenic forcing combined with large natural variability [Serreze et al., 2007]. These changes are widespread and are present in the atmosphere, ocean, sea ice and terrestrial components (e.g., see Serreze et al. [2000] and Overland et al. [2004] for reviews). One of the most striking of these changes is the recent decrease in the summer Arctic sea ice cover. Over the satellite record (1979–2007), a decrease of 0.7 million km² per decade in the September ice cover has occurred, amounting to a loss of almost 30% of the ice cover in 29 years. This change is not linear over the 1979 to present record but instead has shown accelerated retreat in more recent years [Serreze et al., 2003; Stroeve et al., 2005]. The recent minimum for 2007 [Stroeve et al., 2008] was particularly dramatic, being about 1.3 million km² lower than the previous 2005 September record of 5.6 million km² [Fetterer et al., 2002]. This represents a 3 standard deviation excursion from the linear trend. Additionally, the August mean ice extent surpassed the September minimum in 2005. While the sea ice change is largest in summer, a decrease in Arctic ice cover since 1979 is present in all months [Serreze et al., 2007].

[1]National Center for Atmospheric Research, Boulder, Colorado, USA.

[2]Department of Atmospheric Sciences, University of Washington, Seattle, Washington, USA.

[3]Ocean and Atmosphere Sciences, McGill University, Montreal, Quebec, Canada.

Arctic Sea Ice Decline: Observations, Projections, Mechanisms, and Implications
Geophysical Monograph Series 180
Copyright 2008 by the American Geophysical Union.
10.1029/180GM10

Climate models project that decreasing Arctic ice cover will continue into the foreseeable future. An analysis of the models participating in the Intergovernmental Panel on Climate Change fourth assessment report shows that all models simulate reduced Arctic ice cover throughout the 21st century, with about 50% of them reaching seasonally ice-free conditions by 2100 [*Arzel et al.*, 2006; *Zhang and Walsh*, 2006]. As reported by *Holland et al.* [2006a] (hereinafter referred to as HBT), future reductions in September (summer minimum) ice cover occur quite abruptly in some models, with periods of relative stability followed by very rapid ice loss. Model simulations run with a middle-range forcing scenario (Special Report on Emissions Scenario (SRES) A1B) suggest that in the future (generally from 2015 to 2050) short-timescale (5–10 years) September ice loss could occur 3–4 times faster than what has been observed over comparable length time periods through 2005. When observed conditions from 2007 are included, the simulated ice loss over an abrupt event as identified by HBT (see section 3.2) is only from 1.1 to 2.2 times faster (depending on the event) then a comparable length of the observed record. So, while the observed ice cover does not yet exhibit abrupt ice loss as defined by HBT, the observed changes through 2007 are approaching those levels. It is notable that model integrations do not simulate these rates of ice loss until 2015 at the earliest.

HBT found that in the Community Climate System Model, version 3 (CCSM3), there are three factors that contribute to rapid September ice retreat. These include a thinning of the sea ice to a more vulnerable state, pulse-like increases in ocean heat transport to the Arctic that appear to trigger the rapid change in sea ice extent, and the positive surface albedo feedback which accelerates the retreat. One outstanding question from the HBT study is whether the events identified in CCSM3 occur because of the ice cover reaching an unstable threshold ("tipping point") in which the sea ice then rapidly switches to a new, stable (seasonally ice free) equilibrium. Abrupt climate change is often defined in the context of this type of tipping point behavior [e.g., *National Research Council*, 2002]. Here we use a broader definition of "abrupt change" in relation to summer sea ice loss, indicating a change that is fast relative to the forcing but may not constitute a threshold response.

Previous studies using relatively simple models suggest that a threshold instability could exist in the transition to year-round Arctic ice-free conditions [e.g., *North*, 1984; *Gildor and Tziperman*, 2001] and that this may also occur in some GCM simulations [*Winton*, 2006]. It is possible that a transition to seasonally ice-free conditions could also result from such "tipping point" behavior. Alternatively, the abrupt ice loss simulated by CCSM3 may represent the interaction of large intrinsic Arctic variability with increasing forced change because of rising greenhouse gas concentrations. This could occur in the absence of an instability in the ice cover and could have implications for the predictability of such events. The result of an abruptly changing ice cover would be the same, however, with considerable impacts on the socioeconomics, climate, and biological systems in the Arctic. Here we examine the possibility of a critical ice state that leads to the abrupt September ice retreat present in CCSM3 integrations and the role of natural versus forced change in determining simulated rapid ice loss events. The implications for potential predictability of these events are also discussed.

## 2. MODEL INTEGRATIONS

The Community Climate System Model, version 3 (CCSM3) is a state-of-the-art fully coupled climate model, which includes atmosphere, ocean, land, and sea ice components [*Collins et al.*, 2006a]. For the primary integrations considered here, the atmosphere model (CAM3) [*Collins et al.*, 2006b] is run at T85 resolution (approximately 1.4°) with 26 vertical levels. The ocean model [*Smith and Gent*, 2004] includes an isopycnal transport parameterization [*Gent and McWilliams*, 1990] and a surface boundary layer formulation following *Large et al.* [1994]. The dynamic-thermodynamic sea ice model [*Briegleb et al.*, 2004; *Holland et al.*, 2006b] uses the elastic-viscous-plastic rheology [*Hunke and Dukowicz*, 1997], a subgrid-scale ice thickness distribution [*Thorndike et al.*, 1975], and the thermodynamics of *Bitz and Lipscomb* [1999]. Both the ice and ocean models use a nominally 1° resolution grid in which the North Pole is displaced into Greenland. The land component [*Bonan et al.*, 2002] includes a subgrid mosaic of plant functional types and land cover types based on satellite observations. It uses the same spatial grid as the atmospheric model.

We discuss 20th and 21st century CCSM3 simulations that were performed for the Intergovernmental Panel on Climate Change, fourth assessment report (IPCC-AR4) [*Intergovernmental Panel on Climate Change*, 2007]. These will be referred to as the IPCC-AR4 CCSM3 integrations. Eight 20th to 21st century ensemble members are available. The 20th century runs use external forcings based on the observed record and chemical transport models. These include variations in sulfates, solar input, volcanic forcing, ozone, a number of greenhouse gases, halocarbons and black carbon. They were initialized from different years of a multi-century preindustrial (year is 1870) control integration. For the 21st century simulations, we focus on the middle-range SRES A1B forcing which reaches approximately 720 ppm $CO_2$ levels by 2100 [*Intergovernmental Panel on Climate Change*, 2001].

To further evaluate the influence of natural and forced variations we also examine a large 29-member ensemble of CCSM3 21st century model integrations. These runs employ a coarser-resolution atmospheric component with T42 (approximately 2.75°) resolution. The late 20th century Arctic sea ice state that these runs are initialized from is quite different because of the different resolution of the atmospheric model (e.g., see *DeWeaver and Bitz* [2006] for a discussion of resolution and Arctic sea ice in CCSM3 control runs). Of interest here, the early 21st century Arctic sea ice is considerably thicker in these simulations than in their T85 counterparts. This modifies the projected change in ice cover as discussed below. These simulations are also run using the middle-range SRES A1B scenario.

## 3. RESULTS

### 3.1. Simulated Arctic Sea Ice

The ensemble mean IPCC-AR4 CCSM3 sea ice thickness averaged from 1980 to 1999 is shown in Plate 1. In accord with observations [e.g., *Bourke and Garrett*, 1987; *Laxon et al.*, 2003], the thickest ice is present along the Canadian Arctic Archipelago and north of Greenland. However, the model also obtains relatively thick ice throughout the East Siberian Sea in contrast to observations. This is a common problem in many coupled climate models and is likely associated with biases in the wind forcing [e.g., *Bitz et al.*, 2002; *DeWeaver and Bitz*, 2006]. Satellite altimetry measurements suggest an October–March mean ice thickness of 2.73 m for the region south of 81.5N, excluding ice thinner than 1 m [*Laxon et al.*, 2003]. Using a consistent method of averaging, the CCSM3 ensemble mean obtains a value of 2.6 m in good agreement with observations.

The IPCC-AR4 CCSM3 has a reasonable winter ice edge in the Greenland and Bering seas compared to observations, but the winter ice is too extensive in the Labrador Sea and the Sea of Okhotsk (Plate 1). Overall, the CCSM3 integrations obtain a winter ice cover that is from 1 to 1.8 million km² larger than observed (Figure 1). A recent study [*Jochum et al.*, 2008] indicates that substantial improvements in these regions are realized in CCSM3 simulations with modified ocean viscosity. From June to December, the simulated total Arctic ice extent, defined as the area with greater then 15% ice concentration, is in excellent agreement with the observations and the difference between the two is typically smaller than the interannual standard deviation. One small discrepancy from observations during the summer months is that the simulated ice edge is displaced poleward within the Kara Sea.

In any given ensemble member, one does not expect a direct match between a particular year and the corresponding year in the observational record because of the effect of natural variability on the temporal evolution of ice extent. However, we can gauge the consistency of the model simulations and the observed record by comparing statistics of the time series among multiple ensemble members with the observations. An analysis of the September ice extent trends from observations and model simulations (Table 1) confirms that at the Arctic basin scale, the simulated trends are consistent with observations. The observed September ice extent trend from 1979 to 2007 falls within the range of trends from the different ensemble members. The ensemble members differ only in their initial conditions, and as such, the different trends that are present among the ensemble members result from the intrinsic ("natural") model variability. Thus, the simulated trends agree with the observations within these intrinsic variations. As discussed by *Stroeve et al.* [2007], this is in contrast to many other IPCC-AR4 models, which generally simulate smaller trends than observed over the satellite record.

The 1979–2007 interannual standard deviation in September ice extent within the model ensemble members is also generally consistent with observations (Table 1), as the observed value falls within the range of values from different ensemble members. For the detrended 1979–2007 September time series, the observed standard deviation is slightly higher than that simulated in any of the ensemble members. This is due to the very anomalous conditions of 2007, and the simulated values for different ensemble members bracket the observations if only the 1979–2006 time period is considered (Table 1). In general, the IPCC-AR4 CCSM3 integrations appear to simulate realistic year-to-year and long-term variations in the September ice cover compared to observations.

### 3.2. Projected Changes in Ice Cover

Plate 2 shows the projected changes in the September Arctic ice extent within the CCSM3 ensemble members. Abrupt events are indicated by the grey shading. The events are identified following HBT as times when the derivative of the 5-year running mean smoothed September ice extent time series exceeds a loss of 0.5 million km² a⁻¹. The event length is determined by the time around the transition for which the smoothed time series derivative is larger than 0.15 million km² a⁻¹. Only time periods that fit this definition and are longer than 2 years in length are considered "abrupt events." Although subjective, this definition clearly identifies periods of rapid retreat in the September ice cover. These vary from 4 to 10 years in length, and the trends over these events are typically 3–4 times larger than comparable length trends in the smoothed observational time series from 1979 to 2005. If

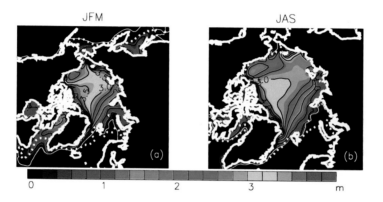

**Plate 1.** (a) January, February, and March (JFM) and (b) July, August, and September (JAS) 1980–1999 ensemble mean ice thickness from the IPCC-AR4 CCSM3 simulations. The ice thickness here is defined as the ice volume per unit grid cell area. The dashed line indicates the observed ice extent. The solid white line is the simulated ice extent.

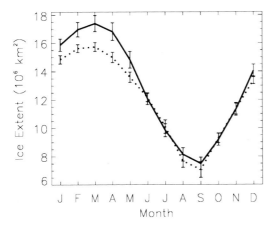

**Figure 1.** Annual cycle of ice extent, defined as the area with greater then 15% ice concentration, averaged from 1980 to 1999 for the IPCC-AR4 CCSM3 ensemble mean (solid line) and the observations (dotted line). The observed values are from the National Snow and Ice Data Center sea ice index [*Fetterer et al.*, 2002], which are obtained from the passive microwave satellite data. The vertical lines indicate the interannual standard deviation for each month.

observed conditions through 2007 are considered, however, the simulated trends over these events are only modestly larger (1.1–2.2 times larger, depending on the event) than the observed ice loss. While the observed ice cover has yet to exhibit abrupt ice loss as defined above, with the extremely low ice conditions in 2007 the observed record is approaching the rates of change of the simulated abrupt events. In the model integrations, these events typically do not occur until after 2020, with the earliest event starting in 2012 (Plate 2g).

If the simulated abrupt transitions are the manifestation of a "tipping point" and result from reaching a certain threshold, then it is likely that they would be preceded by a similar "critical" ice state. Plate 3 shows a sample of the May averaged ice conditions in the year that an abrupt ice loss event is initiated and the region of ice loss over that event. The May ice thickness is shown because this generally represents the conditions at the beginning of the ice melt season. In Plate 3, we only consider the five ice loss events that are initiated between 2020 and 2030. These spatial maps indicate that the ice lost during different abrupt events varies in its initial thickness and the region of loss. The fractional coverage of thin ice in this region also varies (not shown). Quantitative measures of these properties (e.g., the ice thickness in the region where ice is lost during the abrupt event, the ice concentration in this region, and the concentration of thin ice, etc.) have been analyzed and found to differ considerably at the initiation of the different events. The properties over the abrupt ice loss region at the start of the events are instead closely related to the year in which the event begins which varies from 2012 to 2045. Earlier occurring abrupt ice loss events are generally initiated with thicker and more extensive sea ice than those events which occur later in the 21st century. Although the ice conditions vary at the initiation of different events, it is clear that an abrupt ice loss event is only possible if the ice has thinned adequately. No events are present during the 20th century simulations because of a relatively thick ice cover.

The May ice thickness 5 years prior to the time at which a model grid cell transitions from September ice-covered to September ice-free conditions varies regionally but is broadly similar across the different ensemble members (Plate 4). In particular, the thickness 5 years prior to ice-free conditions in the Beaufort, Chukchi, and East Siberian seas

**Table 1.** Trends and Standard Deviation in the 1979–2007 September Ice Extent From Observations and the IPCC-AR4 CCSM3 Ensemble Members[a]

| | | Standard Deviation ($10^6$ km$^2$) | |
| --- | --- | --- | --- |
| | Trend ($10^6$ km$^2$ a$^{-1}$) | Raw Time Series | Detrended Time Series |
| Observations | −0.07 (−0.06) | 0.80 (0.65) | 0.51 (0.42) |
| Run 1 (b.ES01) | −0.08 | 0.80 | 0.43 |
| Run 2 (a) | −0.08 | 0.79 | 0.45 |
| Run 3 (b) | −0.02 | 0.46 | 0.45 |
| Run 4 (c) | −0.04 | 0.56 | 0.44 |
| Run 5 (d) | −0.06 | 0.65 | 0.42 |
| Run 6 (e) | −0.07 | 0.69 | 0.33 |
| Run 7 (f.ES01) | −0.02 | 0.49 | 0.46 |
| Run 8 (g.ES01) | −0.06 | 0.74 | 0.50 |

[a] For the observations, the numbers in parentheses show values for the 1979–2006 time period. The letters in parentheses indicate the given CCSM3 case name and ES01 refers to simulations that were run on the Earth Simulator machine.

is quite thick (2–3 m), whereas for the central ice pack it is only 1–2 m. This is perhaps not too surprising as the shelf regions generally reach September ice-free conditions earlier in the simulation. While the mean ice conditions averaged over the ice loss regions vary considerably across the different events, this could be unduly influenced by the different regions where ice loss occurs. Thus, it is still possible that a common spatially variable "critical ice thickness" is present but that the region and time where this thickness is achieved varies widely from run to run. If true, abrupt loss would be possible if a considerable region of the ice pack reached the appropriate conditions near the same time. The different character of each abrupt ice loss event could then be associated with what region reaches these "critical" conditions.

Figure 2 shows the area of spring sea ice within 25 (and 50) cm of a common "critical ice thickness" for each ensemble member. This "critical ice thickness" is defined by the ensemble mean of the May thickness 5 years prior to the time at which a transition to September ice-free conditions results within each model grid cell (Plate 4, bottom right). Some abrupt events are clearly preceded by anomalously large regions of the ice cover reaching this "critical ice thickness" (e.g., run 4, the second event in run 6, the second event in run 7). However, a number of runs show anomalously low areas of "critical ice thickness" just prior to abrupt September ice loss (e.g., run 1, the first event in run 7). Thus, it is not always the case that an anomalously large region of the ice pack reaches such critical conditions just prior to abrupt ice loss. This suggests that indeed there is no common critical ice state present among the ensemble members that leads to an abrupt ice loss. So, in addition to an adequately thin ice cover, a forcing perturbation appears necessary to initiate the events. This forcing perturbation is unique to the individual ensemble member and results from natural variability. It reinforces changes in external forcing that are applied identically across the members. For this process to occur, a thinned ice cover appears a necessary but not sufficient condition.

*3.2.1. Quantifying natural versus forced change.* There are eight realizations of 20th to 21st century integrations in the suite of IPCC-AR4 CCSM3 scenario runs that are forced with identical external forcing and only vary in their initial conditions. While this is a relatively small number, a comparison between the different realizations can provide insight into the role of natural versus forced change in driving rapid sea ice loss. Figure 3a shows the ensemble mean September sea ice extent. Because of the relatively small ensemble size, this time series exhibits the signature of some individual realizations. However, as averaging across these ensemble members generally removes the uncorrelated natural varia-

tions, we will attribute this ensemble mean response as the "forced" response that results from the changing external forcing. As shown in Figure 3a, this forced response is sizable with a 1.1 million km$^2$ per decade rate of September ice loss from 2000 to 2050. As a result, the ensemble mean reaches near September ice-free conditions by 2050.

Subtracting the ensemble mean from the ice extent time series for each individual realization leaves us with a characterization of the "natural" variations. Figure 3b shows this time series for the extreme ice loss event simulation in ensemble run 1 (Plate 2a). Two things are apparent. First, there is a relatively uniform distribution of positive and negative anomalies about the ensemble mean. Second, the natural variability in September ice extent increases during the early 21st century and then decreases again after 2050 at which time little September ice cover remains. This is quantified in Figure 3c, which shows the 20-year running standard deviation of September ice extent over the 20th to 21st centuries for run 1. Interestingly, while the variance increases, there is little change in skew or kurtosis of the ice extent anomalies as the ice thins (not shown).

The increase in ice extent variability with a thinning ice cover is a consistent feature of the CCSM3 simulations (Plate 5) and indeed of most climate models (not shown) and makes good physical sense. As the ice cover thins, natural variations in atmosphere and/or ocean conditions that modify the sea ice mass budgets translate more readily into a change in ice area as large regions of the thin ice pack can completely melt out. The thinning ice cover is in essence more vulnerable to intrinsic climate perturbations. In turn, the retreating ice cover causes an amplification of the change because of the surface albedo feedback, resulting in still larger ice extent change. Other feedbacks, associated with changing longwave forcing, for example [*Gorodetskaya and Tremblay*, this volume], may also contribute.

*3.2.2. Coarse-resolution results.* One issue with the above analysis is that the separation into natural and forced change relies on a relatively small ensemble member size. To complement these results, we here examine a large 29-member ensemble of simulations runs from 2000 to approximately 2060 using the SRES A1B scenario forcing. These simulations use a coarser-resolution (T42, approximately 2.75°) atmospheric component of the CCSM3 model coupled to the identical ocean and sea ice components as used in the IPCC-AR4 simulations. Aspects of the Arctic climate conditions in control climate integrations with this model are discussed by *Holland et al.* [2006b] and *DeWeaver and Bitz* [2006].

Because of the different atmospheric model resolution, different mean late 20th century sea ice conditions result. In particular, as shown in Plate 6a, the sea ice is considerably

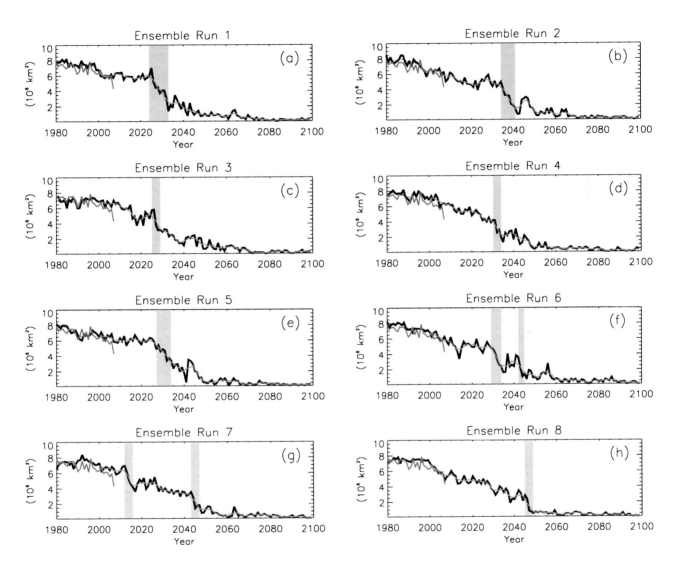

**Plate 2.** Time series of September ice extent from the (a–h) eight IPCC-AR4 CCSM3 ensemble members. The blue line shows the 5-year running mean time series. The observed time series from 1979 to 2007 is shown in red. Grey shading indicates abrupt events as defined in the text.

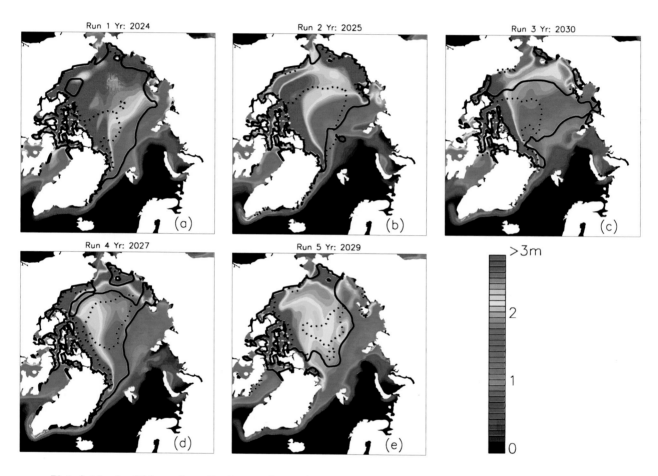

**Plate 3.** May ice thickness immediately preceding an abrupt transition in the ice cover for simulated abrupt transitions occurring between 2020 and 2030. The solid line indicates the September ice extent immediately preceding the event, and the dashed line shows the September ice extent immediately following the event.

thicker in 2000 compared to the IPCC-AR4 integrations. The ice is also distributed differently across the Arctic basin than in the higher-resolution runs, with the thickest ice cover present in the East Siberian Sea in contrast to observations. This is consistent with the sea ice change in response to atmospheric resolution variations in CCSM3 control climate integrations as discussed by *DeWeaver and Bitz* [2006].

The initially thicker ice cover modifies the projection of ice extent. As shown in Plate 6b, the September ice extent loss over the 21st century is delayed in the coarse-resolution runs with similar conditions occurring about 20 years later than in the IPCC-AR4 runs. Plate 6c shows the September ice extent as a function of the May mean Arctic ice thickness. On the basis of this measure, a more consistent picture emerges for the two different sets of model integrations. This suggests that the difference in initial ice thickness is a primary factor responsible for the different projected ice extent loss.

The coarse-resolution ensemble members exhibit rapid September ice loss events as defined above (not shown) but generally do so later in the 21st century, consistent with their initially thicker sea ice. The events typically start around year 2050, but the earliest abrupt transition begins in 2021. The event lengths are almost uniformly distributed between 3 and 7 years. They have typical trends in the smoothed time series of $-0.4$ million km$^2$ per a$^{-1}$ over the event length with the trends varying from $-0.6$ million km$^2$ per a$^{-1}$ to $-0.3$ million km$^2$ a$^{-1}$ for different events. This is similar to the identified trends in the IPCC-AR4 CCSM3 simulations. Of the 29 ensemble members, 17 (about 60%) exhibit an abrupt event. The remaining 12 ensemble members have no periods that meet the defined abrupt event criteria. However, the simulations only run to approximately 2060 and still have about 3 million km$^2$ of ice cover in the 2060 ensemble mean. Additional abrupt events are thus possible if the simulations were extended past 2060.

By subtracting the ensemble mean ice conditions shown in Plate 6b from each individual ensemble member, we are left with information about the intrinsic variations within each simulation. As occurs in the IPCC-AR4 CCSM3 runs, the natural variability in September ice extent (defined as the deviation from the ensemble mean) increases with the thinning ice pack (Plate 6d). Quantitatively, this agrees well with the IPCC-AR4 simulations, suggesting that the small ensemble size in the higher-resolution simulations does not substantially bias those results. Previous work has shown that anthropogenic climate change modifies other aspects of natural variability, such as interannual air temperature and precipitation fluctuations [e.g., *Giorgi and Bi*, 2005].

*3.2.3. Processes contributing to natural variations.* The above analysis suggests that a large and growing intrinsic

variability in September ice extent plays an important role in the simulated abrupt ice loss events. There are numerous factors that can contribute to natural variations in the sea ice cover, including both dynamic and thermodynamic forcing anomalies. As discussed by HBT, pulse-like increases in ocean heat transport to the Arctic play an important role in triggering the simulated ice loss events. These appear in part to be driven by variations in the atmospheric forcing that are suggestive of the North Atlantic Oscillation/Arctic Oscillation [*Hurrell*, 1995; *Thompson and Wallace*, 1998] (Figure 4).

Other intrinsic variations with the modeled climate also likely play a role in the simulated rapid ice loss. However, an analysis of the surface atmosphere-ice heat exchange terms shows little indication of extreme changes preceding the ice loss events (not shown). In contrast, there are indications of feedbacks associated with the changing ice conditions, and many surface flux terms exhibit considerable changes during the ice loss events. As discussed by HBT, the surface albedo feedback and changing net surface shortwave radiation appears to be critically important. Also of interest, there are changes in cloud conditions and cloud radiative forcing at the surface that may play a role in these events. These feedbacks will be examined further in future work.

Of course, natural variations also play an important role in the simulated 20th century Arctic climate. For example, pulse-like ocean heat transport anomalies are present within the 20th century. These have an important effect on simulated ice volume variations but do not translate into an ice area change because of the relatively thick ice that is present. This again highlights that a necessary, if not sufficient, condition for the modeled abrupt ice loss is that the ice cover is adequately thinned.

*3.3. Possible Predictability of the Ice Loss Events*

As discussed above, the abrupt ice loss events appear to result from a growing intrinsic ice extent variability coupled to considerable forced change. There is little indication of a critical ice threshold that causes these events, and the chaotic nature of the natural variations makes the potential predictability of these events difficult. However, as changes in ocean heat transport are implicated as a trigger for the abrupt ice loss and the ocean varies on long timescales, it is possible that knowing the prior ocean state may provide some predictive skill for these events.

Here, we further examine the importance of the prior ice and ocean conditions in the simulation of the events. We focus on the long-lived abrupt ice loss present in ensemble run 1 (Plate 2a) and use a set of sensitivity experiments which are initialized at year 2020 with various ice and ocean

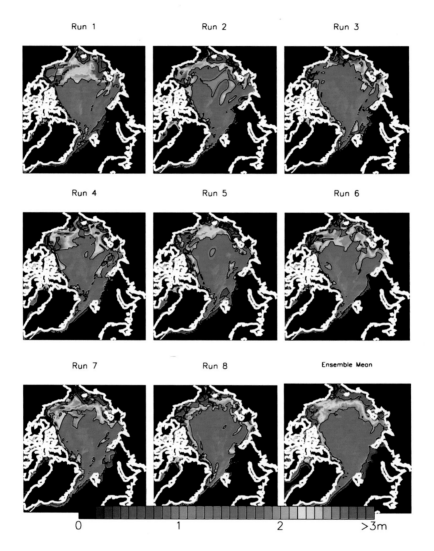

**Plate 4.** May ice thickness that is present 5 years prior to a regional loss in September ice extent. For each ensemble member, the date at which September ice-free conditions first occur is assessed for each model grid cell, and shown here is the May ice thickness that was present 5 years prior to that date for each of the eight ensemble members and for the ensemble mean of the eight runs (bottom right). Black contours indicate 1-m intervals.

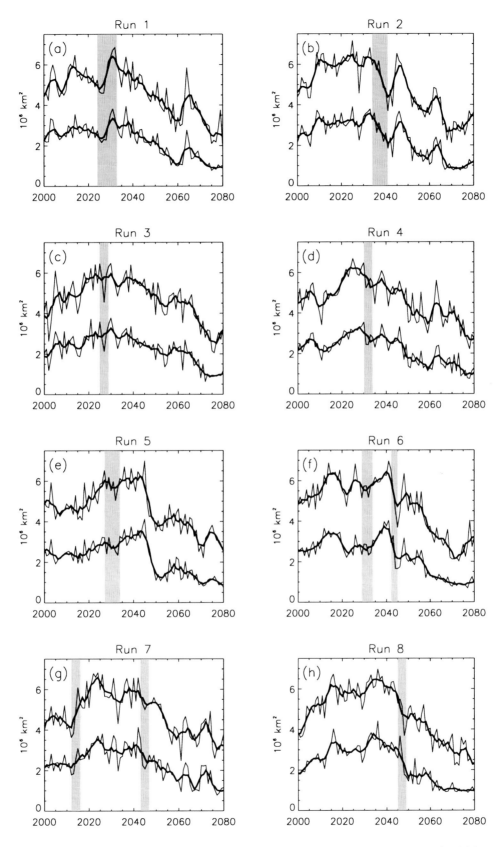

**Figure 2.** Time series of the area of May ice cover that is within 50 and 25 cm of the ensemble mean ice thickness shown in Plate 4, bottom right. Results from each of the eight ensemble members are shown, and each plot has two lines corresponding to the area within 50 cm (top line) and the area within 25 cm (bottom line). The abrupt events for each ensemble member are indicated by grey shading.

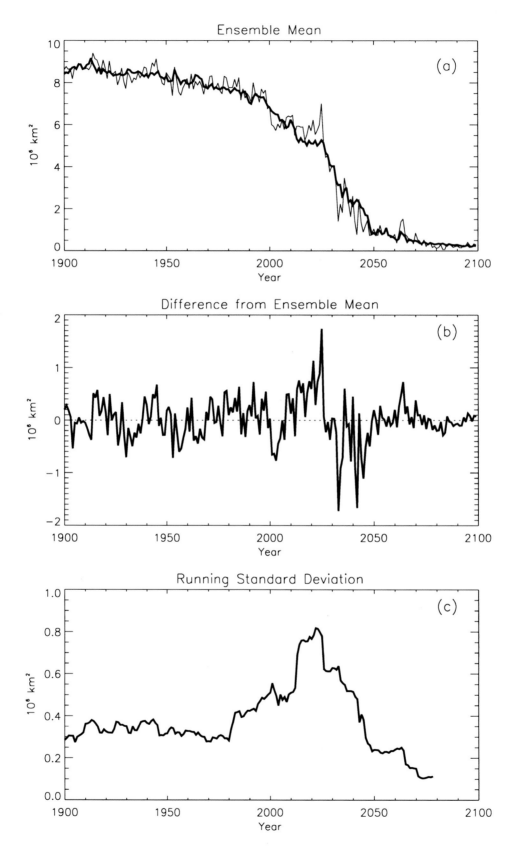

**Figure 3.** (a) Ensemble mean IPCC-AR4 CCSM3 September ice extent time series, (b) deviation from the ensemble mean for run 1, and (c) 20-year running standard deviation of the September ice extent anomaly in run 1. In Figure 3a, the thin line shows the September ice extent time series from run 1.

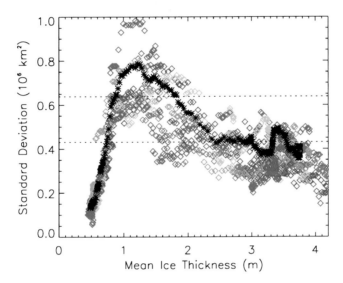

**Plate 5.** Standard deviation of the total Arctic September ice extent anomaly for running 20-year intervals versus the mean Arctic ice thickness from each 20-year interval. The September ice extent anomaly is computed as the September ice extent time series from each ensemble member minus the ensemble mean September ice extent. Each of the eight ensemble members is shown and is indicated by a different color. The black points show the standard deviation computed across all of the ensemble members for a running 20-year time period.

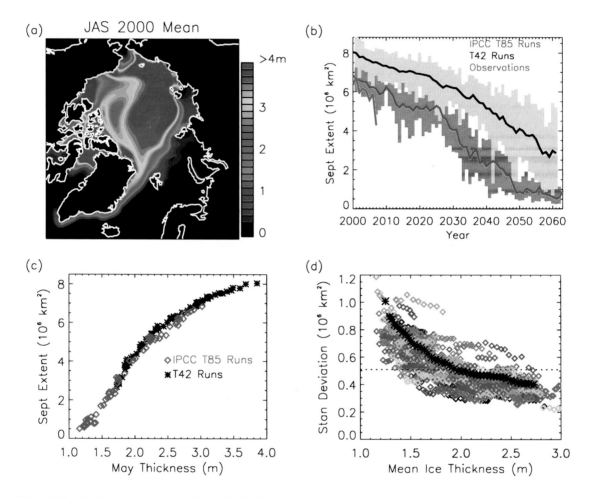

**Plate 6.** Results from the coarse-resolution (T42) simulations. (a) Initial (year 2000) ensemble mean summer (JAS) ice thickness for the T42 simulations. (b) Time series of ensemble mean September ice extent from the IPCC-AR4 CCSM3 simulations (red) and from the 29-member coarse-resolution (T42) CCSM3 simulations (black) with the range in the ensemble members shown by the grey shading and the observed time series shown in green. (c) Ensemble mean September ice extent from these simulations shown as a function of their ensemble mean May ice thickness for each year from 2000 to 2062. (d) The 20-year running standard deviation of the September ice extent anomaly (as computed in Plate 5) versus the mean ice thickness over a running 20-year period from 2000 to 2062 for the T42 simulations. The colored points show values from each individual ensemble member, and the black points show the standard deviation computed across all of the ensemble members for a running 20-year time period.

conditions. The runs are integrated from 2020 to 2040 using identical external forcing as in the IPCC-AR4 CCSM3 integrations. They are initialized at year 2020 with the land and atmospheric conditions from run 1. The initial ice and ocean conditions vary among the sensitivity runs (Table 2). Included among the runs is a simulation, fully initialized with all of the conditions from run 1 (e.g., atmosphere, ocean, sea ice, land state). Because this simulation was run on a different computer from run 1, small (round-off level) perturbations are introduced to the initial state and model computations. This allows us to determine whether the propagation of small errors in this chaotic system contaminates the solution and results in little predictive skill for these events. The remaining four sensitivity runs use either ice or ocean conditions initialized from run 1. For sensitivity runs that apply run 1 initial ice conditions, the ocean initial state is obtained from another of the ensemble members at year 2020 (and similarly for runs initialized with the run 1 ocean state). This results in a set of five sensitivity runs as shown in Table 2. This is too small to provide robust statistics on the predictive skill of these events. However, as illustrated below, they do provide insight into the role of the ice and ocean memory on the simulated rapid ice loss.

The time series of September ice extent from these simulations is shown in Figure 5. Even the simulation with 2020 conditions that are identical to run 1 (Figure 5b) diverges

**Table 2.** Sensitivity Simulations Initialized at Year 2020 From the IPCC-AR4 CCSM3 Simulations[a]

|  | Ice Conditions | Ocean Conditions |
|---|---|---|
| Sensitivity run 1 | run 1 | run 1 |
| Sensitivity run 2 | run 4 | run 1 |
| Sensitivity run 3 | run 3 | run 1 |
| Sensitivity run 4 | run 1 | run 4 |
| Sensitivity run 5 | run 1 | run 3 |

[a] The initial atmosphere and land conditions are taken from run 1. The initial ice and ocean conditions are taken from various ensemble members as shown here.

quite rapidly from the run 1 solution because of the round-off level differences in the integration. None of the sensitivity runs simulate an increase in ice cover from 2020 to 2025 of the same size as run 1, and not all of them simulate an abrupt ice loss. For the three sensitivity simulations that exhibit abrupt September ice loss, the lengths of the events are about half as long as the run 1 event. Also, the timing of the initiation of the event varies among the different sensitivity runs. This suggests that the chaotic nature of the climate system plays a critical role in the year-to-year evolution of the Arctic ice cover and that knowing the ice or ocean conditions several years prior to the event provides little predictive capability. However, this analysis may not be exhaustive, and some predictive skill may be gleaned from longer-timescale atmospheric variations not considered here. Also, to robustly assess potential predictive skill requires a much larger set of sensitivity runs, which is beyond the scope of the present study.

## 4. CONCLUSIONS

The abrupt events present in the IPCC-AR4 CCSM3 simulations occur at different times, over different regions, and with a different resulting sea ice change. There is little indication that large regions of the ice cover reach a threshold in the form of a critical ice thickness or ice area that then results in rapid ice loss. Additionally, the sea ice conditions at the termination of the different events vary from near seasonally ice free to a considerable (>4 million km²) ice cover. While we cannot definitively rule out the presence of a threshold-type behavior that is somewhat masked by climate noise, our results suggest that this is unlikely. Instead, intrinsic variability seems to plays a critical role in the rapid ice loss in these simulations, and they appear to result from an increasing natural variability associated with the thinning ice cover combined with a considerable forced change. This is consistent with the analysis of *Merryfield et al.* [this volume] that considered the bifurcation structure of a simplified model of the CCSM3 events.

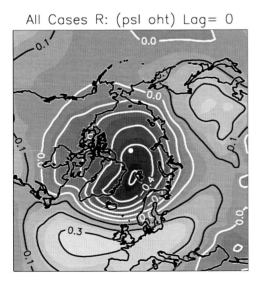

All Cases R: (psl oht) Lag= 0

**Figure 4.** Correlation of detrended annual mean sea level pressure at every point with the detrended time series of annual mean Arctic Ocean heat transport from 2000 to 2049 for all of the ensemble model runs. The detrended time series from each run is concatenated giving a total of 400 temporal data points. The contour interval is 0.1.

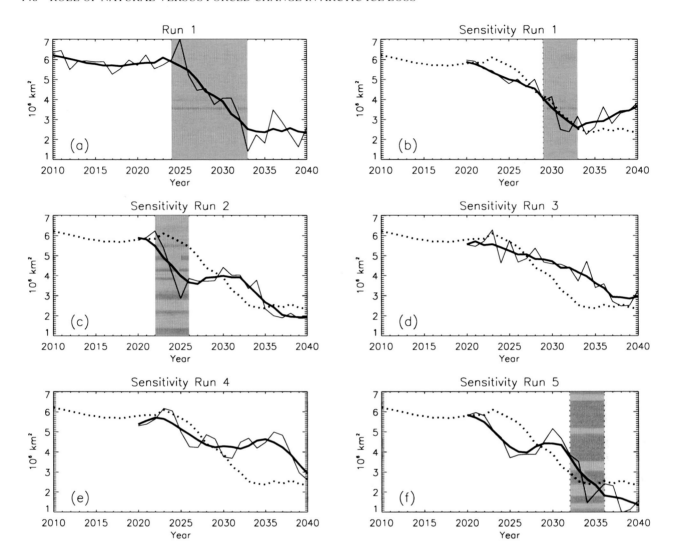

**Figure 5.** Time series of September ice extent for the (a) run 1 ensemble member and (b–f) sensitivity runs listed in Table 2. The 5-year running mean smoothed ice extent from run 1 is shown as the dotted line in Figures 5b–5f. The thin black line indicates the September extent, and the bold black line is the 5-year running mean time series. Abrupt events are indicated by the grey shading.

Our results suggest that the abrupt ice loss events simulated by CCSM3 do not result from threshold or "tipping point" behavior. However, as mentioned above our inability to identify a critical state resulting in rapid ice loss does not necessarily preclude one. In particular, the definition of a critical threshold is somewhat subjective given the complexity of the system. Additionally, only the sea ice state has been considered in this analysis. A threshold could instead exist in other climate variables such as ocean conditions or snow cover. Regardless, the changing aspects of the Arctic sea ice cover indicated by this study, including increas-

ing summer extent variability and the potential for rapid change, could have profound consequences beyond the climate system.

The intrinsic (natural) variability in the September ice extent increases as the ice thins because the natural variations in both dynamic and thermodynamic forcing can more easily produce open water variations. With only a thin ice pack, large regions of the ice cover can be completely melted through or easily converged resulting in a large ice area change. Considerable increases in ice cover still readily occur though since only a thin layer of sea ice needs to survive the melt season in order to result in a large ice extent change. The lower mean ice extent state from which these increases occur is also likely important. More specifically, with smaller summer ice cover and a larger distance between the ice edge and coastline, there is a larger potential for ice edge expansion.

While our analysis and additional sensitivity simulations suggest little predictive capability for these abrupt ice loss events from a knowledge of prior ice and ocean conditions, we can clearly identify some factors that are necessary for them to occur. Probably the most critical is that the ice cover is adequately thin. In the IPCC-AR4 CCSM3 simulations this thinning occurs in response to rising greenhouse gas concentrations and as such, abrupt ice loss events are not present in the 20th century integrations. Another necessary factor is that there is ample natural or intrinsic variability in the modeled system to initiate these events. Finally, large positive feedbacks (such as the surface albedo feedback) are required to accelerate the ice loss. Simulations from various climate models undoubtedly vary in all of these aspects of their simulations, which is likely why they also differ in their simulation of abrupt ice loss.

The thin ice conditions that are necessary for the simulated abrupt ice loss to occur may have some practical implications for understanding observed ice loss. While the large-scale time-varying aspects of Arctic ice thickness are difficult to observe, other related variables (such as ice age) are more easily monitored. There are indications that the Arctic ice age has decreased considerably [*Maslanik et al.*, 2007], which suggests that the Arctic sea ice may well be primed for the type of rapid ice loss simulated by CCSM3. Indeed, the conditions in the summer of 2007 and the synoptic (likely "natural") atmospheric variability implicated in the 2007 ice loss [*Stroeve et al.*, 2008] have some broad similarities to the processes responsible for simulated rapid ice loss in the CCSM3 modeled system. It is possible that the Arctic is currently undergoing rapid sea ice loss not unlike the abrupt ice loss events simulated by CCSM3 for later this century. However, we will only be able to determine this after several more years of observations.

*Acknowledgments.* M.M. Holland would like to acknowledge NASA grant NNG06GB26G. C.M. Bitz was supported by NSF OPP-0454843. B. Tremblay was supported by the Natural Sciences and Engineering Research Council of Canada Discovery Grant Program and by the National Science Foundation under grant OPP-0230325 from the Office of Polar Programs and grant ARC-05-20496 from the Arctic Science Program. D.A. Bailey was supported by NSF OPP-0612388. We thank Eric DeWeaver, Bill Chapman, and an anonymous reviewer for helpful comments that led to improvements in this manuscript. We also thank the community of scientists involved in developing CCSM. Computational facilities have been provided by the National Center for Atmospheric Research (NCAR). Additional simulations were performed by CRIEPI using the Earth Simulator through the international research consortium of CRIEPI, NCAR, and LANL under the Project for Sustainable Coexistence of Human Nature and the Earth of the Japanese Ministry of Education, Culture, Sports, Science and Technology.

## REFERENCES

Arzel, O., T. Fichefet, and H. Goosse (2006), Sea ice evolution over the 20th and 21st centuries as simulated by current AOGCMs, *Ocean Modell., 12*, 401–415.

Bitz, C. M., and W. H. Lipscomb (1999), An energy-conserving thermodynamic model of sea ice, *J. Geophys. Res., 104*, 15,669–15,677.

Bitz, C. M., J. C. Fyfe, and G. M. Flato (2002), Sea ice response to wind forcing from AMIP models, *J. Clim., 15*, 522–536.

Bonan, G. B., et al. (2002), The land surface climatology of the Community Land Model coupled to the NCAR Community Climate Model, *J. Clim., 15*, 3123–3149.

Bourke, R. H., and R. P. Garrett (1987), Sea ice thickness distribution in the Arctic Ocean, *Cold Reg. Sci. Technol., 13*, 259–280.

Briegleb, B. P., C. M. Bitz, E. C. Hunke, W. H. Lipscomb, M. M. Holland, J. L Schramm, and R. E. Moritz (2004), Scientific description of the sea ice component in the Community Climate System Model, version three, *NCAR Tech. Note NCAAR/TN-463+STR*, Natl. Cent. for Atmos. Res., Boulder, Colo.

Collins, W. D., et al. (2006a), The Community Climate System Model: CCSM3, *J. Clim., 19*, 2122–2143.

Collins, W. D., et al. (2006b), The formulation and atmospheric simulation of the Community Atmospheric Model: CAM3, *J. Clim., 19*, 2144–2161.

DeWeaver, E., and C. M. Bitz (2006), Atmospheric circulation and its effect on Arctic sea ice in CCSM3 at medium and high resolutions, *J. Clim., 19*, 2415–2436.

Fetterer, F., K. Knowles, W. Meier, and M. Savoie (2002), Sea ice index, http://nsidc.org/data/seaice_index/, Natl. Snow and Ice Data Cent., Boulder, Colo. (updated 2007)

Gent, P. R. and J. C. McWilliams (1990), Isopycnal mixing in ocean circulation models, *J. Phys. Oceanogr., 20*, 150–155.

Gildor, H., and E. Tziperman (2001), A sea ice climate switch mechanism for the 100-kyr glacial cycles, *J. Geophys. Res., 106*, 9117–9133.

Giorgi, F., and X. Bi (2005), Regional changes in surface climate interannual variability for the 21st century from ensembles of global model simulations, *Geophys. Res. Lett., 32*, L13701, doi:10.1029/2005GL023002.

Gorodetskaya, I. V., and B. Tremblay (2008) Arctic cloud properties and radiative forcing from observations and their role in sea ice decline predicted by the NCAR-CCSM3 model during the 21st century, this volume.

Holland, M. M., C. M. Bitz, and B. Tremblay (2006a), Future abrupt reductions in the summer Arctic sea ice, *Geophys. Res. Lett., 33*, L23503, doi:10.1029/2006GL028024.

Holland, M. M., C. M. Bitz, E. C. Hunke, W. H. Lipscomb, and J. L. Schramm (2006b), Influence of the sea ice thickness distribution on polar climate in CCSM3, *J. Clim.*, *19*, 2398–2414.

Hunke, E. C., and J. K. Dukowicz (1997), An elastic-viscous-plastic model for sea ice dynamics, *J. Phys. Oceanogr.*, *27*, 1849–1867.

Hurrell, J. W. (1995), Decadal trends in the North Atlantic oscillation: Regional temperature and precipitation, *Science, 269*, 676–679.

Intergovernmental Panel on Climate Change (2001), *Climate Change 2001: The Scientific Basis,* edited by J. T. Houghton et al., Cambridge Univ. Press, Cambridge, U. K.

Intergovernmental Panel on Climate Change (2007), *Climate Change 2007: The Physical Science Basis: Contribution of Working Group I to the Fourth Assessment Report of the Intergovernmental Panel on Climate Change,* edited by S. Solomon et al., 996 pp., Cambridge Univ. Press, Cambridge, U. K.

Jochum, M., G. Danabasoglu, M. Holland, Y.-O. Kwon, and W. G. Large (2008), Ocean viscosity and climate, *J. Geophys. Res., 113*, C06017, doi:10.1029/2007JC004515.

Large, W. G., J. C. McWilliams, and S. C. Doney (1994), Ocean vertical mixing: A review and a model with a nonlocal boundary layer parameterization, *Rev. Geophys., 32*, 363–403.

Laxon, S., N. Peacock, and D. Smith (2003), High interannual variability of sea ice thickness in the Arctic region, *Nature, 425*, 947–950.

Maslanik, J. A., C. Fowler, J. Stroeve, S. Drobot, J. Zwally, D. Yi, and W. Emery (2007), A younger, thinner Arctic ice cover: Increased potential for rapid, extensive sea-ice loss, *Geophys. Res. Lett., 34*, L24501, doi:10.1029/2007GL032043.

Merryfield, W. J., M. M. Holland, and A. H. Monahan (2008), Multiple equilibria and abrupt transitions in Arctic summer sea ice extent, this volume.

National Research Council (2002), *Abrupt Climate Change: Inevitable Surprises*, 230 pp., Natl. Acad. Press, Washington, D. C.

North, G. R. (1984) The small ice cap instability in diffusive climate models, *J. Atmos. Sci., 41*, 3390–3395.

Overland, J. E., M. C. Spillane, and N. N. Soreide (2004), Integrated analysis of physical and biological pan-Arctic change, *Clim. Change, 63*, 291–322.

Serreze, M. C., J. E. Walsh, F. S. Chapin III, T. Osterkamp, M. Dyurgerov, V. Romanovsky, W. C. Oechel, J. Morison, T. Zhang, and R. G. Barry (2000), Observational evidence of recent change in the northern high-latitude environment, *Clim. Change, 46*, 159–207.

Serreze, M. C., J. A. Maslanik, T. A. Scambos, F. Fetterer, J. Stroeve, K. Knowles, C. Fowler, S. Drobot, R. G. Barry, and T. M. Haran (2003), A record minimum Arctic sea ice extent and area in 2002, *Geophys. Res. Lett., 30*(3), 1110, doi:10.1029/2002GL016406.

Serreze, M. C., M. M. Holland, and J. Stroeve (2007), Perspectives on the Arctic's shrinking sea-ice cover, *Science, 315*, 1533–1536.

Smith, R., and P. Gent (2004), Reference manual for the Parallel Ocean Program (POP) ocean component of the Community Climate System Model (CCSM2.0 and 3.0), *Rep. LAUR-02-2484,* Los Alamos Natl. Lab., Los Alamos, N. M.

Stroeve, J. C., M. C. Serreze, F. Fetterer, T. Arbetter, W. Meier, J. Maslanik, and K. Knowles (2005), Tracking the Arctic's shrinking ice cover: Another extreme September minimum in 2004, *Geophys. Res. Lett., 32*, L04501, doi:10.1029/2004GL021810.

Stroeve, J., M. M. Holland, W. Meier, T. Scambos, and M. Serreze (2007), Arctic sea ice decline: Faster than forecast, *Geophys. Res. Lett., 34*, L09501, doi:10.1029/2007GL029703.

Stroeve, J., M. Serreze, S. Drobot, S. Gearhead, M. Holland, J. Maslanik, W. Meier, and T. Scambos (2008), Arctic sea ice extent plummets in 2007, *Eos Trans. AGU, 89*(2), 13, doi:10.1029/2008EO020001.

Thompson, D. W. J., and J. M. Wallace (1998), The Arctic oscillation signature in the wintertime geopotential height and temperature fields, *Geophys. Res. Lett., 25*, 1297–1300.

Thorndike, A. S., D. A. Rothrock, G. A. Maykut, and R. Colony (1975), Thickness distribution of sea ice, *J. Geophys. Res., 80*, 4501–4513.

Winton, M. (2006), Does the Arctic sea ice have a tipping point?, *Geophys. Res. Lett., 33*, L23504, doi:10.1029/2006GL028017.

Zhang, X., and J. E. Walsh (2006), Toward a seasonally ice-covered Arctic Ocean: Scenarios from the IPCC AR4 model simulations, *J. Clim., 19*, 1730–1747.

D. A. Bailey and M. M. Holland, National Center for Atmospheric Research, P.O. Box 3000, Boulder, CO 80307-3000, USA. (mholland@ucar.edu)

C. M. Bitz, Department of Atmospheric Sciences, University of Washington, Box 351640, Seattle, WA 98195-1640, USA.

L.-B. Tremblay, Ocean and Atmosphere Sciences, McGill University, 805 Sherbrooke Street West, Montreal, QC H2A 2K6, Canada.

# Multiple Equilibria and Abrupt Transitions
# in Arctic Summer Sea Ice Extent

William J. Merryfield

*Canadian Centre for Climate Modeling and Analysis, Environment Canada, Victoria, British Columbia, Canada*

Marika M. Holland

*National Center for Atmospheric Research, Boulder, Colorado, USA*

Adam H. Monahan

*School of Earth and Ocean Sciences, University of Victoria, Victoria, British Columbia, Canada*

An application of bifurcation theory to the stability of Arctic sea ice cover is described. After reviewing past such efforts, a simple mathematical representation is developed of processes identified as contributing essentially to abrupt decreases in 21st century Arctic summer sea ice extent in climate model simulations of the Community Climate System Model, version 3 (CCSM3). The resulting nonlinear equations admit abrupt sea ice transitions resembling those in CCSM3 and also plausibly represent further gross aspects of simulated Arctic sea ice evolution such as the accelerating decline in summer ice extent in the late 20th and early 21st centuries. Equilibrium solutions to these equations feature multiple equilibria in a physically relevant parameter regime. This enables abrupt changes to be triggered by infinitesimal changes in forcing in the vicinity of the bifurcation or, alternatively, by finite perturbations some distance from the bifurcation, although numerical experiments suggest that abrupt transitions in CCSM3 may arise mainly from the increasing sensitivity of sea ice to fluctuations in ocean heat transport as ice thickness and extent diminish. A caveat is that behavior following a complete seasonal loss of ice cover is sensitive to aspects of the parameterization of ocean shortwave absorption.

> The objective is to illuminate the essential processes and not to embellish them or mix them up with others which are less important. The many simplifications also make it possible to see how the various processes interact, but also make it difficult to assess the quantitative validity of the argument. If the ideas have value, it will be because they serve to develop intuition.
>
> *Thorndike* [1992]

Arctic Sea Ice Decline: Observations, Projections, Mechanisms, and Implications
Geophysical Monograph Series 180
Published in 2008 by the American Geophysical Union.
10.1029/180GM11

## 1. INTRODUCTION

A central question in the study of climate variations is whether climate response is smooth, in the sense of being proportional to small changes in a forcing parameter, or abrupt, so that an infinitesimal change in forcing or a sufficiently large finite perturbation can bring about finite (and possibly large) climate response. Bifurcation theory provides a mathematical framework for representing this dichotomy [e.g., *Guckenheimer and Holmes*, 1983] and for describing sudden changes as shifts between multiple equilibria. While such shifts may sometimes appear to occur in complex models, to apply the tools of bifurcation theory generally requires substantial simplification, with "essential" aspects of the climate system for a particular problem distilled to a relatively simple set of equations. Such an approach can yield physical insights and enables a thorough exploration of parameter dependences, although clearly care must be taken both in formulation and in interpreting results.

One aspect of the climate system that has been thoroughly studied in this manner is the response of the North Atlantic meridional overturning circulation to changes in surface freshwater flux [e.g., *Stommel*, 1961; *Titz et al.*, 2002]. Bifurcation studies of conceptual models, in conjunction with numerical experiments involving circulation models, have indicated that increased freshwater flux into the North Atlantic can cause a sudden "switching off" of the overturning, leading to a rapid cooling in this region [e.g., *Stocker and Wright*, 1991; *Rahmstorf*, 1995; *Manabe and Stouffer*, 1999]. For such abrupt changes to occur requires positive feedback, which in this instance is provided by advective and convective processes [*Rahmstorf*, 1999].

Another component of the climate system with potential to change rapidly is sea ice. In this instance, positive feedback is provided by the absorption of sunlight by low-albedo open water when high-albedo ice cover retreats, promoting further ice melt and inhibiting freezing. The possibility that Arctic sea ice in particular might abruptly retreat under anthropogenic warming has received increasing attention [*Lindsay and Zhang*, 2005; *Winton*, 2006], in part because of the unexpectedly rapid shrinkage of summer ice cover in recent years [e.g., *Stroeve et al.*, 2007], as well as recent climate model results that exhibit very rapid declines in 21st century summer ice extent under projected anthropogenic forcings [e.g., *Holland et al.*, 2006a].

This paper has three main parts. The first, in section 2, is a brief review of several previous investigations that have examined multiple equilibria of sea ice. The second, in section 3, is a case study in which we attempt to represent the complex behavior of the CCSM3 climate model, which includes sudden decreases in 21st century summer Arctic ice extent, with a simple set of equations. The third part, in section 4, demonstrates how seemingly subtle changes in the parameterizations used in section 3 can lead to qualitative differences in behavior, illustrating some of the challenges

inherent in such simplified analyses of the climate system. Discussion and conclusions are provided in section 5.

## 2. MULTIPLE SEA ICE EQUILIBRIA: A BRIEF OVERVIEW

An early quantitative argument that Arctic climate might have two stable regimes, one ice covered and the other ice free, was given by *Budyko* [1966, 1974] based on energy balance considerations. Since than, a number of investigations have examined the possibility that multiple sea ice equilibria can exist because of positive ice-albedo feedback and that the associated bifurcation structure can be obtained from a simplified set of equations. A brief review of these efforts follows.

### 2.1. Planetary Energy Balance Models

*North* [1975, 1984] considered a simple energy balance climate model consisting of an ordinary differential equation for equilibrium surface temperature as a function of latitude, with balances between solar heating, outgoing longwave radiation and poleward atmospheric heat transport represented. (The model was originally devised to study ice sheet inception, although *North* [1984] recognized its applicability to Arctic sea ice.) Surface albedo takes on two values, corresponding to ice or a less reflective earth surface, depending on temperature. As a solar constant parameter $Q$ increases, the equilibrium latitudinal extent of ice cover diminishes until a critical value is reached. At this point a further infinitesimal increase in $Q$ causes the system to jump to a second, ice-free equilibrium. These stable equilibria coexist over a range of $Q$ and are connected by an unstable equilibrium solution branch representing a small ice cap [*North*, 1984, Figure 1]. This instability was described as the small ice cap instability or SICI.

### 2.2. Local Thermodynamic Models

A second approach that has yielded simplified equations whose equilibria can be studied considers the thermodynamics of the atmosphere-ocean-ice system locally, ignoring horizontal nonuniformities. *Thorndike* [1992] examined such balances under seasonal forcing. The warming influences of solar radiation and poleward atmospheric and ocean heat transport and the cooling influence of upward longwave radiation were included, taking albedo changes into account. The resulting equations have two stable equilibria, one representing perennial ice cover and the other ice free, that coexist over a range of poleward heat transports [*Thorndike*, 1992, Figure 9]; a third equilibrium, ice free in summer

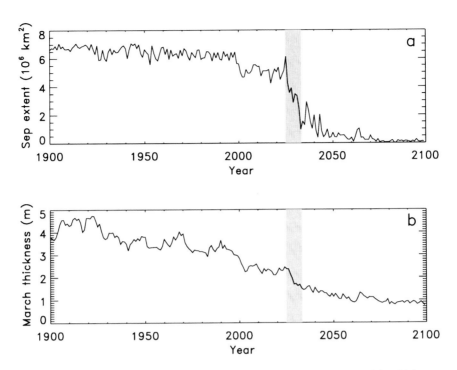

**Figure 1.** Time series of 20th and 21st century (a) September ice extent and (b) mean March ice thickness, as represented by a single realization of CCSM3 under the Special Report on Emissions Scenarios A1B forcing scenario. The gray bands represent an interval of abrupt decrease in ice extent as defined in the text.

only, is unstable. This "all or nothing" situation resembles the SICI phenomenon described by *North* [1984].

*Flato and Brown* [1996] applied a local thermodynamic ice model derived from that of *Maykut and Untersteiner* [1971] to study climate sensitivity of landfast Arctic sea ice. The model contains detailed parameterizations of ice properties and includes effects of snow. When subject to an annual cycle of forcing derived from observations, the model exhibits two distinct equilibria, one consisting of perennial ice having a modest seasonal cycle in thickness and the other seasonally ice free. These solutions are separated by a third, unstable equilibrium having intermediate annual mean thickness [*Flato and Brown*, 1996, Figure 8c].

An intermediate level of complexity is represented by the investigation of *Björk and Söderkvist* [2002]. Their model is closer to a column version of a comprehensive climate model, with ice cover described by a thickness distribution, a prognostic one-dimensional ocean having a dynamic ocean mixed layer and parameterized lateral ocean transports, and an atmospheric model built on that of *Thorndike* [1992]. Forcing is provided by monthly means of observed climatological quantities. Like the simpler model of *Thorndike* [1992], it exhibits dual equilibria having perennial ice in one instance and no ice cover in the other.

Finally, in a more idealized context, *Taylor and Feltham* [2005] treated sea ice as a floating slab subject to surface heating by solar radiation and cooling by other processes. A simple radiative transfer model was used to represent radiative penetration and associated internal heating. The ice thickness equilibria, determined analytically, consist of a stable solution branch that decreases with increasing incident shortwave flux, until a critical flux is reached. At that point an infinitesimal further increase in shortwave flux results in collapse from finite to zero thickness. The disappearance of the stable equilibrium occurs through its merger with an unstable solution branch [*Taylor and Feltham*, 2005, Figure 6].

### 2.3. Climate System Box Models

As a means for studying climate system evolution over very long timescales, *Gildor and Tziperman* [2001] devised a coupled box model comprising a planetary domain divided into four latitude bands, with both sea ice and ice sheets on land included. Their main finding was that growth of sea ice cover in a cooling climate reduces the atmospheric supply of moisture to ice sheets, resulting in a negative feedback that leads, in turn, to glacial cycle-like oscillations on a

100,000-year timescale. These oscillations feature relatively abrupt onsets and collapses of sea ice cover that occur on timescales of a few decades, minute compared to the oscillation period. The model behaved similarly when extended to incorporate seasonal and orbital forcing [*Gildor and Tziperman*, 2000], as well as greater resolution in latitude [*Sayag et al.*, 2004]. Although these studies did not show explicitly that multiple sea ice equilibria exist under fixed climatic forcing, this is strongly implied by the abrupt "switching" or threshold behavior for sea ice in their solutions.

To achieve varying degrees of simplification, each of these studies emphasized particular aspects of sea ice and its relation to climate, as summarized in Table 1; typically, the simpler formulations yield analytical determinations of equilibria, whereas for the more complex ones equilibria were determined numerically. However, one aspect common to all these studies that have found multiple sea ice equilibria is the increase in surface shortwave absorption that results from decreased albedo as ice cover thins or retreats.

In section 3 we describe an attempt to represent the behavior of Arctic sea ice in a complex climate model through a set of equations sufficiently simple that their equilibria can be obtained analytically.

## 3. ANALYSIS OF ABRUPT ARCTIC SEA ICE TRANSITIONS IN CCSM3

After explaining the motivation for this study in section 3.1, simple nonlinear equations describing Arctic sea ice evolution in CCSM3 are formulated in section 3.2. Their bifurcation structure is analyzed section in 3.3, where it is found that multiple equilibria can exist provided changes in ocean shortwave absorption due to ice-albedo feedback are sufficiently strong. Numerical solutions of the simplified equations are compared with CCSM3 behavior in section 3.4; additional solutions exhibiting vacillations between stable equilibria and hysteresis effects are also briefly considered. Finally, the predictability of simulated sea ice abrupt transitions is examined in section 3.5.

### 3.1. Background

In light of the retreat of Arctic sea ice at a rate approaching 10% per decade in recent years [e.g., *Stroeve et al.*, 2005], possibly attributable to anthropogenically induced climatic warming [*Overpeck et al.*, 2005], it is clearly of interest to examine climate model projections of future Arctic sea ice trends. Recently, *Holland et al.* [2006a] (hereinafter referred to as HBT) described such projections from version 3 of the Community Climate System Model (CCSM3), which simulates with reasonable fidelity the observed sea ice extent and trends in the late 20th and early 21st centuries. They considered an ensemble of seven simulations employing the Special Report on Emissions Scenarios (SRES) A1B forcing scenario [*Intergovernmental Panel on Climate Change*, 2001] and found in each that the future Arctic summer sea ice extent undergoes sudden and substantial decreases at rates far exceeding those already observed. These events, which they termed "abrupt transitions," occur as early as 2015 and lead to a nearly ice-free summertime Arctic Ocean

**Table 1.** Studies of Multiple Sea Ice Equilibria[a]

|  | North [1984] | Thorndike [1992] | Flato and Brown [1996] | Taylor and Feltham [2005] | Björk and Söderkvist [2002] | Gildor and Tziperman [2001] | This Study |
|---|---|---|---|---|---|---|---|
| Domain | planetary | local | local | local | local | planetary | Arctic |
| SW heating | • | • | • | • | • | • | • |
| Albedo changes | • | • | • | • | • | • | • |
| Radiative cooling | • | • | • | ∘ | • | • | - |
| SW penetration | - | - | • | • | - | - | - |
| Atmospheric heat transport | • | - | - | - | • | • | - |
| Ocean heat transport | - | • | - | - | - | • | • |
| Seasonality | - | • | • | - | • | - | • |
| Partial ice cover | • | - | - | - | • | • | • |
| Ice thickness | - | • | • | • | • | - | • |
| Ice dynamics/export | - | - | - | - | • | - | - |
| Snow cover | - | - | • | - | • | - | - |
| Analytical solutions | • | ∘ | - | • | - | - | • |

[a] Symbols represent the following: • indicates considered; ∘ indicates partial or incomplete; and - indicates not considered.

by 2050 (Figure 1). (As defined by HBT, abrupt transitions occur when the 5-year running mean of September ice extent decreases more rapidly than $0.5 \times 10^6 \, \text{km}^2 \, \text{a}^{-1}$. If the decrease occurs between years $n$ and $n + 1$, the event is considered to straddle years $n - 2$ to $n + 1$ encompassed by the running averages.)

HBT identified three enabling conditions for the simulated abrupt sea ice retreats: (1) a superlinear increase in the rate of open water formation per centimeter of ice melt as the sea ice cover thins; (2) rapid increases in ocean heat transport into the Arctic, which act as a trigger; and (3) positive ice-albedo feedback, which intensifies the change. These three conditions are essentially thermodynamic, whereas ice dynamics, e.g., through export or consolidation, was found not to play an essential role.

The remainder of this section describes a very simple mathematical representation of these three processes and explores its ability to describe abrupt transitions in Arctic summer sea ice extent. There are two main motivations for this approach. The first is that it enables the HBT hypothesis that these process are the primary drivers for abrupt transitions to be evaluated within a plausible quantitative framework, with minimal additional assumptions. The second motivation is that, to the extent that these equations capture some essential aspects of the modeled Arctic sea ice system, analytical exploration of their solutions and parameter dependences could potentially yield an improved understanding of the response of Arctic sea ice to climatic warming in CCSM3 and possibly in other climate models as well.

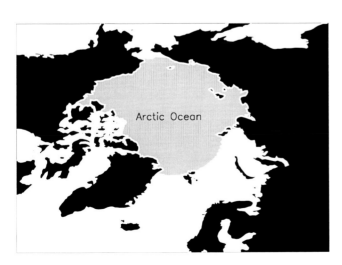

**Figure 2.** Arctic Ocean region defined in section 3.2.

## 3.2. Formulation

We begin by considering an Arctic Ocean region that is nearly perennially ice covered in the absence of anthropogenic forcing (Figure 2). This region includes the Siberian seas eastward of 100°E and the Beaufort Sea but excludes waters equatorward of 80°N between eastern Greenland and 100°E including the Barents and Kara seas, where ice cover is inhibited by strong heat transport from the Atlantic Ocean, even in a preindustrial climate. The ocean area of this region, and hence maximum possible ice extent, is $A_{\text{max}} = 7.23 \times 10^6 \, \text{km}^2$.

For compactness and self-consistency, the evolving quantities considered are restricted to those examined by HBT, namely, the areal mean winter (March) ice thickness, denoted by $T_n$, the summertime (September) ice extent $A_n$, the areal mean cumulative melt $M_n$, and the areal and annual mean ocean heat transport $H_n$ and open water shortwave absorption $Q_n$, where the subscript refers to calendar year $n$. All quantities are defined in terms of the Arctic Ocean region described above.

Winter ice thickness is considered to be determined by a balance between freezing, melt, and ice export. In the absence of ocean heat transport and for perennial ice cover, this balance is described by a "baseline" forcing $F$, approximated as constant. This balance is perturbed by heating of the ocean mixed layer due to a given year's net ocean heat transport $H_n$ and to enhanced shortwave absorption by open water (ice-albedo feedback). Both processes, summed over year $n$, are considered to reduce the following year's winter ice thickness, both through reduced freezing and increased melt:

$$T_{n+1} = \max[F - wH_n - wQ_n, 0], \qquad (1)$$

where $w = 0.104 \, \text{m} \, (\text{W m}^{-2})^{-1}$ is a conversion factor equal to the thickness of pure 0°C ice melted by $1 \, \text{W m}^{-2}$ over 1 year, assuming an ice density of $905 \, \text{kg m}^{-3}$, and a physical lower limit of $T_{n+1} = 0$ has been imposed. This formulation is justified by noting that for pure ice the latent heat of fusion is $334 \, \text{kJ kg}^{-1}$, whereas the specific heat is $2.1 \, \text{kJ kg}^{-1} \, \text{deg}^{-1}$; the energy required to melt a given mass of ice thus exceeds that required to warm it from −15°C to 0°C by an order of magnitude. A more careful accounting of saline influences such as discussed by *Bitz and Liscomb* [1999] is not attempted under the gross approximations adopted here.

It remains to express ocean heat transport $H_n$, considered as a specified forcing, and the open water shortwave absorption $Q_n$ in a manner that is broadly consistent with the CCSM3 model results, and represents with reasonable fidelity the processes considered by HBT. These representations are described in turn in sections 3.2.1–3.2.3.

*3.2.1. Efficiency of open water formation.* HBT define the efficiency of open water formation as the open water formed, as a fraction of $A_{max}$, per meter of areal mean ice melt accumulated over the May–August melt season. In Figure 3, annual values of open water formation efficiency are plotted against mean March ice thickness (representing the wintertime maximum) for the simulation shown in Figure 1.

The rapid increase in open water formation efficiency with decreasing March ice thickness was identified by HBT as playing an important role in the simulated abrupt transitions. This behavior is parameterized here by

$$\text{open water formation efficiency} = T^*/T_n, \quad (2)$$

where $T^*$ is a dimensionless constant and $T_n$ represents mean March ice thickness (in m) for year $n$. Such behavior resembles that obtained by *Hibler* [1979, equations (14)–(16)] under an assumption that ice thickness is uniformly distributed between 0 and twice the mean thickness ($2T$ in our notation). Under such conditions, $M$ cm of ice melt will completely melt an areal fraction $M/2T$ of the original ice cover, which implies an open water formation efficiency of $1/2T$, or $T^* = 0.5$. In representing the more complex behavior in CCSM3, we specify $T^*$ according to a least squares fit to the data in Figure 3. Because the focus here is on the range of ice thicknesses at which the abrupt transitions occur, only thicknesses $T_n < 3$ m are considered in the fit. This procedure yields

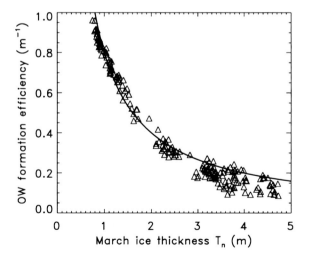

**Figure 3.** Yearly values of open water (OW) formation efficiency, defined as open water gained (as a fraction of Arctic regional area $A_{max}$) per meter of mean accumulated May–August ice melt, plotted against mean March thickness. The curve represents the fit discussed in section 3.2.1.

$T^* = 0.8$ (solid curve in Figure 3), a value used subsequently except where noted.

*3.2.2. Ocean heat transport.* The second factor identified by HBT as contributing to abrupt transitions are rapid transient increases in ocean heat transport (OHT). These "pulses" resemble observed sudden warmings in Atlantic waters entering the Arctic Ocean [e.g., *Polyakov et al.*, 2005] and are superimposed on a more gradual increase in OHT qualitatively similar to that seen in other climate models [*Holland and Bitz*, 2003]. The transient OHT increases tend to lead the sea ice abrupt transitions by 1–2 years (HBT). In Figure 4a, the solid curve represents OHT for the CCSM3 realization depicted in Figure 1. This time series features a sudden increase near 2025 which appears to be associated with the sea ice abrupt transition.

The more gradual increase in OHT associated with climatic warming was examined by *Bitz et al.* [2006] in a CCSM3 run subject to 1% year$^{-1}$ increase in $CO_2$. They showed the OHT increase to be attributable both to warmer water entering from the subpolar Atlantic and increased volume transport into the Arctic basin (primarily the latter). This increased ventilation peaks at around the time of $CO_2$ doubling, and Bitz et al. argued that it is driven by an increase in buoyancy forcing caused by increased sea ice production that results from more open water, thinner summer ice, and the inverse dependence of winter sea ice growth on ice thickness. (This effect could be represented schematically in Figure 8 as an arrow directed from "increased open water" to "increased OHT.")

The evolution of OHT is represented here according to a statistical model which is fit to the OHT time series from CCSM3. The synthetic OHT time series, denoted $H_n$, consist of a smoothly evolving ensemble mean $\overline{H}_n$, to which is added a more rapidly varying stochastic component $H'_n$. For the evolving ensemble mean we take

$$\overline{H}_n = H_0 + \left(\frac{H_1 - H_0}{2}\right)\left[1 + \tanh\left(\frac{t_n - t^*}{\delta_t}\right)\right], \quad (3)$$

where $t_n$ denotes year $n$, and take $H_0 = 1$ W m$^{-2}$, $H_1 = 14$ W m$^{-2}$, $t^* = 2025$, and $\delta_t = 50$, which yields the dashed curve in Figure 4a. The stochastic component $H'_n$ is modeled as an AR(1)-like process having a smoothly evolving ensemble variance $\sigma_n^2$, so that

$$H_n = \overline{H}_n + \phi(H_{n-1} - \overline{H}_{n-1}) + (1 - \phi^2)^{1/2}\sigma_n\varepsilon_n, \quad (4)$$

[e.g., *von Storch and Zwiers*, 1999] which has the desired mean and variance. Here $\varepsilon_n$ is a white noise process having

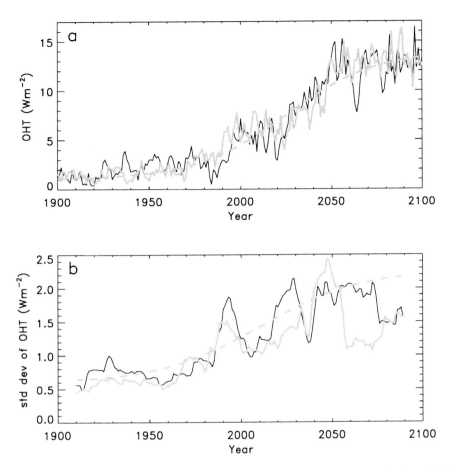

**Figure 4.** Ocean heat transports entering Arctic region: (a) annual means and (b) standard deviations. Solid curves correspond to the CCSM3 realization shown in Figure 1. Shaded dashed curves denote the ensemble mean of the statistical model, and shaded curves denote a single realization of the statistical model. The single-realization standard deviations in Figure 4b are computed using a 21-year sliding window.

unit variance, and $\phi = \exp(-1/t_A) \approx 0.61$ for an assumed autocorrelation time of $t_A = 2$ years, similar to that of the OHT time series from CCSM3. The ensemble variance is taken to evolve as $\sigma_n^2 = \sigma_0^2(\bar{H}_n/H_0)$ with $\sigma_0 = 0.6$ W m$^{-2}$. This yields an ensemble standard deviation, described by the dashed curve in Figure 4b, which corresponds reasonably well to the standard deviation of the CCSM3 time series within a running 21-year window (solid curve in Figure 4b); for comparison, a similar running standard deviation applied to the synthetic time series in Figure 4a is also shown (shaded curve in Figure 4b).

*3.2.3. Ice-albedo feedback.* The decrease in surface albedo as sea ice retreats in a warming climate leads to increased surface absorption of shortwave radiation and is believed to constitute a significant positive feedback which amplifies polar warming relative to other regions [e.g., *Holland and*

*Bitz*, 2003]. HBT and *Winton* [2006] discuss the likely role of ice-albedo feedback in accelerating ice retreat.

To parameterize this feedback, we consider the shortwave radiation $Q_{SW}$ absorbed by the ocean in the Arctic region, which is approximately equal to $Q_{SW} = (1 - a_{ice})(1 - \alpha_{ocn})F_{SW}$, where $\alpha_{ocn}$ is the ocean albedo, $a_{ice}$ is the ice-covered fraction, and $F_{SW}$ is the downwelling shortwave radiation at the surface. Penetration of solar radiation through the ice into the ocean is neglected. Changes in $Q_{SW}$ are dominated by decreasing ice cover, with changes in $F_{SW}$, which decreases by approximately 18% over the 21st century, playing a small compensating role. (Such a decrease, seen also in other current generation climate models, is associated with increased cloudiness [*Sorteberg et al.*, 2007].) This suggests approximating changes in annually averaged ocean shortwave absorption $Q_n$ as proportional to September open water area $A_{max} - A_n$:

$$Q_n = b(A_{\max} - A_n). \tag{5}$$

The plausibility of such a relation is supported by Figure 5, which plots values of annual and areal mean $Q_{SW}$ against annual values of $A_{\max} - A$ for the CCSM3 run whose summertime sea ice retreat is shown in Figure 1a. These two time series are closely correlated ($r^2 = 0.98$). The linear regression (solid curve in Figure 5) has an offset $Q_0 \approx 5.5$ W m$^{-2}$ at $A_{\max} - A = 0$, which is taken to represent ocean shortwave absorption in leads and through ice and is considered part of the baseline forcing $F$ in (1). By approximation to the regression slope $1.86 \times 10^{-12}$ W m$^{-4}$, the shortwave absorption parameter for CCSM3 is assigned the value $b = 2 \times 10^{-12}$ W m$^{-4}$.

*3.2.4. Simple equations describing sea ice evolution.* Substituting (5) describing open water ocean shortwave absorption into (1) leads to winter ice thickness evolution being described by

$$T_{n+1} = \max[F - wH_n - wb(A_{\max} - A_n), 0]. \tag{6}$$

The September ice extent $A_n$ for the previous calendar year is taken to be governed by the open water formation efficiency relation (2):

$$A_n = A_{\max}[1 - (T^*/T_n)M_n], \tag{7}$$

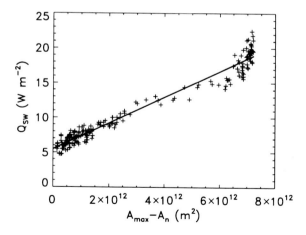

**Figure 5.** Symbols denote annual and areal mean shortwave radiation $Q_{SW}$ absorbed in the Arctic Ocean region shown in Figure 2, as a function of September open water area $A_{\max} - A$, for the CCSM3 simulation whose Arctic ice evolution is shown in Figure 1. The solid line is a linear regression on these values.

where $M_n$ is the accumulated March–September ice melt in meters.

To close these equations, $M_n$ must be specified as a function of known quantities. The total melt consists of the sum of the surface melt, $M^{(s)}$, due primarily to radiative heating of the ice surface, and the basal melt, $M^{(b)}$, due to heat transferred from ocean to ice, where lateral melting is neglected as it is a small term. While both $M^{(s)}$ and $M^{(b)}$ increase in a warming climate, the increase in total melt in CCSM3 arises primarily from $M^{(b)}$. Therefore, as a further gross approximation we take $M^{(s)}$ to be constant, assigning a representative value of $M_0^{(s)} = 0.4$ m. This is considered adequate, given that the values simulated by CCSM3, low-pass filtered by a sliding 21-year window, differ by less than 0.1 m a$^{-1}$ from this value over the simulated period 1900–2099 (not shown).

Because increased OHT into the Arctic Ocean should tend to warm Arctic waters and because HBT observed that OHT pulses appear to act as a trigger for sea ice abrupt transitions, we take the basal melt term $M^{(b)}$ to depend on $H_n$. In order to maintain as simple a set of physical assumptions as is feasible, it is assumed that there is a perfect transfer of heat input from OHT to melting of sea ice where ice is present, which implies $M^{(b)} = M_0^{(b)} + wH(\bar{A}/A_{\max})$, where $\bar{A}$ is the annual mean ice extent. Drawing from available information, we specify the latter as the average of winter ice extent, assumed equal to $A_{\max}$, and summer ice extent $A_n$, so that $\bar{A}_n = (A_{\max} + A_n)/2$. On physical grounds $M^{(b)}$ cannot increase indefinitely with increasing OHT because total mean melt is limited by the winter mean ice thickness $T$, i.e., $M^{(s)} + M^{(b)} \leq T$ for any particular year. However, applying this constraint to (7) would prevent $A$ from decreasing below $(1 - T^*)A_{\max}$, or approximately $1.45 \times 10^6$ km$^2$ for the chosen parameters. Therefore, to enable zero summer ice extent to be attained we instead limit $M$ by $M^{(s)} + M^{(b)} \leq T/T^*$. The melt parameterization thus becomes

$$M_n = \min[M_0^{(s)} + M_0^{(b)} + wH_n(1 + A_n/A_{\max})/2, T_n/T^*], \tag{8}$$

with $M_0^{(s)} = 0.4$ m as discussed above and $M_0^{(b)} = 0.2$ m, which provides a reasonable fit to the initial, approximately linear increase of $M$ with $H$. The modeled variation of $M_n$ with $H_n$ in the CCSM3 simulation considered previously, low-pass filtered by a sliding 21-year window, is indicated in Figure 6 by the symbols, whereas the solid curve represents (8) as computed for similarly low-passed CCSM3 $T_n$, $A_n$, and $M_n$.

We note that the albedo feedback, which as discussed by HBT may contribute to ice melt, is not included in the melt parameterization (8), and justify this on the somewhat ad hoc grounds that this heat input is by nature concentrated in ice-free regions, and in any case its inclusion would worsen

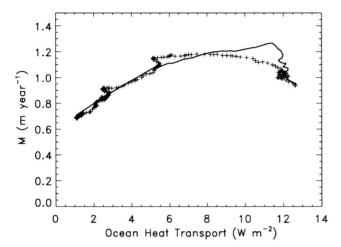

**Figure 6.** Symbols denote accumulated annual melt $M$ versus annual mean ocean heat transport $H$ for the Arctic region, low-pass filtered by a sliding 21-year window, from CCSM3 simulation. Solid curve represents melt as parameterized by equation (8), using low-pass filtered CCSM3 ocean heat transport $H$ and March ice thickness $T$ as input.

the tendency for overestimation of melting implied by (8). The albedo effect is instead viewed as warming the ocean mixed layer in ice-free regions, thus delaying the onset of seasonal ice growth, as suggested by its role in suppressing ice growth in (6).

*3.2.5. Calibration to CCSM3.* In addition to the time-dependent winter ice thickness $T_n$ and summer ice extent $A_n$, both predicted, and OHT forcing $H_n$ specified as input, equations (6)–(8) contain several time-independent coefficients (Table 2). Two of these, the maximum Arctic regional ice extent $A_{max}$ and the factor $w$ relating thickness of ice melted to a given heat input, are physical constants. Several others, including the ocean shortwave absorption parameter $b$, surface melt $M_0^{(s)}$ and basal melt $M_0^{(b)}$ in the absence of ocean

heat transport, have been set to CCSM3-relevant values. Only the coefficient $F$ representing baseline climatic forcing of ice growth in the absence of ocean heat transport and the open water shortwave absorption remain to be specified.

To fix $F$, we apply as input the actual CCSM3 time series for $H_n$ in Figure 4a and adjust $F$ to provide a qualitatively reasonable match between output $T_n$ and $A_n$ and the actual CCSM3 time series in Figures 1a and 1b. A value of $F = 3.2$ m yields the time series in Figure 7, which shows (1) a slight, gradual decrease in $A_n$ and $T_n$ throughout most of the 20th century; (2) a modest but more sudden decrease in both quantities near 2000, followed by 2 decades or so that feature fluctuations but no further decrease; (3) an abrupt transition centered near 2027, during which $A_n$ decreases by more than 50% in less than a decade; (4) minimal summer ice cover after 2040 or so, with a brief but minor recovery near 2065, and (5) winter ice thicknesses of less than 1 m in the late 21st century, with a local maximum coinciding with the minor recovery in summer extent around 2065. There are also some notable differences between the CCSM3 and synthetic time series; in particular, winter thickness and summer extent are significantly underestimated throughout much of the 20th century. (A contributing factor for this discrepancy is the overestimation of open water formation efficiency for $T_n$ greater than 3 m or so that is apparent in Figure 3.) In addition, the CCSM3 time series exhibit significantly more high-frequency variability, which may arise from atmospheric influences, such as the Arctic Oscillation, that modulate ice export and summer ice extent [e.g., *Rigor and Wallace*, 2004; *Lindsay and Zhang*, 2005] but are not included in the simple representation considered here.

*3.2.6. Summary.* Equations (6)–(8) describe evolution of annual values $T_n$ and $A_n$ of Arctic sea ice winter thickness and summer extent under changes in forcing, considered to result from time dependence of ocean heat transport and changing ocean shortwave absorption associated with the

**Table 2.** Parameters in Equations (6)–(8) Describing Artic Sea Ice Evolution

| Parameter | Description | CCSM3 Value |
|---|---|---|
| $F$ | baseline winter ice thickness | 3.2 m |
| $b$ | ocean shortwave absorption parameter | $2 \times 10^{-12}$ W m$^{-4}$ |
| $T^*$ | open water formation efficiency coefficient | 0.8 |
| $M_s^{(0)}$ | annual surface ice melt | 0.4 m |
| $M_b^{(0)}$ | baseline annual basal ice melt | 0.2 m |
| $\sigma_0$ | ocean heat transport variability | 0.6 W m$^{-2}$ |
| $A_{max}$ | ocean area in Arctic region[a] | 7.23 x 10$^6$ km$^2$ |
| $w$ | ice thickness melted by 1 W m$^{-2}$ in 1 year[a] | 0.104 m |

[a] Physical parameters; values are model-independent.

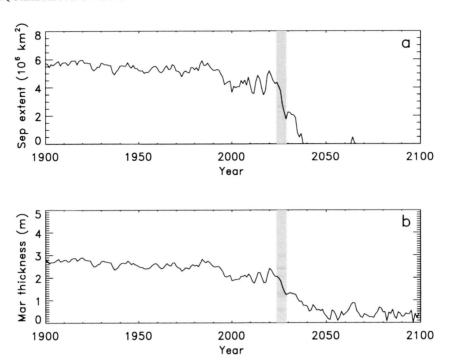

**Figure 7.** Time series of 20th and 21st century (a) September ice extent $A_n$ and (b) mean March ice thickness $T_n$, as in Figure 1, but generated by equations (6)–(8) with the CCSM3 time series of ocean heat transport shown in Figure 4a as input. The gray bands represent an abrupt transition according to the same criterion as in Figure 1.

surface albedo effect. The modeled feedbacks and the way in which the system responds to an increase in OHT are illustrated schematically in Figure 8; here the key nonlinearity resulting from the inverse dependence of summer open water area on the previous winter's ice thickness as discussed in 3.2.1 is represented by the thick arrow.

In obtaining the above expressions major simplifications have been made, and clearly such a representation does not capture all the complexity of the Arctic climate system as simulated by CCSM3. Nonetheless, by considering in isolation a small number of processes thought to be relevant to sea ice abrupt transitions, further insights into this phenomenon can potentially be gained, and the possible range of behaviors, under differing climatic conditions and in different models, can plausibly be explored. As an initial step, equilibrium solutions to these equations are examined next.

### 3.3. Equilibrium Solutions

Equations (6)–(8) describing Arctic sea ice evolution contain nonlinearities arising from the inverse dependence of open water production efficiency on $T_n$ in (7) and from clipping to prevent unphysical negative values of $T_n$ and $A_n$ in

(6) and (8). Such nonlinearity can potentially give rise to multiple equilibria.

Equilibria of (6)–(8) can be obtained as a function of $H$ by setting $T_n$ and $A_n$ to constant equilibrium values $T_e$ and $A_e$. Substituting (8) into (7) and (7) into (6) then leads to

$$T_e(H) = F - wH - wbA_{max}\left[\frac{(T^*/T_e)(M_0^{(s)} + M_0^{(b)} + wH)}{1 + (T^*/T_e)wH/2}\right],$$
(9a)

$$A_e(H) = A_{max}\left[\frac{1 - (T^*/T_e)(M_0^{(s)} + M_0^{(b)} + wH/2)}{1 + (T^*/T_e)wH/2}\right], \quad (9b)$$

in instances where (8) yields $M \leq T_e/T^*$. When forcing is sufficiently strong that $M = T_e/T^*$ according to (8), the equilibrium solution is

$$T_e(H) = \max[F - wH - wbA_{max}, 0], \quad (10a)$$

$$A_e(H) = 0. \quad (10b)$$

For $H > w^{-1}F - bA_{max}$ this equilibrium solution implies zero winter ice thickness, i.e., a perennially ice-free Arctic, as is approached under quadrupling of $CO_2$ in CCSM3 [*Winton, 2006*].

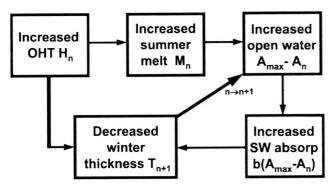

**Figure 8.** Schematic illustration of the web of feedbacks represented by equations (6)–(8), indicating how the system responds to an increase in ocean heat transport (OHT). The thick arrow represents the key nonlinearity, an inverse dependence of summer open water area on the previous winter's ice thickness, whose basis is discussed in section 3.2.1.

Because (9a) is quadratic in $T_e$, there are potentially three physically realizable (i.e., real valued and nonnegative) solutions to (9) and (10). Expressions for these solutions and for various critical values of the parameters are obtained in Appendix A.

Equilibrium solutions for the CCSM3 parameter values listed in Table 2 are illustrated in Figure 9, where the solid curves indicate stable (attracting) solutions and the dashed curves indicate unstable solutions, as determined in Appendix A. It is seen that, for values of $H$ smaller than a critical value $H_c^-$, there is a single equilibrium, described by one of the solutions to (9) and denoted $A_e^+$ in Figure 9a, for which $A_e$ and $T_e$ decrease smoothly with increasing $H$. For values of $H$ larger than a second critical value $H_c^+$, there is again a single equilibrium, described by (10) and denoted $A_e^0$, having $A_e = 0$ and $T_e$ declining gradually with increasing $H$ until $T_e = 0$ is attained for $H \geq F/w - bA_{max}$. In the intermediate regime $H_c^- \leq H \leq H_c^+$, these two stable equilibria coexist and are connected by an unstable equilibrium branch $A_e^-$ arising from the second solution of (9).

At least three aspects of this result bear upon abrupt transitions like those in Figures 1 and 7:

1. If $H$ increases gradually from relatively small values to values exceeding $H_c^+$, $A$ will track $A_e^+$ and hence decrease smoothly until $H = H_c^+$. At this point, any further increase in $H$ will cause $A$ to undergo an abrupt, hysteretic transition from finite $A_e^+$ to $A_e^0 = 0$, as indicated by the downward arrow in Figure 9b.

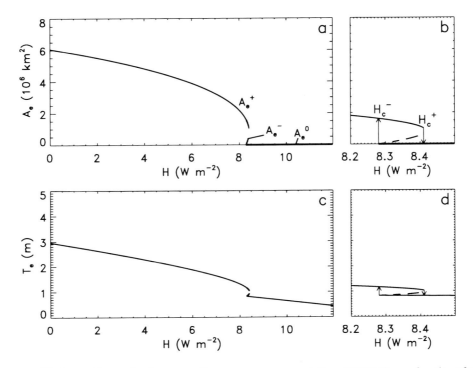

**Figure 9.** (a) Equilibria $A_e$ of September ice extent, for parameters representative of CCSM3, as a function of ocean heat transport $H$. (b) Close-up of multiple equilibrium regime for $A_e$, showing hysteretic transitions at $H_c^-$ and $H_c^+$ (arrows). (c) Equilibria $T_e$ of March ice thickness. (d) Close-up of the multiple equilibrium regime for $T_e$. Solid curves denote stable equilibria, and dashed curves denote unstable equilibria.

2. For lesser values of $H$ that lie within the multiple equilibrium regime $H_c^- \leq H \leq H_c^+$, a finite perturbation that draws $A$ below the unstable equilibrium $A_e^-$ also will result in an abrupt transition to $A = 0$. Conversely, a sufficiently large positive perturbation from $A_e^0$ will induce a transition to $A_e^+$. Hence, in this regime, sufficiently large fluctuations can result in "switching" between the two stable equilibria, a phenomenon examined further in section 3.4.2.

3. As $H$ increases toward $H_c^+$, the impact on $A$ of fluctuations in $H$ steadily increases. This tendency can be quantified by considering the sensitivity of $A$ to changes in $H$. If $(T_n, A_n)$ initially are in equilibrium on the upper stable branch at some $H < H_c^+$, then the response of the following year's summer sea ice extent to an incremental OHT increase $dH$ is

$$\frac{dA}{dH}\bigg|_{A_e^+(H)} \equiv \lim_{dH \to 0} \frac{A_{n+1} - A_e^+(H)}{dH}$$

$$= -A_{\max} \frac{T^* w}{T_e^{+2}} \left[ \frac{M_0^{(s)} + M_0^{(b)} + (T_e^+ + wH)(1 + A_e^+ / A_{\max})/2}{1 + (T^*/T_e^+)wH/2} \right],$$

$$H < H_c^+, \tag{11}$$

where $T_e^+(H)$ and $A_e^+(H)$ are given by (A1), (A2), and (A4). This sensitivity, plotted as a function of $H$ in Figure 10, becomes increasingly negative as $H$ approaches $H_c^+$, implying that sensitivity to fluctuations in $H$ amplifies as the hysteretic transition is approached. This result, together with the tendency for OHT fluctuations to increase as $H$ increases (Figure 4b), suggests that particularly large fluctuations in

$A$ should occur in the neighborhood of abrupt transitions as appears to be the case in Figures 1 and 7. This phenomenon and the potential for its observational detection are discussed further in section 3.5.

The dependence of the equilibrium solution structure on the parameters in (6)–(8) is now considered, as this may bear on the widely differing responses of Arctic sea ice to climatic warming among different climate models, which is discussed briefly by HBT, as well as the model intercomparisons of *Flato and Participating CMIP Modelling Groups* [2004], *Arzel et al.* [2006], and *Zhang and Walsh* [2006].

*3.3.1. Dependence on F and b.* A key feature of the equilibrium solutions to (6)–(8) with parameter values fit to CCSM3 behavior as described above is the multiple equilibrium regime that occurs for $H_c^- < H < H_c^+$ and the attendant hysteretic transitions, as illustrated in Figure 9. The transition relevant to a warming climate is the one from $A_e^+$ to $A_e^0$ that occurs as $H$ increases through $H_c^+$. Two key parameters characterizing this transition are the critical ocean heat transport $H_c^+$, given by (A3), and the minimum stable sea ice area $A_e^+(H_c^+)$, which determines the magnitude of the transition.

Figure 11 illustrates the dependence of these quantities on the baseline ice formation parameter $F$ and ocean shortwave absorption parameter $b$, with remaining parameters fixed at CCSM3 values. (The strength of ice-albedo feedback may be sensitive to the type of ice model employed, for example multiple versus single thickness category [*Holland et al.*, 2006b], whereas winter ice thickness is known to differ widely among climate models [*Flato and Participating CMIP Modelling Groups*, 2004]). The CCSM3 values for $F$ and $b$ are indicated by the asterisk, and letters a–d sig-

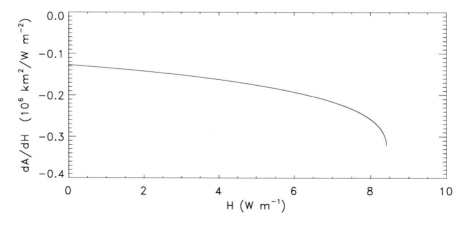

**Figure 10.** Sensitivity $dA/dH$ of summer ice extent $A$ to fluctuations in $H$ when the system is perturbed from equilibrium on the upper stable branch $A_e^+$ for the case illustrated in Figure 9.

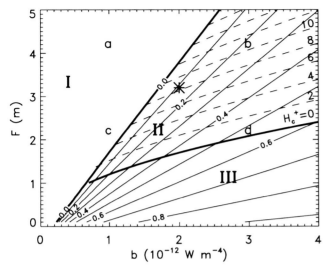

**Figure 11.** Properties of the hysteretic transition from $A_e^+ > 0$ to $A_e^0 = 0$ as in Figure 9b, as functions of parameters $F$ and $b$ with other parameters assigned CCSM3 values listed in Table 2. Thin solid contours denote summer ice extents $A_e^+ > 0$ (scaled by $A_{max}$) at which the transition occurs (contour interval 0.1), and thin dashed contours denote the corresponding critical ocean heat transports $H_c^+$. The (thick) zero contours of both quantities divide the plane into three regimes, as discussed in the text. The asterisk indicates $(F, b)$ characterizing CCSM3, and the letters a–d indicate values of $(F, b)$ in Figures 12a–12d, respectively.

nify cases shown in Figures 12a–12d, respectively. The thin solid lines contour $A_e^+(H_c^+)$ in increments of $0.1\, A_{max}$, whereas the dashed contours indicate positive $H_c^+$ in increments of $2\ \mathrm{W\ m^{-2}}$. Zero values for both quantities, indicated by the heavy solid contours, divide the parameter space into three regimes. In regime I, the bifurcation at $H_c^+$ occurs formally at $A_e < 0$ but is not realized because of clipping at the minimum physical value $A_e = 0$. Thus, as $H$ increases, $A_e^+$ approaches zero continuously, without an abrupt transition, as in Figures 12a and 12c. This regime is realized under high values of $F$ and thus could be viewed (within the limitations of this model) as representing behavior under colder climatic conditions such as the Last Glacial Maximum. By contrast, in regime II, an abrupt transition from $A_e^+(H_c^+) > 0$ to $A_e^0 = 0$ occurs as $H$ increases through $H_c^+ > 0$ as is evident in Figure 12b and Figure 12d. For CCSM3 parameter values the system lies within this regime, so that hysteretic transitions between multiple equilibria occur for $H > 0$ as shown previously in Figure 9.

Regime III characterizes a situation in which the transition from $A_e^+$ to $A_e^0$ occurs for $H < 0$, so that for the $H \geq 0$ regime that is of interest here only the $A_e^0 = 0$ equilibrium is realized. This case is not considered further.

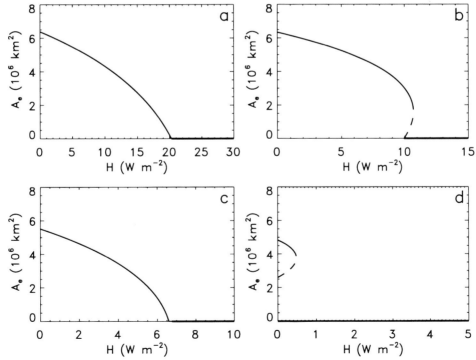

**Figure 12.** Equilibria $A_e$ of summer ice extent, as a function of $H$, for parameter values $F$ and $b$: (a) (4.2 m, $1 \times 10^{-12}$ W m$^{-4}$), (b) (4.2 m, $3 \times 10^{-12}$ W m$^{-2}$), (c) (2.2, $1 \times 10^{-12}$ W m$^{-2}$), and (d) (2.2, $3 \times 10^{-12}$ W m$^{-2}$), indicated by corresponding letters a–d in Figure 11.

*3.3.2. Dependence on $M_s^{(0)}$, $M_b^{(0)}$, and $T^*$.* The dependence of hysteretic transition properties $A_e^+(H_c^+)$ and $H_c^+$ on $M_s^{(0)} + M_b^{(0)}$ and $T^*$ is illustrated in Figure 13, where $F$ and $b$ are fixed at CCSM3 values and, as in Figure 11, the solid contours indicate $A_e^+(H_c^+)$ and the dashed contours $H_c^+$. Again, the zero contours of these quantities divide the plane into three regimes, denoted as I', II', and III'. Regime II', which contains the CCSM3 case (indicated again by the asterisk), is analogous to regime II in Figure 11; that is, a hysteretic transition from positive $A_e^+$ to $A_e^0 = 0$ occurs for a critical $H_c^+ > 0$. Similarly, regime I' is analogous to regime I in Figure 11, in that as $H$ increases $A_e^+$ approaches $A_e^0$ in a continuous manner. Finally, regime III' is analogous to III in Figure 11 in that the transition from $A_e^+$ to $A_e^0$ occurs at negative $H$, so only the $A_e^0$ equilibrium is present when $H > 0$. In regime II', the magnitude of the abrupt transition increases as $T^*$ and $M_s^{(0)} + M_b^{(0)}$ decrease, as does the critical OHT value $H_c^+$.

### 3.4. Numerical Experiments

In this section numerical solutions to (6)–(8), forced by synthetic OHT time series $H_n$ obtained from (3) and (4), are described. Objectives are (1) to examine statistical properties of sea ice evolution and abrupt transitions under climatic warming as simulated by (6)–(8), with a view toward determining the extent to which multiple equilibria may play a

role, and (2) to illustrate possible additional consequences of multiple equilibria for Arctic sea ice evolution, including vacillation between multiple stable equilibria, as well as hysteretic behavior that occurs when a period of warming is followed by a gradual return to a cooler climate.

*3.4.1. Abrupt transitions in a warming climate.* To obtain a statistical description of Arctic sea ice evolution under climatic warming as simulated by (6)–(8), ensembles consisting of $10^4$ realizations of the random process $\varepsilon_n$, which governs the fluctuating component of the synthetic OHT time series $H_n$ described by (3) and (4), are considered. For default values of the parameters, these synthetic time series resemble statistically the CCSM3 OHT time series, as demonstrated in Figure 4. To assist in determining the relative importance of OHT fluctuations and multiple equilibria in contributing to abrupt transitions, sensitivity is considered to the scale factor $\sigma_0$ of OHT fluctuations and to the ocean shortwave absorption parameter $b$ that substantially governs whether multiple equilibria are present (see Figure 11).

Results for these various cases are illustrated in Plate 1. We consider first that for which $\sigma_0$ and $b$ are set to their CCSM3 default values, shown in the middle panel of the topmost row in Plate 1. Here and in the other panels, ensemble mean $A_n$ is indicated by the thick blue curve, and a range of 1 ensemble standard deviation $\sigma_A$ above and below the mean is indicated by the thin blue curves. The thick red curve indicates the probability $p_{abrupt}$ that a given year lies within an interval of abrupt transition in $A$ as defined in section 1, and the thin red curve the probability of abrupt increase as defined by an analogous criterion. In this panel, the probability of an abrupt decrease occurring in the ensemble of seven CCSM3 21st century simulations subject to the SRES A1B forcing scenario is additionally indicated by the green histogram. Overall, the temporal distribution and frequency of the abrupt decreases in CCSM3 and in the synthetic time series agree reasonably well. One notable difference is that the latter feature some abrupt increases associated with large swings in $A_n$ driven by OHT fluctuations, whereas in CCSM3 no such increases exceeding the abrupt transition threshold occur. These increases tend to be paired with subsequent decreases, particularly after 2050 or so, which together compose partial, temporary recoveries in ice extent. One such event is visible near year 2065 in Figure 7a, although in this instance the abrupt transition threshold is not exceeded.

A further result pertaining to the synthetic time series in this panel is that, as warming progresses, the decrease in $A_n$ below a particular threshold can be hastened or delayed by a decade or more for realizations lying within $\pm\sigma_A$ of the mean. It is also seen that $\sigma_A$ tends to be largest at approximately

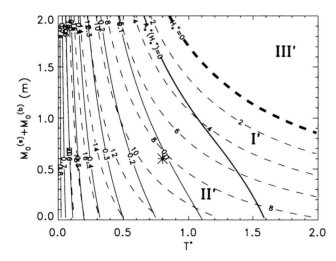

**Figure 13.** Properties of the hysteretic transition from $A_e^+ > 0$ to $A_e^0 > 0$, as functions of parameters $M_0^{(s)} + M_0^{(b)}$ and $T^*$, with $F$ and $b$ fixed at CCSM3 values. Contour definitions and intervals are as in Figure 11. The asterisk denotes CCSM3 values, and regimes I', II', and III' are described in the text.

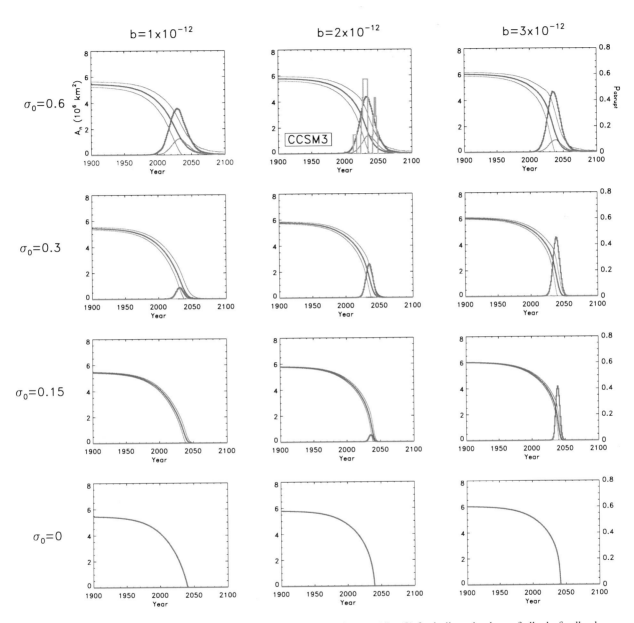

$b=1\times10^{-12}$          $b=2\times10^{-12}$          $b=3\times10^{-12}$

$\sigma_0=0.6$

$\sigma_0=0.3$

$\sigma_0=0.15$

$\sigma_0=0$

CCSM3

**Plate 1.** Statistical properties of $10^4$ member ensembles of solutions to (6)−(8) for indicated values of albedo feedback parameter $b$ (in W m$^{-4}$) and OHT variability parameter $\sigma_0$ (in W m$^{-2}$). Thick blue curves represent ensemble mean $A_n$, and thin blue curves represent ranges of 1 standard deviation about these values (left-hand scale). Thick red curves indicate probability that a given year lies within an interval of abrupt decrease in $A_n$, and thin red curves indicate probability of an abrupt increase (right-hand scale). For the case with CCSM3 parameter values (top middle panel), the probability of abrupt decrease as determined from an ensemble of seven CCSM3 simulations is indicated in green.

the time when mean $A$ is decreasing most rapidly and abrupt decreases occur most frequently, a result that is revisited in the next section.

Dependence on $b$ is illustrated by considering values $b = 1 \times 10^{-12}$ W m$^{-4}$ and $3 \times 10^{-12}$ W m$^{-4}$, in addition to the CCSM3 value of $2 \times 10^{-12}$ W m$^{-4}$. To facilitate comparison, $F$ is adjusted in the non-CCSM3 cases so that, in the absence of OHT fluctuations, the transition from finite $A$ to $A = 0$ (with or without hysteresis) occurs near year 2040 as in the default case. These values are $F = 2.55$ m in the low-$b$ case, which lies well within the single-equilibrium regime I in Figure 11, and $F = 3.8$ m in the high-$b$ case, which lies well within the multiple-equilibrium regime II.

With $\sigma_0 = 0.6$ W m$^{-2}$, the CCSM3 default, it is seen in the top row of Plate 1 that the probability and timing of abrupt transitions is comparable for all three values of $b$, even though the decrease in mean $A$ occurs somewhat more rapidly for higher $b$ than for lower $b$. Thus, it appears that in this instance the primary influence governing abrupt transitions is the relatively large OHT fluctuations, together with the increased sensitivity of $A$ to changes in $H$ as the transition to $A_e = 0$ is approached, as discussed in section 3.3 and illustrated in Figure 10. However, for reduced OHT variability as simulated using lesser values of $\sigma_0$ (second and third rows of Plate 1), the frequency of abrupt transitions becomes much more sensitive to $b$. For example, for $\sigma_0 = 0.15$ W m$^{-2}$, 4 times smaller than the value characterizing CCSM3, abrupt transitions are entirely absent for $b = 1 \times 10^{-12}$ W m$^{-4}$, whereas for $b = 3 \times 10^{-12}$ W m$^{-4}$, $p_{\text{abrupt}}$ exceeds 0.4 near year 2040. Not surprisingly, as $\sigma_0$ decreases, the range of times over which abrupt decreases occur becomes increasingly narrow. Also of note is the even greater sensitivity to $\sigma_0$ of the abrupt increases, which have become very infrequent even for $\sigma_0 = 0.3$ W m$^{-2}$.

Finally, the bottom row of panels, for which $\sigma_0 = 0$, shows that in our simple model some OHT variability is essential for abrupt decreases to occur for the parameters considered. (For the case $b = 3 \times 10^{-12}$ W m$^{-4}$, however, the abrupt decrease threshold is nearly met, so that for slightly larger $b$ an abrupt decrease would occur near 2040.)

*3.4.2. Vacillation between equilibria in a warmer control climate.* In a nonlinear system having multiple equilibria that is forced stochastically, fluctuations in forcing can potentially trigger transitions between stable equilibria, even when the ensemble mean forcing is stationary. This phenomenon has been discussed in the context of simple models of oceanic thermohaline circulation, e.g., by *Monahan* [2002a, 2002b], and in the sea ice context by *Flato and Brown* [1996]. In the present context, one might consider a situation in which warming has stabilized at some future date, and climate, though warmer than at present, is stationary. Such a situation can be represented by assigning a fixed value ensemble mean $\overline{H}$ in equation (4).

Figure 14 illustrates such a scenario with constant $\overline{H} = 8$ W m$^{-2}$ with other parameters, including the standard deviation $\sigma_0 = 0.6$ W m$^{-2}$ of the stochastic component of forcing, set to their CCSM3 default values. This value for $\overline{H}$ is realized in CCSM3 at around 2030 (Figure 4a) and lies within the range of $H$ for which multiple equilibria are present (Figure 9b). In the 200-year time series shown in Figure 14a, $A_n$ appears to vacillate between very small or zero values and somewhat larger values in the range $2 \times 10^6$ km$^2 \leq A_n \leq 4 \times 10^6$ km$^2$. This impression is borne out by the probability density for $A_n$, computed from a $10^4$-year time series, which is clearly bimodal (Figure 14b).

Such bimodality persists when $\overline{H}$ and $\sigma_0$ are varied somewhat about the values assigned in Figure 14. With $\overline{H}$ fixed, for example, the value of $A_n$ at which the upper lobe of $p(A_n)$ peaks increases as $\sigma_0$ increases, exceeding $5 \times 10^6$ km$^2$ for $\sigma_0 = 3$ W m$^{-2}$. Conversely, when $\sigma_0$ is reduced at constant $\overline{H}$, the value of $A_n$ characterizing the peak of the upper lobe of $p(A_n)$ decreases toward the $A_n$ equilibrium value of about $2 \times 10^6$ km$^2$. If $\sigma_0 = 0.6$ W m$^{-2}$ is fixed instead, this bimodality persists for $\overline{H}$ in the range 6.3 W m$^{-2} \leq \overline{H} \leq 8.8$ W m$^{-2}$,

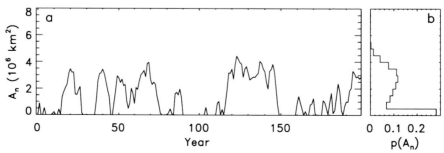

**Figure 14.** (a) A 200-year time series of $A_n$ for CCSM3 parameter values and fixed $\overline{H} = 8$ W m$^{-2}$ characterizing CCSM3 climate near 2030. (b) Probability density of $A_n$ for a $10^4$-year continuation of this time series.

which considerably exceeds the range in $H$ over which multiple equilibria are present. Such vestigial influences of multiple equilibria have been discussed by *Monahan* [2002b].

*3.4.3. Hysteresis under reversal of climatic change.* It is clear from Figures 9 and 12 that when multiple stable equilibria are present within a given range in $H$, hysteresis will occur if $H$ gradually increases and subsequently decreases across this range, as might occur, for example, if climatic warming were to be arrested and then reverse. Such effects

are illustrated in Figure 15, which shows $A_n$ as a function of $\overline{H}$, with $\overline{H}_n$ given by (2) for years 1901 to 2100 and by the reverse of this time series from 2101 to 2300. As in Plate 1, three cases are considered: $b = 1 \times 10^{-12}$, $2 \times 10^{-12}$, and $3 \times 10^{-12}$ W m$^{-4}$, with $F$ adjusted in each case to ensure that the transition between equilibria $A_e^+ > 0$ and $A_e^0 = 0$ occurs at similar $\overline{H} \approx 8.4$ W m$^{-2}$ as in the default case. Ensembles of $10^4$ realizations are again considered, with $H$ variability specified by the CCSM3 default value $\sigma_0 = 0.6$ W m$^{-2}$. Resulting ensemble mean $A_n$ are indicated by the solid curves:

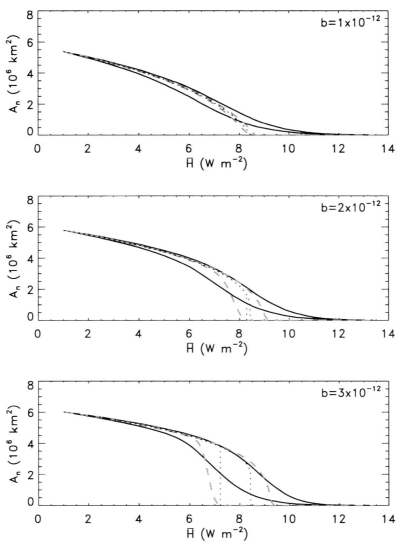

**Figure 15.** Hysteretic behavior of $A_n$ when a CCSM3-like increase in ensemble mean OHT described by (3) and (4) is followed by a similar decrease. Solid curves denote ensemble mean $A_n$ obtained from $10^4$ realizations of fluctuating component of OHT; dashed curves indicate evolution under the same time-dependent mean OHT but with fluctuations absent; dotted curves denote ideal hysteresis cycles described by equilibria $A_e$. As in Plate 1, three values for $b$ are considered, with $F$ adjusted so that in each case transition from $A_e > 0$ to $A_e > 0$ occurs at comparable $\overline{H}$.

the upper is described during the period of increasing $\overline{H}$, and the lower is described during the period of decreasing $\overline{H}$. For comparison the dashed curves describe evolution of $A_n$ for the same time-dependent $\overline{H}$ but without fluctuations ($\sigma_0 = 0$), and the dotted curves describe ideal hysteresis cycles determined from the equilibrium solutions.

One aspect of Figure 15 that is immediately evident is that the numerical solutions exhibit a separation between the ascending-$\overline{H}$ and descending-$\overline{H}$ branches in all three cases, even though multiple equilibria are present only for the larger two values of $b$. This is because of the finite time required for solutions of (6)–(8) to adjust to changes in $H$. Such a separation depends on the rate of change of $\overline{H}$ and is reduced if $\overline{H}$ varies more slowly than in the case illustrated here. The overall separation of the two branches does, however, increase with $b$, in accordance with the increasingly hysteretic character of the equilibria.

A second aspect of Figure 15 that bears mention is that the $H$ fluctuations greatly broaden the range in $\overline{H}$ over which the ascending and descending branches separate. This is because in the presence of fluctuations the transitions between equilibria become "blurred," an effect which is also evident in Plate 1. An alternative view is that in the presence of fluctuations multiple equilibria can exert a vestigial influence beyond the parameter range in which they are formally present through ephemeral trapping of solution trajectories, as discussed by *Monahan* [2002b].

### 3.5. Are the Simulated Abrupt Transitions Predictable?

When multiple equilibria are present, as in Figure 9, for example, rapid transitions from relatively large $A$ to $A \approx 0$ can result from an incremental increase in forcing $H$. In such cases, there is relatively little prior indication from the general magnitude of $A$ that a seasonally ice-free Arctic is imminent. This raises the question of whether other aspects of the time series for $A$ might indicate more clearly that such a threshold is being approached.

One potential candidate for such an indicator is the ensemble standard deviation of $A$,

$$\sigma_n^e = \left[ (N-1)^{-1} \sum_{i=1}^{N} (A_{n,i} - \overline{A}_n)^2 \right]^{1/2}, \quad (12)$$

where $N$ is the number of realizations, $i$ denotes an individual realization, and the overbar denotes an ensemble average. This is suggested by the result, noted in section 3.3, that when multiple equilibria are present the modeled sea ice response to OHT fluctuations amplifies dramatically as the bifurcation point $H_c^+$ is approached, as discussed by *Held and Kleinen* [2004] and *Carpenter and Brock* [2006]. In the case of the actual climate system, of course, only a single

realization is available, and other fluctuation measures must be considered. Figure 16a plots for each year a single-realization standard deviation computed from $A$ for that year and the 10 preceding years, i.e.,

$$\sigma_n^i = \left[ M^{-1} \sum_{m=n-M}^{n} (A_{m,i} - <A>_{n,i})^2 \right]^{1/2}, \quad (13)$$

with $M = 10$, where $<>$ denotes an average over years $n - M$ to $n$, for $b = 1, 2$, and $3 \times 10^{-12}$ W m$^{-4}$. Other parameters are set as in the top row of Plate 1, so that in each case $A$ vanishes near 2040 if $H$ fluctuations are absent; recall that multiple equilibria occur for the larger two values of $b$. To damp statistical noise, this fluctuation measure is averaged over each realization of a $10^4$-member ensemble; that is, we plot $\sigma_n^i$. The resulting values (thick curves) increase significantly, by a factor between 5 and 10, from the beginning of the 20th century to the interval 2020 to 2050 where most abrupt transitions occur (see Plate 1). However, the peak values are only weakly sensitive to $b$, in contrast to the ensemble standard deviations $\sigma_n^e$ (not shown), which increase by $\approx 40\%$ between the lowest and highest values of $b$. (A similar result is obtained if the temporal standard deviation (13) is computed with the linear trend removed.) Thus there is no clear signal that a bifurcation is being approached, although in all cases the temporally increasing variance does point to an accelerating decline in sea ice cover which may or may not be characterized by a transition between multiple equilibria.

For comparison, the thin curve in Figure 16a shows $\sigma_n^i$ calculated from six realizations of CCSM3. Its general behavior is similar to that of the thick curves, except that the result is statistically noisier, as expected with many fewer realizations. (Results obtained using six realizations of the simple model exhibit comparable noise.) The CCSM3 standard deviations noticeably exceed those of the simple model prior to about 1980. This is likely due to atmospheric forcing variability, which is not accounted for in the simple model and is comparatively most important in this epoch when OHT forcing is relatively small (Figure 4).

A second potential indicator of an impending bifurcation is the integral time scale $\tau_n$, which can be obtained from the correlation matrix via

$$\tau_n = \frac{1}{2} \sum_m \text{cor}(A_n, A_m) \quad (14)$$

and is a measure of the temporal persistence of fluctuations in $A$. (For an AR(1) process, $\tau$ is equal to the characteristic autocorrelation time scale.) An increase in this quantity as a random process approaches a saddle-node bifurcation is a generic behavior, as discussed by *Kleinen et al.* [2003] and *Held and Kleinen* [2004] in the context of the thermohaline

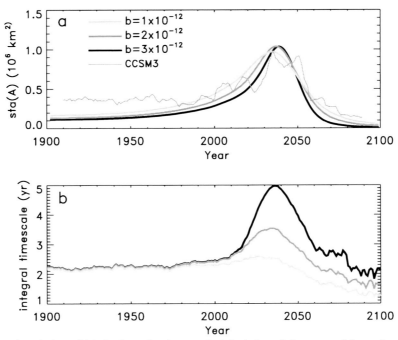

**Figure 16.** Temporal evolution of (a) single-realization standard deviation of $A$, computed for each year $n$ from values $A_{n-10}$ to $A_n$ as in (13) and subsequently averaged over $10^4$ realizations, for the three parameter sets considered in the top row of Plate 1 (thick curves), or six CCSM realizations (thin curve) and (b) integral time scale $\tau_n$ for the same three-parameter sets, also computed from $10^4$ realizations.

circulation and by *Carpenter and Brock* [2006] in the context of ecosystem regime shifts, and thus could provide a potential "early warning mechanism" for the approach of a regime threshold. As an idealized test of this approach, $\tau_n$ was computed from $10^4$-realization ensembles, summing from $m = n - 10$ to $m = n + 10$, for each of the three cases considered in Figure 16a. The results, plotted in Figure 16b, show that $\tau \approx 2$ years throughout the 20th century, reflecting the autocorrelation time scale of the forcing $H_n$. In the cases $b = 3 \times 10^{-12}$ W m$^{-4}$ and $b = 2 \times 10^{-12}$ W m$^{-4}$ in which the saddle-node bifurcation is realized, $\tau_n$ increases substantially over the 2 decades or so preceding the bifurcation point, whereas little effect is seen for the case $b = 1 \times 10^{-12}$ W m$^{-4}$ in which the saddle-node bifurcation is not realized. (This increase in integral time scale with increasing $b$ could explain why the single-realization standard deviation computed for a moving window according to (13) is less sensitive to $b$ than the ensemble standard deviation: in the former case, increased $\tau$ leads to a larger variance fraction at periods excluded by the 11-year window. This will tend to compensate any tendency for overall variance to increase with $b$.)

This result suggests that similar early warning signals for the collapse of summertime Arctic ice extent via such a bifurcation could exist, although the detection of trends

in integral time scale from a single realization of the random process is a challenging problem of time series analysis (computing $\tau_n$ as above even from six realizations of CCSM3 or our stochastic model does not yield an obvious signal). An interesting direction of future research would be the development of statistical measures of proximity to a bifurcation in which trends are most easily detected [*Held and Kleinen*, 2004].

## 4. SENSITIVITY TO PARAMETERIZATIONS

The major simplifications adopted in section 3 enable analytical tractability but are imperfect, as is inevitable in studies of this type. In order to assess sensitivity to these simplifications, this section examines the effects of changes that make the parameterizations arguably more realistic, though more complex.

A significant simplification in section 3 is the "two-season approximation" implied by selecting March ice thickness and September ice extent as dependent variables. Consider in particular equation (5), which posits that annual mean ocean shortwave absorption increases proportionally to September open water area. While such a relation is plausible based on Figure 5, it does not allow for ice cover in other sunlit months to continue to decline after September ice extent $A_n$ has saturated at zero. To examine the importance of this effect, we first express annual mean ocean shortwave absorption due to open water in terms of monthly open water fractions $\Omega_{n,m}$ for year $n$ as

$$Q_n = (1 - \alpha_{\text{ocn}}) \frac{1}{12} \sum_m \Omega_{n,m} F_{SW,m}, \qquad (15)$$

where $\alpha_{\text{ocn}}$ is ocean albedo as in section 3, and $F_{SW,m}$, the monthly incident shortwave fluxes, are assigned values in Table 3 obtained from mean CCSM3 fluxes in the Arctic region for 1950–1999. The summation is over all sunlit months, including minor February and October contributions not listed in Table 3.

To close this expression, we first examine the extent to which $\Omega_{n,m}$ can be deduced from September open water fraction,

$$\Omega_n \equiv (A_{\max} - A_n)/A_{\max}, \qquad (16)$$

when $\Omega_n < 1$. Table 3 shows the correlations between $\Omega_n$ and March–August open water fractions in CCSM3 for years 1900–2049 prior to $A_n$ saturating near zero. The correlations range from approximately 0.83 to 0.99, indicating that monthly open water fractions $\Omega_{n,m}$ are to a reasonable approximation proportional to $\Omega_n$. The constants of proportionality are taken to be the coefficients of regression $a_m$ between CCSM3 September open water fraction and that in other months, indicated in Table 3 (the constant terms in the regressions are negligible). Obtaining monthly open water fractions from $\Omega_{n,m} = a_m \Omega_n$, annual mean ocean shortwave absorption continues to be given by (5), but with

$$b = \frac{(1 - \alpha_{\text{ocn}})}{A_{\max}} \frac{1}{12} \sum_m a_m F_{SW,m}. \qquad (17)$$

Assigning $\alpha_{\text{ocn}} = 0.1$ gives $b = 2.06 \times 10^{-12}$ W m$^{-4}$, similar to the value determined in section 3.

To obtain $Q_n$ when September is ice free ($A_n = 0$, $\Omega_n = 1$), we continue to assert a proportionality $\Omega_{n,m} = a_m \tilde{\Omega}_n$, where $\tilde{\Omega}_n$ is determined in the same manner as $\Omega_n$ from March thick-

ness $T_n$ and annual melt $M_n$ via (7) and (16), but without truncating at the physical maximum value unity for $\Omega_n$. For consistency we replace expression (8) for $M_n$, which was based on a two-season approximation of annual mean ice cover, with

$$M_n = M_0^{(s)} + M_0^{(b)} + wH_n(1 - \overline{\Omega}_{n,m}), \qquad (18)$$

where $\overline{\Omega}_{n,m}$ is the annual mean open water fraction. (The two-season approximation adopted in section 3 corresponds to $\overline{\Omega}_{n,m} = (1 - A_n/A_{\max})/2$.) From (7) and (16), we thus have

$$\tilde{\Omega}_n = (T^*/T_n)M_n, \qquad (19)$$

with $M_n$ given by (18). The monthly open water fractions, truncated to physical values, are then given by

$$\Omega_{n,m} = \min[a_m \tilde{\Omega}_n, 1]. \qquad (20)$$

Equations (18)–(20) form an implicit relation which must be solved for the $\Omega_{n,m}$ to obtain $Q_n$ through (15).

Under ice-free September conditions $\tilde{\Omega}_n \geq 1$, the above relations become complicated enough to impede analytical progress. To illustrate differences between the approach described here and that of section 3 numerically, we consider $H_n$ increasing according to (3) over 1900–2100 and decreasing in the same manner over the subsequent two centuries (Figure 17a). For simplicity, the stochastic component of $H_n$ is ignored. According to the section 3 formulation (solid curves in Figures 17b and 17c), $A_n$ and $T_n$ decrease until $A_n$ collapses to zero shortly before 2040. After $H_n$ decreases back through its critical value, $A_n$ recovers (with some hysteresis) around 2170. The section 4 formulation (dashed curves in Figure 17) yields comparable behavior prior to 2040 (the decline occurs a little more rapidly because (18) implies slightly more melt than (8)). At this stage a significant difference is that $T_n$ also collapses, implying perennial, as opposed to seasonal, loss of ice. In addition, the hysteresis is more extreme, so much so that ice cover fails to reappear even when $H_n$ has returned to 1900 values. This phenom-

**Table 3.** Quantities Relating to Parameterization of Monthly Ocean Shortwave Absorption

| | March | April | May | June | July | August | September |
|---|---|---|---|---|---|---|---|
| Mean ↓ SW $F_{SW,m}^a$ (W m$^{-2}$) | 33.2 | 130.8 | 219.2 | 231.5 | 174.7 | 99.8 | 37.9 |
| Correlation coefficients[b] | 0.874 | 0.866 | 0.826 | 0.828 | 0.940 | 0.990 | 1 |
| Regression coefficients $a_m$[b] | 0.012 | 0.012 | 0.021 | 0.078 | 0.287 | 0.814 | 1 |

[a] CCSM3 mean monthly incident surface shortwave fluxes for 1950–1999.
[b] Correlation and regression coefficients relating CCSM3 monthly open water areas to September open water area $A_{\max} - A_n$ for years 1900–2049.

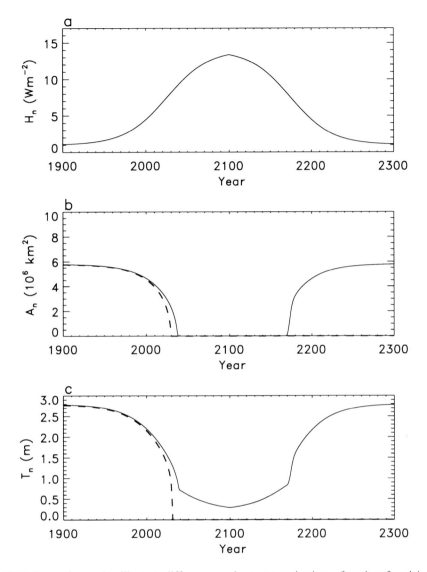

**Figure 17.** (a) OHT time series used to illustrate differences under parameterizations of sections 3 and 4; (b) time series of September ice area under section 3 parameterizations (solid curve) versus section 4 parameterizations (dashed curve); (c) time series of March ice thickness, curves similarly labeled.

enon resembles the small ice cap instability found in other simplified representations of sea ice described in section 2.

What is the source of these differences? The answer lies in the parameterized ocean shortwave absorption. As long as $A_n > 0$, parameterizations (5) and (15) behave similarly (solid and dashed curves in Figure 18) because of their close relation through (17). In particular, the forms of the two equilibrium solution branches connected by the saddle-node bifurcation remain much the same. However, once September ice cover is lost ($A_n = 0$), (5) implies a saturation of shortwave absorption, whereas under (15) $Q_n$ can

continue to increase as ice cover in the other sunlit months declines. The additional shortwave absorption becomes sufficient to prevent freezing in winter, and $Q_n$ saturates at its value for perennial absence of ice. In actuality, the importance of this effect is likely to be diminished by processes not represented explicitly in our simple model, namely, the strong sensible, latent, and longwave heat losses that will occur from open water in winter. The loss of a much larger fraction of the absorbed shortwave energy from the sea surface under ice-free winter conditions than when ice is present should tend to mitigate the large differences

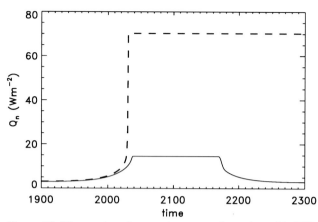

**Figure 18.** Time series of ocean shortwave absorption with OHT forcing as in Figure 17 under section 3 parameterizations (solid curve) versus section 4 parameterizations (dashed curve).

in Figure 17 between behavior under the two shortwave parameterizations.

Whether an abrupt loss of perennial ice can, in fact, occur in a climate model was investigated by *Winton* [2006], who found that this phenomenon does occur in one model (Max Planck Institute ECHAM5), though not in CCSM3.

## 5. DISCUSSION AND CONCLUSIONS

By quantifying in a simple manner three effects identified by HBT as contributing crucially to abrupt declines in summer Arctic sea ice extent in CCSM3 21st century climate simulations, a nonlinear set of equations (6)–(8) was obtained that appears able to reproduce some aspects of the CCSM3 results. The key nonlinearity is the inverse dependence of open water formation on winter ice thickness in (7), whereas the impact of this effect on the following winter's ice thickness is crucially enabled by the ice-albedo feedback term in (6), which serves as an amplifier; if the coefficient $b$ in this expression is sufficiently large, then abrupt disappearances of summer ice cover can occur under infinitesimal increases in OHT. This is in contrast to the SICI phenomenon discussed in section 2.1, for which albedo contrast provides the essential nonlinearity and the minimum size of a viable ice cap is determined by a length scale characterizing diffusive atmospheric heat transport.

The main utility of the approach outlined here is its extreme simplification: to the extent that the most important physical processes are plausibly represented, it enables the phenomenology of abrupt transitions to be examined in a context removed from the massive computational expenditure involved in running and analyzing results from a comprehensive climate model. This reduction has enabled us to (1) obtain statistics encompassing many realizations of ran-

dom forcing, (2) examine sensitivity to various parameters, and (3) obtain an analytical description of the equilibria and their stability.

It was found in section 3.3 that multiple stable equilibria exist for physically relevant values of the parameters, including those representative of CCSM3. It is thus tempting to speculate that transitions between equilibria triggered by OHT increases play a key role in the abrupt transitions in CCSM3. However, in the numerical experiments described in section 3.4.1 and illustrated in Plate 1, when OHT fluctuations are as large as in CCSM3, the abrupt transitions occur primarily as a consequence of the fluctuations themselves, together with the increased sensitivity of sea ice to perturbations as summer ice extent decreases. Multiple equilibria play a larger role in abrupt transitions when OHT fluctuations are smaller.

These results suggest that the physical processes identified by HBT as being associated with abrupt sea ice transitions in CCSM3 are, in fact, sufficient to trigger such behavior. However, a significant caveat is that behavior following abrupt transitions is sensitive to parameterization changes, described in section 4, that may represent more plausibly the effect of complete seasonal loss of ice cover on ocean shortwave absorption. Abrupt transitions then lead to a sudden perennial disappearance of ice cover that is strongly hysteretic. Such behavior is reminiscent of the small ice cap instability that occurs in certain other simplified representations of sea ice but is not seen in CCSM3.

This case study thus illustrates both some of the promises and potential pitfalls of imposing major simplifications in order to study fundamental climate system behavior. On one hand, as climate models become more complex, a more complete and detailed simulation of the climate system is produced. Arguably, this may not by itself enable simulated phenomena to be fundamentally understood; for this purpose, simplified models and conceptual frameworks can provide valuable tools [*Held*, 2005]. On the other hand, while such simplifications may enable solving for and identifying key influences on climate system phenomena (analytically in some cases as here), results may be sensitive to which processes are included and how the included processes are represented, as, for example, in box models of ocean circulation and simple conceptual models of ENSO [e.g., *Olbers*, 2001].

## APPENDIX A: STRUCTURE AND STABILITY OF MULTIPLE EQUILIBRIA

To obtain equilibria $T_e$ and $A_e$ as functions of $H$, (6)–(8) are first combined to obtain closed expressions for $T_e$, given by (9a) and (10a). The quadratic equation (9a) has solutions

$$T_e^{\pm}(H) = \frac{1}{2}[F - (1 + T^*/2)wH]\left[1 \pm \sqrt{1 - Z(H)}\right], \quad (A1)$$

where

$$Z(H) \equiv \frac{4wbA_{max}T^*(M_0^{(s)} + M_0^{(b)} + wH) - 2(F - wH)T^*wH}{[F - (1 + T^*/2)wH]^2}. \quad (A2)$$

Clearly, $Z \geq 0$ for $H \geq 0$, while $Z \leq 1$ is necessary for either branch to be a physical solution. This implies a critical value for $H$, corresponding to a saddle node bifurcation, above which (A2) ceases to be real valued:

$$H_c^+ = \frac{F}{w\zeta} + \frac{2bA_{max}T^*}{\zeta^2}\left[1 - \sqrt{1 + \frac{F\zeta + (M_0^{(s)} + M_0^{(b)})\zeta^2}{wbA_{max}T^*}}\right], \quad (A3)$$

where $\zeta \equiv 1 - T^*/2$. Corresponding values $A_e^{\pm}$ are obtained by substituting (A1) into (9b):

$$A_e^{\pm}(H) = A_{max}\left[\frac{1 - (T^*/T_e^{\pm})(M_0^{(s)} + M_0^{(b)} + wH/2)}{1 + (T^*/T_e^{\pm})wH/2}\right]. \quad (A4)$$

Conditions for these solutions to be physically realizable, in addition to $H \leq H_c^+$, are $T_e^{\pm}$ and $A_e^{\pm} \geq 0$. If $A_e^+(H_c^+) = A_e^-(H_c^+) > 0$, then the saddle node bifurcation is realized as in Figures 9, 12b, and 12d. In this case, $A_e^-$ is nonnegative on the interval $H_c^- \leq H \leq H_c^+$, where

$$H_c^- = \frac{F - (M_0^{(s)} + M^{(b)})T^* - wbA_{max}}{w(1 + T^*/2)} \quad (A5)$$

is obtained by setting $A_e^- = 0$. Otherwise, if $A_e^+(H_c^+) < 0$, the saddle node bifurcation is not realized, as in Figures 12a and 12c. In this instance, $A_e^-$ is always negative and hence unphysical, and $A_e^+$ is nonnegative for $H \leq H_c^0$, with $H_c^0$ given by the right-hand side of (A5).

The third equilibrium solution $T_e^0$, $A_e^0$, given by (10a) and (10b), exists for $H_c^- \leq H \leq \infty$ when $A_e^+(H_c^+) > 0$ (multiple equilibrium case), and for $H_c^0 \leq H \leq \infty$ when $A_e^+(H_c^+) \leq 0$ (single equilibrium case).

The local stability of these solutions can be deduced by considering the response to small perturbations from equilibrium, e.g., by letting $A_n = A_e^{\pm}(1 + \varepsilon)$ with $|\varepsilon| \to 0$; the solution is asymptotically stable (i.e., attracting) if $-2 < (A_{n+1} - A_n)/\varepsilon A_e^{\pm} < 0$ and unstable otherwise. Carrying out this procedure analytically leads to

$$\frac{A_{n+1} - A_n}{\varepsilon A_e^{\pm}} = \mp\frac{F - (1 + T^*/2)wH}{T_e^{\pm} + T^*wH/2}\sqrt{1 - Z}, \quad (A6)$$

where $T_e^{\pm}$ and $Z$ are given by (A1) and (A2). From (A1) and (A6) it can be deduced that in the physical regime where $T_e^{\pm} \geq 0$ and $0 \leq Z \leq 1$, $A_e^+$ is always stable, and $A_e^-$ is always unstable. The third branch $A_e^0$ is deduced to be stable by similar means.

*Acknowledgments.* Greg Flato, Slava Kharin, Eric DeWeaver, and three anonymous reviewers are thanked for suggesting improvements to earlier versions of this paper.

## REFERENCES

Arzel, O., T. Fichefet, and H. Goosse (2006), Sea ice evolution over the 20th and 21st centuries as simulated by current AOGCMs, *Ocean. Modell.*, *12*, 401–415.

Bitz, C. M., and W. H. Liscomb (1999), An energy-conserving thermodynamic model of sea ice, *J. Geophys. Res.*, *104*, 15,669–15,677.

Bitz, C. M., P. R. Gent, R. A. Woodgate, M. M. Holland, and R. Lindsay (2006), The influence of sea ice on ocean heat uptake in response to increasing $CO_2$, *J. Clim.*, *19*, 2437–2450.

Björk, G., and J. Söderkvist (2002) Dependence of the Arctic Ocean ice thickness distribution on the poleward energy flux in the atmosphere, *J. Geophys. Res.*, *107*(C10), 3173, doi:10.1029/2000JC000723.

Budyko, M. L. (1966), Polar ice and climate, in *Proceedings of the Symposium on the Arctic Heat Budget and Atmospheric Circulation, Rand Corp. Res. Memo.*, RM-5233NSF, edited by J. O. Fletcher, pp. 3–22, Rand Corp., Santa Monica, Calif.

Budyko, M. L. (1974), *Climate and Life, Int. Geophys. Ser.*, vol. 18, 508 pp., Academic, New York.

Carpenter, S. R., and W. A. Brock (2006), Rising variance: A leading indicator of ecological transition, *Ecol. Lett.*, *9*, 311–318.

Flato, G. M., and R. O. Brown (1996), Variability and climate sensitivity of landfast Arctic sea ice, *J. Geophys. Res.*, *101*, 25,767–25,777.

Flato, G. M., and Participating CMIP Modelling Groups, (2004), Sea-ice and its response to $CO_2$ forcing as simulated by global climate models, *Clim. Dyn.*, *23*, 229–241.

Gildor, H., and E. Tziperman (2000), Sea ice as the glacial cycles' climate switch: Role of seasonal and orbital forcing, *Paleoceanography*, *15*, 605–615.

Gildor, H., and E. Tziperman (2001), A sea ice climate switch mechanism for the 100 kyr glacial cycles, *J. Geophys. Res.*, *106*, 9117–9133.

Guckenheimer, J., and P. Holmes (1983), *Nonlinear Oscillations, Dynamical Systems, and Bifurcations of Vector Fields*, 459 pp., Springer, New York.

Held, I. M. (2005), The gap between simulation and understanding in climate modeling, *Bull. Am. Meteorol. Soc.*, *86*, 1609–1614.

Held, H., and T. Kleinen (2004), Detection of climate system bifurcations by degenerate fingerprinting, *Geophys. Res. Lett.*, *31*, L23207, doi:10.1029/2004GL020972.

Hibler, W. D. (1979), A dynamic thermodynamic sea ice model, *J. Phys. Oceanogr.*, *9*, 815–846.

Holland, M. M., and C. M. Bitz (2003) Polar amplification of climate change in coupled models, *Clim. Dyn.*, *21*, 221–232.

Holland, M. M., C. M. Bitz, and B. Tremblay (2006a), Future abrupt transitions in the summer Arctic sea ice, *Geophys. Res. Lett.*, *33*, L23503, doi:10.1029/2006GL028024.

Holland, M. M., C. M. Bitz, E. C. Hunke, W. H. Liscombe, and J. L. Schramm (2006b), Influence of the sea ice thickness distribution on polar climate in CCSM3, *J. Clim.*, *19*, 2398–2414.

Intergovernmental Panel on Climate Change (2001), *Climate Change 2001: The Scientific Basis: Contribution of Working Group 1 to the Third Assessment Report of the Intergovernmental Panel on Climate Change*, edited by J. T. Houghton et al., Cambridge Univ. Press, Cambridge, U. K.

Kleinen, T., H. Held, and G. Petschel-Held (2003) The potential role of spectral properties in detecting thresholds in the Earth system: Application to the thermohaline circulation, *Ocean Dyn.*, *53*, 53–63.

Lindsay, R. W., and J. Zhang (2005), The thinning of Arctic sea ice 1988–2003: Have we passed a tipping point?, *J. Clim.*, *18*, 4879–4894.

Manabe, S., and R. J. Stouffer (1999), Are two modes of thermohaline circulation stable?, *Tellus, Ser. A*, *51*, 400–411.

Maykut, G. A., and N. Untersteiner (1971), Some results from a time-dependent thermodynamic model of sea ice, *J. Geophys. Res.*, *76*, 1550–1575.

Monahan, A. H. (2002a), Stabilization of climate regimes by noise in a simple model of the thermohaline circulation, *J. Phys. Oceanogr.*, *32*, 2072–2085.

Monahan, A. H. (2002b), Correlation effects in a simple stochastic model of the thermohaline circulation, *Stochastics Dyn.*, *2*, 437–462.

North, G. R. (1975), Analytical solution to a simple climate model with diffusive heat transport, *J. Atmos. Sci.*, *32*, 1301–1307.

North, G. R. (1984), The small ice cap instability in diffusive climate models, *J. Atmos. Sci.*, *41*, 3390–3395.

Olbers, D. (2001), A gallery of simple models from climate physics, in *Stochastic Climate Models, Prog. Probab.*, vol. 49, edited by P. Imkeller and J.-S. von Storch, pp. 3–63, Birkhäuser, Boston, Mass.

Overpeck, J. T., et al. (2005), Arctic system on trajectory to new, seasonally ice-free state, *Eos Trans. AGU*, *86*(34), 309.

Polyakov, I. V., et al. (2005), One more step toward a warmer Arctic, *Geophys. Res. Lett.*, *32*, L17605, doi:10.1029/2005GL023740.

Rahmstorf, S. (1995), Bifurcations of the Atlantic thermohaline circulation in response to changes in the hydrological cycle, *Nature*, *378*, 145–148.

Rahmstorf, S. (1999), Rapid transitions of the thermohaline ocean circulation—A modelling perspective, in *Reconstructing Ocean History: A Window Into the Future*, edited by F. Abrantes and A. C. Mix, pp. 139–149, Kluwer Acad., New York.

Rigor, I. G., and J. M. Wallace (2004), Variations in the age of Arctic sea-ice and summer sea-ice extent, *Geophys. Res. Lett.*, *31*, L09401, doi:10.1029/2004GL019492.

Sayag, A., E. Tziperman, and M. Ghil (2004), Rapid switch-like sea ice growth and land ice–sea ice hysteresis, *Paleooceanography*, *19*, PA1021, doi:10.1029/2003PA000946.

Sorteberg, A., V. Kattsov, J. E. Walsh, and T. Pavlova (2007), The Arctic surface energy budget as simulated with the IPCC AR4 AOGCMs, *Clim. Dyn.*, *29*, 131–156.

Stocker T. F., and D. G. Wright (1991), Rapid transitions of the ocean's deep circulation induced by changes in the surface water fluxes, *Nature*, *351*, 729–732.

Stommel, H. (1961), Thermohaline convection with two stable regimes of flow, *Tellus*, *13*, 224–230.

Stroeve, J. C., M. C. Serreze, F. Fetterer, T. Arbetter, W. Meier, J. Maslanik, and K. Knowles (2005), Tracking the Arctic's shrinking ice cover: Another extreme September minimum in 2004, *Geophys. Res. Lett.*, *32*, L04501, doi:10.1029/2004GL021810.

Stroeve J., M. M. Holland, W. Meier, T. Scambos, and M. Serreze (2007), Arctic sea ice decline: Faster than forecast, *Geophys. Res. Lett.*, *34*, L09501, doi:10.1029/2007GL029703.

Taylor, P. D., and D. L. Feltham (2005), Multiple stationary solutions of an irradiated slab, *J. Cryst. Growth*, *276*, 688–697.

Thorndike, A. S. (1992), A toy model linking atmospheric thermal radiation and sea ice growth, *J. Geophys. Res.*, *97*, 9401–9410.

Titz, S., T. Kuhlbrodt, S. Rahmstorf, and U. Feudel (2002) On freshwater-dependent bifurcations in box models of the inter-hemispheric thermohaline circulation, *Tellus, Ser. A*, *54*, 89–98.

von Storch, H., and F. W. Zwiers (1999), *Statistical Analysis in Climate Research*, 484 pp., Cambridge Univ. Press, New York.

Winton, M. (2006), Does Arctic sea ice have a tipping point?, *Geophys. Res. Lett.*, *33*, L23504, doi:10.1029/2006GL028017.

Zhang, X., and J. E. Walsh (2006), Toward a seasonally ice-covered Arctic Ocean: Scenarios from the IPCC AR4 model simulations, *J. Clim.*, *19*, 1730–1747.

M. M. Holland, National Center for Atmospheric Research, Boulder, CO 80307-3000, USA.

W. Merryfield, Canadian Centre for Climate Modeling and Analysis, University of Victoria, P.O. Box 1700, Victoria, BC V8W 2Y2, Canada. (Bill.Merryfield@ec.gc.ca)

A. H. Monahan, School of Earth and Ocean Sciences, University of Victoria, Victoria, BC V8W 3P6, Canada.

# What Is the Trajectory of Arctic Sea Ice?

Harry L. Stern and Ronald W. Lindsay

*Polar Science Center, Applied Physics Laboratory, University of Washington, Seattle, Washington, USA*

Cecilia M. Bitz and Paul Hezel

*Department of Atmospheric Sciences, University of Washington, Seattle, Washington, USA*

We consider the trajectory in phase space of the Arctic sea ice thickness distribution, in which each dimension or component is the time series of sea ice area for a given ice thickness bin. We analyze the trajectory as determined by output from an ice-ocean model, finding that the first two principal components account for 98% of the variance. Simplifying the ice thickness distribution into thin ice, thick ice, and open water, we construct a simple empirical linear model that converges to a stable annual cycle from any initial state. When we include a quadratic nonlinearity to simulate a crude ice-albedo feedback, the model exhibits two stable states, one with perennial ice and one with ice-free summers, resembling the projections of some climate models for the late 21st century. We discuss the interplay between external forcing, internal dynamics, and "tipping points" in the decline of Arctic sea ice.

## 1. INTRODUCTION

Significant changes in the Arctic sea ice, ocean, atmosphere, ice sheets, and freshwater cycle over the past few decades are well documented [*Intergovernmental Panel on Climate Change*, 2007]. Several recent articles refer to the "trajectories" of these components, in which the word trajectory is used in a figurative sense [e.g., *Overpeck et al.*, 2005; *Peterson et al.*, 2006]. The literal meaning of a trajectory is a path in physical space or phase space. In this work we consider the trajectory of sea ice in the ice thickness phase space. We then analyze that trajectory as determined by model sea ice thickness distributions. Our use of the word trajectory does not refer to the physical motion of ice floes.

Arctic Sea Ice Decline: Observations, Projections, Mechanisms, and Implications
Geophysical Monograph Series 180
Copyright 2008 by the American Geophysical Union.
10.1029/180GM12

The sea ice thickness distribution $g(h)$ over some region R gives the fractional area of R covered by ice of thickness h. It is the fundamental description of the Arctic sea ice cover [*Thorndike et al.*, 1975]. The processes controlling $g(h)$ are ice growth, melt, divergence, ridging, and import/export (into or out of R). Models that simulate the evolution of $g(h)$ often use discrete bins of ice thickness. Let $g_k$ represent the fractional area covered by ice with thickness in the range $h_k < h < h_{k+1}$, for k = 1 to n, where n is the number of bins. The area with $h < h_1$ is considered open water. The sum of the fractional areas $g_1 + \ldots + g_n$ is the total ice concentration. Now consider $(g_1, \ldots, g_n)$ as a point in n-dimensional space. As time progresses, the point moves in the n space, tracing out a trajectory. That is what we mean by the trajectory of Arctic sea ice. A point on the trajectory gives the ice thickness distribution at a particular time. Each component $g_k(t)$ is a time series of the fractional ice area in thickness bin k. The region R over which this description applies may be chosen to be as small as one model grid cell or as large as the entire Arctic Ocean.

We analyze the trajectory of Arctic sea ice from two models. One is a retrospective coupled ice-ocean model that assimilates sea ice concentration data and is run for the years 1958–2005. Using the output of this model, we construct a pair of equations for the evolution of thin ice and thick ice that exhibits two stable annual cycles: one with ice-free summers and one with sea ice year-round. The second model is a coupled global general circulation climate model that is forced with increasing greenhouse gas and aerosol concentrations and is run for the years 1870–2099. This model attains ice-free summers by the second half of the 21st century. We use these models to contrast the influence of external forcing versus internal dynamics. We end with a brief discussion of tipping points and the transition from one state to another.

## 2. MODELS

The first model used in this study is the Pan-Arctic Ice-Ocean Modeling and Assimilation System (PIOMAS), which has been used in a wide range of retrospective climate studies [*Zhang and Rothrock*, 2005]. It is based on the parallel ocean and ice model of *Zhang and Rothrock* [2003]. It consists of the Parallel Ocean Program (POP) ocean model developed at Los Alamos National Laboratory coupled to a multicategory ice thickness and enthalpy distribution sea ice model [*Zhang and Rothrock*, 2001; *Hibler*, 1980]. The POP model is a Bryan-Cox-Semtner type ocean model [*Bryan*, 1969; *Cox*, 1984; *Semtner*, 1976] with numerous improvements. The sea ice model consists of five main components: (1) a momentum equation that determines ice motion, (2) a viscous-plastic rheology that determines the internal ice stress, (3) a heat equation that determines ice temperature profiles and ice growth or decay, (4) an ice thickness distribution equation that conserves ice mass, and (5) an enthalpy distribution equation that conserves ice thermal energy. The model has seven ice thickness categories plus open water (Table 1), and a latitude/longitude grid with mean spacing 22 km, with the

**Table 1.** PIOMAS Sea Ice Thickness Bins

| Bin | Lower Limit (m) | Upper Limit (m) | Designation |
|---|---|---|---|
| 0 | 0.00 | 0.10 | open water |
| 1 | 0.10 | 0.66 | thin ice |
| 2 | 0.66 | 1.93 | thin ice |
| 3 | 1.93 | 4.20 | thick ice |
| 4 | 4.20 | 7.74 | thick ice |
| 5 | 7.74 | 12.74 | thick ice |
| 6 | 12.74 | 19.31 | thick ice |
| 7 | 19.31 | 27.51 | thick ice |

**Table 2.** CCSM3 Sea Ice Thickness Bins

| Bin | Lower Limit (m) | Upper Limit (m) | Designation |
|---|---|---|---|
| 0 | 0.00 | 0.00 | open water |
| 1 | >0.00 | 0.65 | thin ice |
| 2 | 0.65 | 1.39 | thin ice |
| 3 | 1.39 | 2.47 | thick ice |
| 4 | 2.47 | 4.57 | thick ice |
| 5 | 4.57 | infinity | thick ice |

pole displaced over Greenland. The model assimilates open-ocean sea surface temperature (SST) data using a nudging method with a 15-day time constant, and sea ice concentration data using a nudging method that emphasizes the ice extent and minimizes the effect of observational errors in the interior of the pack ice [*Lindsay and Zhang*, 2006]. The assimilated data (HadISST [*Rayner et al.*, 2003]) consists of monthly averaged values on a 1° grid. The effect of the assimilation of total ice concentration is that the sum of the model ice concentrations over all the bins ($g_1 + ... + g_7$) is constrained to match (approximately) the observations. The assimilation of ice concentration also improves the modeled ice thickness compared to submarine and moored upward looking sonar data [*Lindsay and Zhang*, 2006]. Thus the PIOMAS ice concentration and ice thickness are in reasonable agreement with observations. The model is run for 48 years (1958–2005), forced with daily fields of 10 m surface wind velocities, 2 m air temperature and humidity, precipitation, and downwelling longwave and shortwave radiative fluxes. These are obtained from the ERA-40 reanalysis for 1958–2001, and from the ECMWF operational analysis for 2002–2005. The daily sea ice thickness distributions computed by the model are then temporally averaged over each month, and spatially averaged over $6.6 \times 10^6$ km$^2$ of the central Arctic Ocean shown in Figure 1. Thus the end result is one sea ice thickness distribution (with seven bins) for each month, 1958 through 2005.

The second model used in this study is the Community Climate System Model version 3.0 (CCSM3) [*Collins et al.*, 2006]. It is a state-of-the-art fully coupled climate model composed of four separate models simultaneously simulating the Earth's atmosphere, ocean, land surface, and sea ice. The sea ice model has five ice thickness categories plus open water (Table 2), and a nominal 1° grid with the pole displaced over Greenland. The ocean model (POP) is the same as that used in PIOMAS. The 20th century simulation is forced with observed $CO_2$, $CH_4$, $NO_2$, solar variability, volcanic sulfate aerosols, and industrial aerosols. The 21st century simulation is an Intergovernmental Panel on Climate Change Special

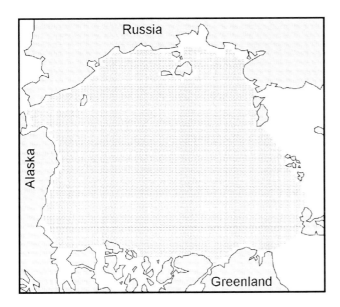

**Figure 1.** The sea ice thickness distribution from PIOMAS is averaged over the 6.6 million km² of the central Arctic Ocean shown by the gray cells.

Report on Emissions Scenarios A1B scenario [*Meehl et al.*, 2006] which assumes a moderate level of conservation and technical advance to limit emissions (i.e., greenhouse gas levels rise more slowly after about 2050). The distribution and thickness of Arctic sea ice have improved in CCSM3 relative to earlier versions. The model simulates sea ice extent, thickness, and trends reasonably well in the late 20th and early 21st centuries [*Holland et al.*, 2006a]. The model is run for 230 years, of which we analyze the last 150 years (1950–2099). The sea ice thickness distributions are averaged in a manner similar to those from PIOMAS: monthly in time and Arctic-wide in space. The spatial domain is essentially the same as that shown in Figure 1 (although slightly more area along the northern edge of the Canadian Archipelago is included).

## 3. ANALYSIS

### 3.1. Principal Component Analysis

Consider the trajectory of sea ice from the PIOMAS model, which is a curve in seven-dimensional space. It is natural to ask whether this curve occupies the full seven dimensions, or whether it is primarily confined to a lower-dimensional subspace. Thus we look for linear combinations of the components $g_1$ through $g_7$ that account for as much variance as possible, i.e., principal component analysis. After removing

the mean from each time series $g_k(t)$, we form the $7 \times 7$ covariance matrix and compute its eigenvalues and eigenvectors. The sum of the eigenvalues is the total variance of all the components, which includes the seasonal cycle. It turns out that the first two eigenvalues account for 98% of the variance; the trajectory of Arctic sea ice is essentially two-dimensional. The first two principal components are given by

$$PC_1 = -0.56g_1 + 0.76g_2 + 0.33g_3 + 0.03g_4$$
$$PC_2 = 0.35g_1 + 0.56g_2 - 0.69g_3 - 0.29g_4 \qquad (1)$$
$$- 0.07g_5 - 0.02g_6 .$$

The two-dimensional trajectory in principal component space is shown in Figure 2. The annual cycles go clockwise, drifting slowly over time toward increasing values of $PC_2$. Notice from equation (1) that $PC_2$ is roughly "thin ice minus thick ice." Thus the trajectory indicates increasing thin ice and decreasing thick ice.

To investigate the effect of the number of ice thickness bins on the principal component analysis, we also analyzed a 56-year run (1948–2003) of a version of the PIOMAS model with 11 ice thickness bins. We found that the first two principal components account for 91% of the total variance.

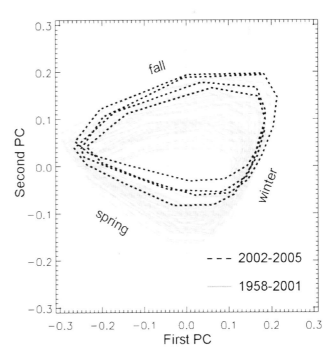

**Figure 2.** Trajectory of Arctic sea ice in principal component (PC) space. The annual cycles go clockwise, drifting slowly over time toward increasing values of the second principal component.

Thus the two-dimensionality of the trajectory is fairly robust. Nevertheless, it is not straightforward to give physical interpretations to the principal components in equation (1). Therefore we take a slightly different approach, as follows.

### 3.2. Thin Ice and Thick Ice

Motivated by the results in the previous section, we form linear combinations of the bins $g_1$ through $g_7$ that have simple physical interpretations, at the expense of accounting for less than the maximum possible variance. The combinations are simply thin ice ($g_1 + g_2$) and thick ice ($g_3 + g_4 + g_5 + g_6 + g_7$). Unlike the principal components, which are uncorrelated, thin ice and thick ice have nonzero covariance, and thus the total variance cannot be neatly partitioned into thin and thick contributions. The sum of $\text{var}(g_1) + \ldots + \text{var}(g_7)$ is $300 \times 10^{-4}$. The covariance matrix of the first two principal components is

$$\begin{pmatrix} 69 & 0 \\ 0 & 224 \end{pmatrix} \times 10^{-4},$$

while the covariance matrix of thin ice (first row/column) and thick ice (second row/column) is

$$\begin{pmatrix} 69 & -50 \\ -50 & 113 \end{pmatrix} \times 10^{-4}.$$

Thus thin ice and thick ice are negatively correlated, but their variances (69 and $113 \times 10^{-4}$) capture a reasonably large fraction (61%) of the total variability.

Figure 3 shows the trajectory of Arctic sea ice in thin/thick space. The axes give the concentration of each ice type. All years (1958–2005) are shown in gray. The annual cycles go counterclockwise. Monthly values for January to December are shown for the years 1966 and 2005. The distance from a point on the trajectory to the line thin + thick = 1 is the fraction of open water. Consider the physical sources and sinks of thin ice and thick ice throughout the year. During the winter, thin ice grows thicker (sink of thin ice, source of thick ice). This can be seen in January through April as the total ice concentration remains close to 1 but the trajectory moves toward the upper left. May is a transition month, when thick ice reaches a maximum. From June through September, thick ice decreases. This melting of thick ice is a source of thin ice, which is offset to some extent by the melting of the thin ice itself. In 1966 the thin ice increased from June to September (being replenished by melting thick ice more quickly than its own melt rate), whereas in 2005 the thin ice decreased from June to September (being replenished by melting thick ice more slowly than its own melt rate).

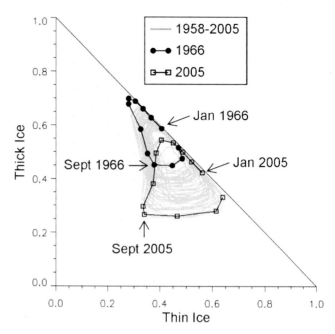

**Figure 3.** Trajectory of Arctic sea ice in thin/thick space from PIOMAS model output. Axes give the concentration of each ice type. All years (1958–2005) are shown in gray. The annual cycles go counterclockwise. Monthly values are shown for the years 1966 (circles) and 2005 (squares).

Notice that the total ice concentration in September 1966 was about 0.8 while in September 2005 it was 0.6. October and November are freeze-up months, with rapid growth of thin ice and hardly any increase in thick ice. December is another transition month, when full winter conditions are reached, with a total ice concentration close to 1 again.

The gradual shift over time toward less thick ice and lower total ice concentration in September is clearly evident in Figure 3 (more so than in the principal component trajectory of Figure 2). What are the factors driving this transition? They can be separated into two broad categories: external forcing and internal dynamics. External forcing refers to (for example) increasing air temperature that melts more ice; internal dynamics refers to the response of the system to perturbations in its state. For example, if the system has multiple stable equilibrium states, a small perturbation could knock the system from one basin of attraction into another, sending it on a course toward less ice even when normal forcing conditions are restored. Several investigators have found multiple stable states in sea ice models [e.g., *Flato and Brown*, 1996; *Hibler et al.*, 2006; *Merryfield et al.*, this volume], which we discuss in more detail later. These were physical models, in which processes such as ice growth and melt were explicitly included. Physical models often have to be "tuned" to reproduce observed behavior. Alternatively, empirical models reproduce observed behavior by design.

### 3.3. Empirical Models

While *Merryfield et al.* [this volume] constructed a simple physical model to mimic the essential behavior of the sea ice component of CCSM3, we adopt an empirical approach toward "modeling the model" in which we postulate a simple form for the evolution equations of thin ice and thick ice that includes undetermined coefficients, use the output of PIOMAS or CCSM3 to find the coefficients that best reproduce the thin/thick trajectory, and then study the stability properties of the resulting equations.

*3.3.1. One-dimensional linear model.* We start with a one-dimensional formulation. Let $z$ equal the total sea ice concentration in the Arctic. We postulate $dz/dt = p(t)(1 - z)$, where $p(t)$ is an annually periodic function of the form $p(t) = A \cos(\omega(t - \tau))$. With $t$ measured in months, the frequency $\omega$ is $2\pi/12$. The constant $\tau = 10.3$ months is a phase shift to align the minima of $z(t)$ with the end of summer. The motivation for the form of this equation is that it is linear in $z$, the solutions are periodic, and it gives the correct shape of the annual cycle (as shown presently). If $A$ is a constant then the amplitude of each annual cycle is the same. If we allow a different value of $A$ each year, we can reproduce interannual variability. Figure 4 shows the solution of this one-dimensional equation (solid line), together with the monthly Arctic sea ice concentration from PIOMAS (black dots). The annual values of $A$ used in this 10-year interval are [0.98, 0.96, 0.97, 0.99, 0.98, 0.83, 0.81, 0.90, 0.84, 1.19]. The fit is remarkably good: the bias between $z$ and the PIOMAS ice

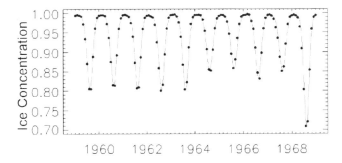

**Figure 4.** Solid curve is the solution of $dz/dt = p(t)(1 - z)$, where $p(t)$ is a cosine function with a period of 1 year and a different amplitude for each year shown by solid curve. Black dots indicate monthly Arctic sea ice concentration from PIOMAS.

concentration is less than 0.002, with a standard deviation of 0.008. Apparently the annual cycle has only one degree of freedom per year, an amplitude that determines the September minimum. The success of this one-dimensional fit motivates the following two-dimensional case.

*3.3.2. Two-dimensional linear model.* Let $x$ equal thin ice concentration and $y$ equal thick ice concentration. We wish to formulate simple equations for $x$ and $y$ that reproduce, in an average sense, the annual trajectories of Figure 3. Based on the one-dimensional case, we choose linear equations with annually periodic coefficients:

$$dx/dt = a(t)(1 - x - y) + b(t)x + c(t)y$$
$$dy/dt = d(t)(1 - x - y) + e(t)x + f(t)y. \tag{2}$$

The term $1 - x - y$ is the open water concentration. The six functions $a$, $b$, $c$, $d$, $e$, and $f$ are determined from the 48-year time series of PIOMAS thin ice and thick ice concentrations using a least squares method. Let time be measured in months, and replace the continuous $dx/dt$ with $\Delta x$, which represents the change in thin ice concentration from one month to the next. Consider the change from, say, January to February. We write

$$\Delta x_k = a(1 - x_k - y_k) + bx_k + cy_k + error_k, \tag{3}$$

where $x_k$ and $y_k$ are the concentrations of thin ice and thick ice in January of year $k$ and $\Delta x_k$ is the change in thin ice concentration from January to February of year $k$. This is 48 equations (48 years, $k = 1$ to 48) with three unknowns: $a$, $b$, and $c$. We find $a$, $b$, and $c$ by using a standard least squares procedure that minimizes the variance of the error term. The process is repeated for the change from February to March, March to April, etc. Thus $a(t)$ consists of 12 values, one for each month, and similarly for $b(t)$ and $c(t)$. Finally, the same procedure is applied to the thick ice equation (for $\Delta y_k$) to obtain the 12 values of $d(t)$, $e(t)$, and $f(t)$. The equations (2) can then be integrated in time, starting from some initial state $(x_0, y_0)$. Figure 5 shows the resulting equilibrium trajectory that is obtained, no matter what the initial state. The full 48-year thin ice/thick ice trajectory from PIOMAS is shown in gray. The equilibrium trajectory is clearly an average of the 48 years, by construction. The coefficients $a(t)$, $b(t)$, $c(t)$, $d(t)$, $e(t)$, and $f(t)$ parameterize the annual cycle of forcing fields such as air temperature (i.e., cold in winter and warm in summer). If the coefficients were constants instead of periodic functions, the equilibrium would be a single point.

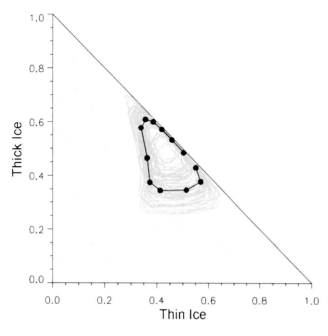

**Figure 5.** Monthly values of the equilibrium trajectory (circles) obtained by integrating equations (2). The 48-year PIOMAS trajectory is shown in gray.

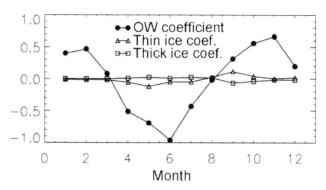

**Figure 6.** Coefficients of the sum of equations (2): open water (OW) coefficient $a(t) + d(t)$ (circles), thin ice coefficient $b(t) + e(t)$ (triangles), and thick ice coefficient $c(t) + f(t)$ (squares).

The periodicity of the coefficients allows for an annual cycle of thin ice and thick ice. There is no interannual variability in this empirical model. The point of this two-dimensional exercise is that a simple linear model can reproduce the annual cycle of thin ice and thick ice fairly well, and it has one stable state: the trajectory always converges to that of Figure 5 no matter what the initial condition. It will never evolve to an ice-free Arctic.

If we add together equations (2), we obtain a single equation for the total ice concentration, $z = x + y$, of the form $dz/dt = (a + d)(1 - z) + (b + e)x + (c + f)y$. On the basis of the one-dimensional case above, we would expect $a+d$ to be approximately a cosine function with amplitude close to 1, and $b + e$ and $c + f$ should be zero or small. This is indeed the case, as shown in Figure 6. This confirms that the two-dimensional model is consistent with the one-dimensional model.

A trend in external forcing (such as increasing atmospheric $CO_2$) is capable of driving the trajectory of Arctic sea ice toward a state with ice-free summers, according to the recent predictions of several models [*Zhang and Walsh*, 2006]. Figure 7 shows the trajectory of the CCSM3 model for 150 years (1950–2099). The annual cycles through 2005 are similar to those from PIOMAS (Figure 3). From 1950 through about 2005, the amount of thick ice in September decreases but the amount of thin ice in September is fairly constant. After 2005 the September trajectory swings toward

(0,0), with decreasing amounts of thin ice nearly every year. The annual cycle for 2044 is shifted downward relative to 2005, i.e., less thick ice, but there is also a noticeable change in the shape of the annual cycle: (1) In 2044 there is a loss of thin ice from June to July, whereas in 2005 and earlier years there is a gain of thin ice. (2) In 2044 the loss of thin ice from July to August is far greater than in the earlier years. This might be understood as follows:

1. When there is a relatively large amount of thick ice in June, and it begins to melt, it creates thin ice faster than the thin ice itself can melt. But when there is relatively little thick ice in June, its conversion to thin ice cannot keep up with the melting of the thin ice itself.

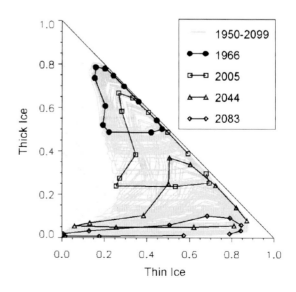

**Figure 7.** Trajectory of the CCSM3 model for 150 years, 1950–2099 (gray). Monthly values are shown for the years 1966, 2005, 2044, and 2083.

2. With the creation of more open water in July, the ice-albedo feedback may be exerting more influence. The fraction of open water in July goes from 0.19 (1966) to 0.27 (2005) to 0.51 (2044). That is an increase of 42% in the first 39-year period and 89% in the second 39-year period. With more open water in July, more solar radiation is absorbed by the ocean, leading to more melting of thin ice and the creation of more open water.

These processes can also be understood in terms of open water formation efficiency (OWFE) [*Holland et al.*, 2006b; *Merryfield et al.*, this volume], which is the change in open water fraction per meter of ice melt, over the course of the melt season (May to September). OWFE increases dramatically in CCSM3 simulations as the mean winter ice thickness decreases; that is, it becomes much easier to create open water for a given amount of melt if one starts with thinner ice. The strength of the OWFE depends on the ice thickness distribution, i.e., on the internal model dynamics. The ice-free summers in CCSM3 toward the end of the 21st century are undoubtedly due to both the trend in forcing and the internal model dynamics, but the relative contributions of these factors are unknown.

What role might internal dynamics play in the transition to an ice-free Arctic summer? *Flato and Brown* [1996] constructed a one-dimensional thermodynamic model to study landfast sea ice. The model successfully reproduced seasonal and interannual variability of ice thickness at two locations in the Canadian Arctic for which data were available. The model was also shown to have two stable states: thin seasonal ice and thick perennial ice. A relatively modest change in climate was sufficient to cause the model to jump from the thick perennial state to the thin seasonal state. An unstable equilibrium existed between the two stable states (as is always the case) which arose because of the contrast in albedo between thick ice and thin ice/open water. *Merryfield et al.* [this volume] studied the phenomenon of abrupt sea ice reductions in CCSM3 (as documented by *Holland et al.* [2006b]) by constructing a simplified physical model of the essential processes with just two dependent variables: winter sea ice thickness and summer sea ice extent. They found that pulses of ocean heat transport into the Arctic, along with increased sensitivity of summer sea ice to declining ice extent, were the likely causes of the abrupt drops in summer sea ice. Their nonlinear model also exhibited multiple stable states in a physically relevant parameter regime. These examples suggest that internal dynamics could play a role in guiding the trajectory of sea ice toward a new equilibrium state after the system crosses from one basin of attraction into another, and that the ice-albedo feedback and increasing summer open water are likely to be important factors.

*3.3.3. Two-dimensional nonlinear model.* We investigate the above ideas by making a small modification to the equations (2). We add a quadratic term to the equation for thin ice:

$$
\begin{aligned}
dx/dt &= a(t)(1 - x - y) + b(t)x + c(t)y \\
&\quad + \alpha(t)(1 - x - y)^2 \\
dy/dt &= d(t)(1 - x - y) + e(t)x + f(t)y ,
\end{aligned}
\tag{4}
$$

where $\alpha(t)$ is negative in summer and zero otherwise. The quadratic term simulates enhanced melting of thin ice in summer as the fraction of open water increases. Equation (3) is therefore modified by adding a corresponding quadratic term:

$$
\begin{aligned}
\Delta x_k &= a(1 - x_k - y_k) + bx_k + cy_k \\
&\quad + \alpha(1 - x_k - y_k)^2 + \text{error}_k ,
\end{aligned}
\tag{5}
$$

where we specify a negative value of $\alpha$ for the June-July and July-August equations, and $\alpha = 0$ otherwise. We then solve for $a$, $b$, and $c$ as before using the PIOMAS model output and a standard least squares procedure. We need only do this for the June-July and July-August thin ice equations; all the other coefficients remain the same. Having found the new coefficients, we then integrate the equations (4), with specified initial conditions for thin ice and thick ice in September. (If a negative value of $x$ or $y$ is obtained for a given month, it is reset to zero before continuing the integration.) The results are shown in Figure 8, using ad hoc values of $\alpha = -3$ for June-July and $\alpha = -6$ for July-August. There are now two equilibrium cycles, one with perennial ice and one with seasonal ice, depending on the initial conditions. If we start the integration in September within the gray triangle (bottom left), the trajectory evolves to the seasonal ice cycle. If we start outside the gray triangle, the trajectory evolves to the perennial ice cycle. Comparison of the perennial ice cycle with that of Figure 5 obtained from the linear equations (2) shows that the thin ice and thick ice fractions match within 0.02 in every month: the perennial ice cycle is nearly unchanged. The quadratic nonlinearity has allowed the emergence of a seasonal ice cycle, which can only be sustained if the summer open water is sufficiently large. The seasonal ice cycle resembles the annual cycle for the year 2044 in CCSM3 (Figure 7). Comparing the size of terms in equations (2) and (4), we find that when the open water fraction $1-x-y$ becomes greater than about 0.2, the nonlinearity becomes important, with increasing dominance as $1-x-y$ increases.

Figure 9 shows the approach to equilibrium. The black dots are different September initial conditions from throughout the domain. Black and gray paths trace subsequent Septembers obtained by integrating equations (4) with $\alpha = -3$ for June-July and $\alpha = -6$ for July-August. Black trajectories lead to the equilibrium September value for perennial ice; gray trajectories lead to the equilibrium September value for seasonal ice. Notice how the September trajectories all rapidly approach a curving "attractor" and then continue along the attractor to the equilibrium point (gray circle). The triangles are the 48 Septembers of the PIOMAS model, clustered loosely around the perennial ice equilibrium point and along the attractor. Although the black and gray paths are derived from equations (4) with coefficients fit to the PIOMAS model output, we also show the 150 Septembers of the CCSM3 model (squares and pluses). The CCSM3 Septembers for 1950–2005 (squares) do not match those of PIOMAS because of the different thin ice/thick ice cutoffs (Tables 1 and 2) and because of differences between the two models. However, after about 2025 the CCSM3 Septembers follow a path close to the curving attractor toward the seasonal ice equilibrium near (0,0).

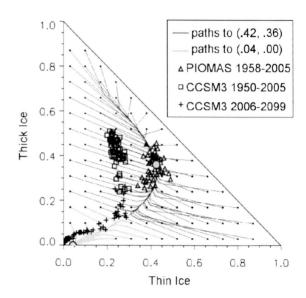

**Figure 9.** Approach to equilibrium. The black dots, with spacing 0.07, are initial conditions in September. The black and gray paths are the September trajectories obtained by integrating equations (4). Black trajectories lead to the equilibrium September value at (0.42, 0.36) (gray circle). Gray trajectories lead to the equilibrium September value at (0.04, 0.00) (gray circle). Triangles are the 48 Septembers of the PIOMAS model (1958–2005). Squares and pluses are the 150 Septembers of the CCSM3 model (1950–2099).

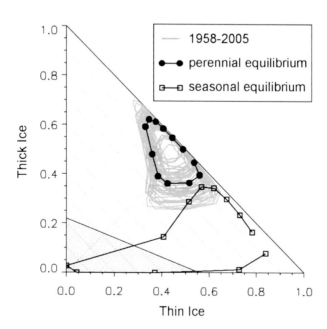

**Figure 8.** Perennial ice equilibrium cycle (circles) and seasonal ice equilibrium cycle (squares) obtained by integrating equations (4). Initial conditions for September within the gray triangle (bottom left) lead to the seasonal ice cycle. Initial conditions for September outside the gray triangle lead to the perennial ice cycle. The 48-year PIOMAS trajectory is shown in gray.

## 4. DISCUSSION

We reiterate that the equations (2) and (4) do not explicitly model physical processes. The annually periodic coefficients, derived empirically, are meant to simulate the annual cycles of the mean external forcing fields, with no interannual variability. The linear equations (2) are perhaps the simplest possible nontrivial evolution equations for thin ice and thick ice, and the quadratic term in equation (4) is a simple nonlinearity. Nevertheless, these empirical models illustrate that (1) it is not difficult to formulate a system with multiple stable states and (2) the inclusion of a quadratic nonlinearity that mimics the ice-albedo feedback leads to the emergence of a stable annual cycle with ice-free summers (August–September). *Serreze et al.* [2007] raised the question of how a seasonally ice-free Arctic Ocean might be realized: through gradual decline or through a rapid transition once the ice thins to a more vulnerable state? Our empirical model would allow the second scenario to exist if there were a mechanism to nudge the system across the "tipping point" from one stable state to another.

The concept of a tipping point has recently gained popularity in the press [*Walker*, 2006]. A tipping point is an unstable equilibrium between two stable ones. (In our non-linear empirical model, the tipping point for September is actually the entire hypotenuse of the gray triangle in Figure 8). The transition across a tipping point could be precipitated by a particular event, or it could be gradual. *Lindsay and Zhang* [2005] hypothesized that the thinning of Arctic sea ice since 1988 was triggered by the export of older, thicker ice out of the Arctic basin in the late 1980s and early 1990s, driven by atmospheric circulation patterns associated with high values of the Arctic Oscillation (AO) and the Pacific Decadal Oscillation (PDO). The thinning continued in the following years, even though the AO and PDO returned to near normal levels, because the ice-albedo feedback was then exerting a larger influence because of the increase in summer open water and thin ice. The abrupt reductions in future Arctic sea ice in CCSM3 simulations were triggered by "pulse-like" events of ocean heat transport into the Arctic Ocean [*Holland et al.*, 2006b, this volume], followed by a similar increase in the influence of the ice-albedo feedback due to more open water. *Winton* [2006] examined two climate models that both became seasonally ice free gradually as the air temperature increased, but further warming caused an abrupt loss of sea ice year-round in one of the models (MPI) because of the ice-albedo feedback. The other model (CCSM3) lost all of its sea ice more gradually, driven primarily by the ocean heat flux. Thus we have examples of both abrupt and gradual changes between states.

A simple model that exhibits a tipping point is $dz/dt = -kz(z-a)(z-b) + F(t)$ where $k > 0$ is a constant, $0 < a < b$ are constants, and $F(t)$ is the external forcing. (We are thinking of $z$ as the September sea ice extent). First consider $F = 0$. The two stable equilibria are $z = 0$ and $z = b$. They are separated by the unstable equilibrium point (tipping point) $z = a$. Any trajectory that starts with $z > a$ converges to $z = b$; any trajectory that starts with $z < a$ converges to $z = 0$. The constant $k$ (with units 1/time) determines the rate of convergence. Large $k$ means strong internal dynamics (short response time of the system); small $k$ means weak internal dynamics (long response time). Now turn on the external forcing F. Clearly F can be formulated to kick the system back and forth between its two regions of attraction, either randomly or cyclically, suddenly or gradually. If F is of the form $A\sin(\omega t)$ and we nondimensionalize time by $t' = kt$ then the nondimensional external forcing is $F' = (A/k)\sin((\omega/k)t')$. The ratio $A/k$ is the strength of the external forcing relative to the strength of the internal dynamics. The ratio $\omega/k$ is the response time of the system relative to the period of the external forcing. If $A/k$ is small, the system will not cross the tipping point; if $A/k$ is large, it will cross back and forth regularly. Suppose instead that $F'$ is a random process with mean zero and standard deviation $S$. Then $S/k$ is analogous

to $A/k$, and the autocorrelation time scale of F' is analogous to the period $2\pi/\omega$. If $S/k$ is small, the system will fluctuate about one of the stable equilibrium points. At larger values of $S/k$, the system will make occasional, gradual transitions across the tipping point, sometimes flipping back and forth several times before approaching one of the stable equilibrium points. At still larger values of $S/k$, the transitions are abrupt. In the limiting case of large $A/k$ or $S/k$, the system is dominated by the external forcing, and the internal dynamics become negligible.

We are led to the following thoughts about tipping points: (1) A tipping point may be approached or crossed suddenly because of an externally forced event, or the transition may be more gradual as the external forcing changes gradually. (2) When the system is near a tipping point, it is more sensitive or susceptible to being nudged into a new state by small perturbations. (3) Crossing a tipping point is not an irreversible event. The system can be driven back into the previous state if the external forcing changes course. (4) The trajectory of the system is determined by a balance between the amplitude and period of the external forcing and the response time of the system (internal dynamics).

## 5. SUMMARY AND CONCLUSIONS

We have defined the trajectory of sea ice to be the path in phase space of the ice thickness distribution, $(g_1(t),\ldots, g_n(t))$, where $g_k(t)$ is the fractional area of sea ice in bin k at time t. We analyzed the 48-year monthly output of an ice-ocean model with seven bins and found that the first two principal components of the trajectory account for 98% of the variance: the trajectory is essentially two-dimensional. Simplifying the ice thickness distribution into thin ice, thick ice, and open water, we constructed a linear model with empirically determined periodic coefficients that matches the mean annual cycle of the 48-year ice-ocean model output. The linear model was found to be stable. We then modified the linear model by adding a quadratic term to simulate enhanced melting of thin ice in summer when the amount of open water is large, i.e., a crude ice-albedo feedback. The nonlinear model was found to have two stable annual cycles, one with ice-free summers and one with perennial summer ice, qualitatively similar to the results of *Flato and Brown* [1996]. The annual cycle with ice-free summers resembles the late 21st century annual cycles of the CCSM3 model projection.

The projection of an ice-free Arctic in summer is not new, nor is the idea that Arctic sea ice may have multiple stable states. Twenty-eight years ago, *Parkinson and Kellogg* [1979] ran a sea ice model forced by atmospheric warming commensurate with a doubling of $CO_2$. They found that a 5°C increase in surface air temperature led to the disappearance of sea ice in August and September. Also in 1979, *Kellogg* [1979, p. 85] wrote, "There are good reasons to believe that the Arctic Ocean may have just two stable states, a largely frozen-over one (as at present) and an ice-free one." What is new today compared to 1979 is the sustained downward trend in Arctic sea ice, capped by record-shattering losses in August and September 2007. Using data from the National Snow and Ice Data Center [*Fetterer et al.*, 2002], we find that the best linear fit of September sea ice extent versus year (1979–2006) leaves residuals with standard deviation $4.2 \times 10^5$ km$^2$, in terms of which the September 2007 sea ice extent falls 4 standard deviations below the trend line. This is the type of abrupt loss of summer sea ice simulated by CCSM3 starting in the year 2024 [*Holland et al.*, 2006b, this volume], and it continues the pattern of observed ice extent declining faster than model predictions [*Stroeve et al.*, 2007]. One could argue that Arctic sea ice is now entering a new regime. In any case, it seems likely that increasing greenhouse gases, driving increasing air temperatures, will eventually lead to summers without sea ice in the Arctic Ocean. The interplay between the internal dynamics of the climate system and the external forcing will determine the extent to which this outcome is achieved sooner rather than later. Whether or not CCSM3 or the real climate system actually has a "tipping point" is still unknown.

Several extensions of this work are possible. One could divide the Arctic Ocean and peripheral seas into regions, and construct and analyze the trajectory of sea ice in each region. One could add a third category of ice thickness to the present thin ice and thick ice (e.g., medium thick ice), thereby capturing more of the variance. A more interesting extension would be to construct a simplified physical model of the evolution of thin ice and thick ice, rather than an empirical model. This was proposed by *Stern et al.* [1995], based on the framework of *Thorndike et al.* [1975] and *Hibler* [1980]. One could then analyze the stability of the physical model and attempt to attribute the observed changes in sea ice to changes in the external forcing and to the influence of internal dynamics.

*Acknowledgments.* We thank J. Zhang for providing the PIO-MAS model output, and we acknowledge the computational facilities of the National Center for Atmospheric Research (NCAR) for the CCSM3 model output. We thank the Editor, Eric DeWeaver, for his helpful comments. This work was supported by the NASA Cryospheric Sciences Program (grant NNG06GA84G) (H.S. and R.L.) and the NSF Office of Polar Programs (grant 0454843) (C.B. and P.H.).

## REFERENCES

Bryan, K. (1969), A numerical method for the study of the circulation of the world oceans, *J. Comput. Phys.*, *4*, 347–376.

Collins, W. D., et al. (2006), The Community Climate System Model version 3 (CCSM3), *J. Clim.*, *19*, 2122–2143.

Cox, M. D. (1984), A primitive equation, three-dimensional model of the oceans, *GFDL Ocean Group Tech. Rep. 1*, Geophys. Fluid Dyn. Lab., Princeton, N. J.

Fetterer, F., K. Knowles, W. Meier, and M. Savoie (2002), Sea ice index, http://nsidc.org/data/seaice_index/, Natl. Snow and Ice Data Cent., Boulder, Colo. (updated 2007)

Flato, G. M., and R. D. Brown (1996), Variability and climate sensitivity of landfast Arctic sea ice, *J. Geophys. Res.*, *101*(C10), 25,767–25,777.

Hibler, W. D., III (1980), Modeling a variable thickness sea ice cover, *Mon. Weather Rev.*, *1*, 943–973.

Hibler, W. D., III, J. K. Hutchings, and C. F. Ip (2006), Sea-ice arching and multiple flow states of Arctic pack ice, *Ann. Glaciol.*, *44*, 339–344.

Holland, M. M., C. M. Bitz, E. C. Hunke, W. H. Lipscomb, and J. L. Schramm (2006a), Influence of the sea ice thickness distribution on polar climate in CCSM3, *J. Clim.*, *19*, 2398–2414.

Holland, M. M., C. M. Bitz, and B. Tremblay (2006b), Future abrupt reductions in the summer Arctic sea ice, *Geophys. Res. Lett.*, *33*, L23503, doi:10.1029/2006GL028024.

Holland, M. M., C. M. Bitz, B. Tremblay, and D. A. Bailey (2008), The role of natural versus forced change in future rapid summer Arctic ice loss, this volume.

Intergovernmental Panel on Climate Change (2007), *Climate Change 2007: The Physical Science Basis: Contribution of Working Group I to the Fourth Assessment Report of the Intergovernmental Panel on Climate Change*, edited by S. Solomon et al., 996 pp., Cambridge Univ. Press, Cambridge, U. K.

Kellogg, W. W. (1979), Influences of mankind on climate, *Annu. Rev. Earth Planet. Sci.*, *7*, 63–92.

Lindsay, R. W., and J. Zhang (2005), The thinning of Arctic sea ice, 1988–2003: Have we passed a tipping point?, *J. Clim.*, *18*, 4879–4894.

Lindsay, R. W., and J. Zhang (2006), Assimilation of ice concentration in an ice-ocean model, *J. Atmos. Oceanic Technol.*, *23*, 742–749.

Meehl, G. A., W. A. Washington, B. D. Santer, W. D. Collins, J. M. Arblaster, A. Hu, D. M. Lawrence, H. Teng, L. E. Buja, and W. G. Strand (2006), Climate change projections for the twenty-first century and climate change commitment in the CCSM3, *J. Clim.*, *19*, 2597–2616.

Merryfield, W. J., M. M. Holland, and A. H. Monahan (2008), Multiple equilibria and abrupt transitions in Arctic summer sea ice extent, this volume.

Overpeck, J. T., et al. (2005), Arctic system on trajectory to new, seasonally ice-free state, *Eos Trans. AGU*, *86*(34), 309, doi:10.1029/2005EO340001.

Parkinson, C. L., and W. W. Kellogg (1979), Arctic sea ice decay simulated for a $CO_2$-induced temperature rise, *Clim. Change*, *2*, 149–162.

Peterson, B. J., J. McClelland, R. Curry, R. M. Holmes, J. E. Walsh, and K. Aagaard (2006), Trajectory shifts in the Arctic and subarctic freshwater cycle, *Science*, *313*, 1061–1066.

Rayner, N. A., D. E. Parker, E. B. Horton, C. K. Folland, L. V. Alexander, D. P. Rowell, E. C. Kent, and A. Kaplan (2003), Global analyses of sea surface temperature, sea ice, and night marine air temperature since the late nineteenth century, *J. Geophys. Res.*, *108*(D14), 4407, doi:10.1029/2002JD002670.

Semtner, A. J., Jr. (1976), A model for the thermodynamic growth of sea ice in numerical investigations of climate, *J. Phys. Oceanogr.*, *6*, 379–389.

Serreze, M. C., M. M. Holland, and J. Stroeve (2007), Perspectives on the Arctic's shrinking sea-ice cover, *Science*, *315*, 1533–1536.

Stern, H. L., D. A. Rothrock, and R. Kwok (1995), Open water production in Arctic sea ice: Satellite measurements and model parameterizations, *J. Geophys. Res.*, *100*(C10), 20,601–20,612.

Stroeve, J., M. M. Holland, W. Meier, T. Scambos, and M. Serreze (2007), Arctic sea ice decline: Faster than forecast, *Geophys. Res. Lett.*, *34*, L09501, doi:10.1029/2007GL029703.

Thorndike, A. S., D. A. Rothrock, G. A. Maykut, and R. Colony (1975), The thickness distribution of sea ice, *J. Geophys. Res.*, *80*(33), 4501–4513.

Walker, G. (2006), The tipping point of the iceberg, *Nature*, *441*, 802–805.

Winton, M. (2006), Does the Arctic sea ice have a tipping point?, *Geophys. Res. Lett.*, *33*, L23504, doi:10.1029/2006GL028017.

Zhang, J., and D. A. Rothrock (2001), A thickness and enthalpy distribution sea-ice model, *J. Phys. Oceanogr.*, *31*, 2986–3001.

Zhang, J., and D. A. Rothrock (2003), Modeling global sea ice with a thickness and enthalpy distribution model in generalized curvilinear coordinates, *Mon. Weather Rev.*, *131*, 681–697.

Zhang, J., and D. A. Rothrock (2005), The effect of sea ice rheology in numerical investigations of climate, *J. Geophys. Res.*, *110*, C08014, doi:10.1029/2004JC002599.

Zhang, X., and J. E. Walsh (2006), Toward a seasonally ice-covered Arctic Ocean: Scenarios from the IPCC AR4 model simulations, *J. Clim.*, *19*, 1730–1747.

C. M. Bitz and P. Hezel, Department of Atmospheric Sciences, University of Washington, Seattle, Washington 98195-1640, USA.

R. W. Lindsay and H. L. Stern, Polar Science Center, Applied Physics Laboratory, University of Washington, Seattle, Washington 98105, USA. (harry@apl.washington.edu)

# Analysis of Arctic Sea Ice Anomalies in a Coupled Model Control Simulation

Richard I. Cullather

*Lamont-Doherty Earth Observatory of Columbia University, Palisades, New York, USA*

L.-Bruno Tremblay

*Department of Atmospheric and Oceanic Sciences, McGill University, Montreal, Quebec, Canada*

In a 600-year control run simulation of the Community Climate System Model, version 3 (CCSM3) of the National Center for Atmospheric Research, three events of extraordinary minimum September sea ice extent are identified, from which the system recovers within a few years. The control simulation is run in a stable climate mode using constant 1990 solar, trace gas, and aerosol forcing. The identified events are all of similar magnitude to the observed 2007 record low Arctic ice cover. In the first event (simulation year 451), the record low ice extent coincides with an extreme, cyclonic phase of atmospheric circulation and follows a 10-year period of steady decline in Northern Hemisphere sea ice volume that is associated with increased winter ice export. The ice extent recovered the following year when both the ice export and summer melt returned to more normal values. However, Arctic ice volume did not fully recover until 20 years after the event. In the second event (model year 490), the system was again preconditioned by a 10-year decline in Arctic sea ice volume because of larger than normal ice export. The event itself was associated with a coincident peak in early summer melt and ice export. During the recovery phase, the ice extent again rapidly returned to normal while the sea ice volume continued to decline for another 10 years, to its lowest value of the entire simulation. The recovery in Northern Hemisphere sea ice volume is associated with periods of reduced wintertime ice export through Fram Strait. The third event (model year 556–558) is characterized by a longer-lived anomaly (3 years). The system was not preconditioned by low ice volume, indicating that redistribution of sea ice (rather than ice mass loss) played an important role. Recovery for the third event is associated with a period of low ice export associated with a rapid return of an expanded Beaufort Gyre, which serves to recirculate sea ice within the Arctic Ocean. Notwithstanding known climate model deficiencies, the control

Arctic Sea Ice Decline: Observations, Projections, Mechanisms, and Implications
Geophysical Monograph Series 180
Copyright 2008 by the American Geophysical Union.
10.1029/180GM13

simulation suggests that Arctic sea ice extent anomalies of the magnitude observed in the September 2007 minimum are plausible under 1990 forcing conditions. It is reasonable to conclude that more severe anomalies are achievable with more acute forcing conditions that are in excess of those of 1990.

## 1. INTRODUCTION

The Arctic Ocean Basin (Figure 1) has received increased attention in recent years because of observed, rapid changes in its climate [*Serreze et al.*, 2000]. Its perennial sea ice cover is a signature characteristic. As shown in Plate 1, Northern Hemisphere sea ice extent has historically varied seasonally from a maximum cover of over $15.3 \times 10^6$ km$^2$ in

March to a minimum of $6.8 \times 10^6$ km$^2$ in September [*Serreze et al.*, 2007]. For the observational record, a sustained ice pack averaging about 2 to 3 m in thickness [e.g., *Bourke and Garrett*, 1987] has continuously covered the central Arctic Ocean and portions of the surrounding marginal seas. Over the record of satellite passive microwave observations of sea ice cover beginning in 1979, the average sea ice extent has trended downward at a rate of about 3% per decade [*Par-*

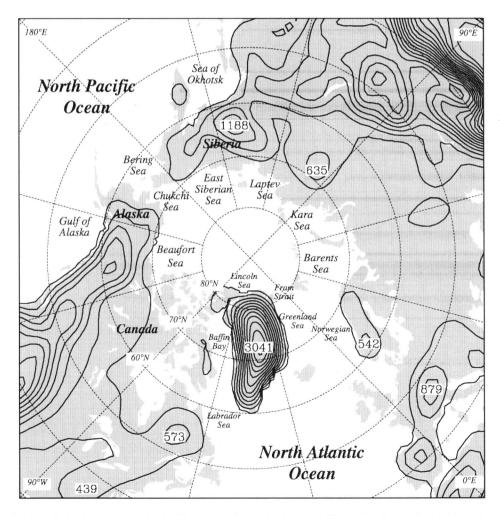

**Figure 1.** Map of the Arctic Ocean basin. The orography of the National Center for Atmospheric Research (NCAR) CCSM3 is contoured every 250 m with local maxima indicated.

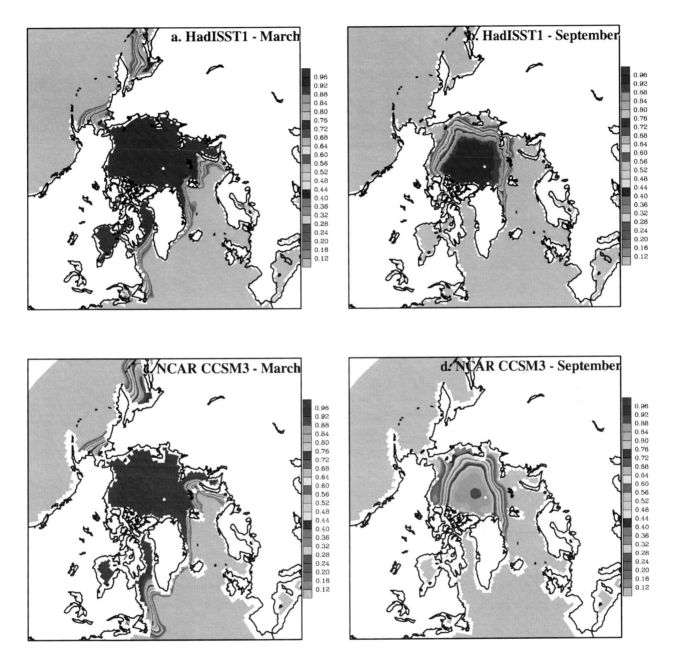

**Plate 1.** HadISST1 [*Rayner et al.*, 2003] 1979–2006 average sea ice concentration for (a) March and (b) September and average sea ice extent for the NCAR CCSM3 for (c) March and (d) September for IPCC 1990 control simulation. The ice fraction is shaded every 4% beginning with 12% coverage.

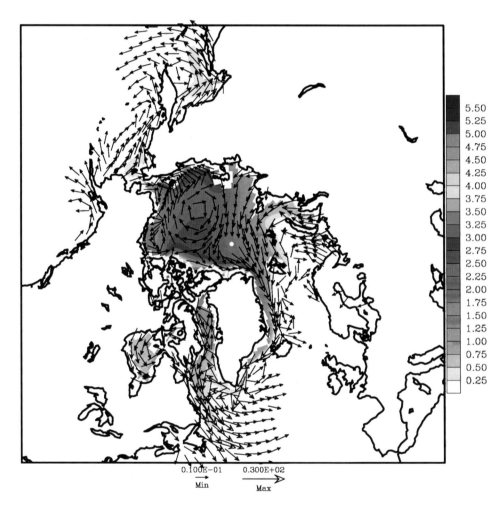

**Plate 2.** Average simulated ice thickness (shaded) and ice velocity for the NCAR CCSM3 IPCC 1990 control simulation averaged for the months of November through April. Ice thickness is shaded every 0.25 m. The maximum ice velocity vector shown corresponds to 0.3 m s$^{-1}$.

*kinson*, 2006], with greater losses of more than 7% per decade for the September minimum [*Lemke et al.*, 2007]. Since 2002, the Arctic Ocean has experienced an unprecedented period of continuously low September ice cover, with larger reductions in more recent years [*Stroeve et al.*, 2008, 2005; *Serreze et al.*, 2003]. As the September minimum roughly denotes the ice cover that has survived the melt season, it may be seen that the downward trend in September ice cover corresponds to a reduction in thick, multiyear ice, which is being replaced by a thinner, seasonal ice cover. Global, coupled models used in prognostic simulations have also indicated that the Arctic perennial ice cover is in jeopardy. The most recent survey by the Intergovernmental Panel on Climate Change (IPCC) has indicated that many models project a seasonally ice-free Arctic Ocean by the end of the 21st century under the high-emission A2 scenario for anthropogenic forcing [*Meehl et al.*, 2007; *Zhang and Walsh*, 2006].

The manner in which an increasing anthropogenic forcing becomes manifest in the climate system is the subject of some discussion within the literature. The Arctic multiyear ice pack may be abated either through an increase in ice export out of the basin, a decrease in wintertime ice growth, an increase summertime melting, or some combination of the three [*Tremblay et al.*, 2007]. These processes may be mapped onto aspects of the Arctic climate system which include the atmospheric circulation, radiative processes, and oceanic poleward energy transport.

The overlying atmosphere has a significant influence on the sea ice cover by imparting a wind stress on the surface, thus affecting the spatial distribution and export, and as a medium for the poleward transport of energy from lower latitudes. In documenting the September 2002 minimum, at the time a record low for the satellite era, *Serreze et al.* [2003] noted anomalies in the summertime atmospheric circulation including unusually low pressure over the central Arctic Basin. Through Ekman transport, low surface pressure tends to cause the ice pack to diverge, which would produce openings and increased melting via the ice-albedo feedback mechanism after dispersal. Subsequent years have not followed this pattern [*Stroeve et al.*, 2005]. Rather, high pressure at the surface has generally been noted in summer, which would result in consolidation and the rapid removal of ice from marginal sea embayments [*Ogi and Wallace*, 2007], as well as a reduction in cloud cover, leading to an increased shortwave radiative flux absorbed at the surface. In contrast to the summertime focus of *Serreze et al.* [2003], there has been considerable interest in the literature of the last decade in changes to the mean wintertime atmospheric circulation toward the positive index phase, or the "Greenland Below" condition of the North Atlantic Oscillation (NAO) [*Hurrell*, 1995]. The NAO is an atmospheric teleconnection pattern

that is characterized by variability in surface westerlies over the North Atlantic. This positive index phase was found to result in a decreased flux of ice that traverses around the Beaufort Gyre from the western to the eastern Arctic and an increased ice export through Fram Strait [*Kwok*, 2000; *Rigor et al.*, 2002]. In disrupting the Beaufort Gyre circulation, thick multiyear ice in the pack is spatially confined to the Canadian Archipelago and away from the Alaskan and Eurasian coastlines. The increased export through Fram Strait produces a direct loss of multiyear ice from the basin. The predominant wintertime mode of atmospheric circulation reached extreme positive values in the late 1980s and early 1990s but has trended toward neutral in more recent years. In this context, the recent reduction in summer ice cover since 2002 is then seen as a dilatory impact of anomalous wintertime atmospheric circulation [*Rigor et al.*, 2002].

An examination of the sea ice cover time series in comparison to observed radiative fluxes and other variables was conducted by *Francis et al.* [2005]. Satellite-derived downwelling longwave flux was found to account for 40% of the variability in the September sea ice extent. *Francis et al.* [2005] suggest that longwave radiatively driven melting may have eroded the perennial sea ice cover in more recent years after dynamically forced reductions in ice volume. The result is consistent with observed increases in springtime cloudiness over the Arctic [*Wang and Key*, 2005].

Finally, oceanographic research in the Arctic in the 1990s was highlighted by the discovery of warming in the Atlantic Layer, located at about 300 m in depth [*Quadfasel et al.*, 1991; *Carmack et al.*, 1995; *Grotenfendt et al.*, 1998]. This layer of inflow water from relatively lower latitudes of the North Atlantic is insulated from the surface sea ice pack by a cold halocline layer, which itself was found to be in retreat in the early 1990s [*Steele and Boyd*, 1998]. Large-scale basal melting of the ice pack would occur if the halocline was significantly weakened or removed. The inflow of Atlantic warm water into the Arctic Basin is, in turn, influenced by atmospheric and oceanic processes local to the North Atlantic, including trends in atmospheric circulation and the oceanic meridional overturning circulation. Changes in the temperature of the Atlantic Layer have been found to be well correlated with atmospheric indices such as the NAO [*Dickson et al.*, 2000; *Jones*, 2001]. Submarine data [*Boyd et al.*, 2002] have suggested that the cold halocline layer recovered over the period 1998–2000 after declining in the early 1990s. More recent studies have begun to comprehensively assess the influence of the warm Atlantic layer on the overlying ice pack, even in the presence of a robust cold halocline. For example, observations in the Arctic using ice-ocean buoys have indicated that weather systems can produce mixing that extends through the halocline [*Yang et al.*, 2004]. A study of observations made north of Severnaya Zemlya in the 1970s and 1980s estimated upward heat fluxes of 4 to 6 W m$^{-2}$ which would affect the overlying sea ice cover [*Walsh et al.*, 2007]. The mechanisms through which warm Atlantic and Pacific water inflows into the Arctic may affect the sea ice cover have been identified as known gaps in the present understanding (Arctic Research Consortium of the United States, Arctic observation integration workshops report, 55 pp., Study of Environ. Arct. Change Proj. Off., Fairbanks, Alaska, 2008, available at http://www.arcus.org/SEARCH/ meetings/2008/aow/report.php).

A goal of this study is to examine the relative roles of the various climate components in coupled model simulations of the Arctic climate. Numerical general circulation models represent an integration of our knowledge of the processes that are associated with the Arctic climate system and that are essential for prognostic assessments. These processes are simulated in a manner that is largely absent of a presupposed prioritization and is physically consistent in conserving mass, energy, and momentum. Global, coupled models also incorporate remote forcings from outside the Arctic Basin. Understanding the output of a climate model may shed additional light on the forcing mechanisms that are affecting the contemporary Arctic. Currently, there is great interest in prognostic simulations of the 21st century because of the intermodel variability in the timing and conditions that are associated with the transition from perennial to seasonal ice cover. For example, *Holland et al.* [2006a] found a multi-staged transition in which different mechanisms dominated at different periods of time, producing a stair-stepped time series of late summer ice concentration. These mechanisms may be viewed to occur as a result of increased interannual variability that is superimposed on a forced signal as the sea ice thins [*Holland et al.*, this volume].

The approach here is to examine a control simulation from a state-of-the-art climate model using constant 1990 trace gas forcing over an extended period of time. Extended control simulations on the order of 1000 years enable a comparison of the relation between variables over a period time of time that is significantly longer than the observational record. In particular, we focus on a 350-year segment of the time series in which possible spin-up effects from the beginning of the simulation may be discounted. Over this time period, the average conditions closely resemble those found in observations of the contemporary Arctic climate, both in the annual mean and in the seasonal cycle. However, this simulation shows drastic declines in sea ice extent occurring 3 to 4 times in the time series. These episodic events are comparable to the recent dramatic decrease in ice extent observed in September 2007, and yet the simulated ice recovers to "normal" sea ice extent conditions within a few years. This contrasts

with anthropogenically forced simulations of the 21st century from the same climate model, which produces virtually ice-free summer Arctic conditions around 2040–2055 from which the system never recovers [*Holland et al.*, 2006a]. An understanding of the mechanisms that are responsible for the decline and recovery of the sea ice cover in control simulations when compared with those of the forced simulations is of particular interest and is timely. Indeed this gives a different perspective on the question of what fraction of the observed recent change in sea ice extent is forced and what fraction is coming from natural variability. Some questions of interest to be examined in this study are as follows:

- Do these events of substantial sea ice cover loss all occur in a similar manner?
- How do anomalous events evolve? Does preconditioning play a role?
- Do these anomalies in ice cover coincide with extreme conditions in atmospheric or oceanic circulation?
- How rapidly does the ice pack recover?

This paper is divided as follows: Section 2 describes the Community Climate System Model, version 3 (CCSM3), which is used in this study. A brief evaluation of this model in comparison to the contemporary Arctic climate is also presented in this section. A description of the mechanisms responsible for the decline and recovery in Northern Hemisphere sea ice cover is given in section 3. Finally, a discussion of the findings of this study and the implications of these results for the contemporary Arctic is given in section 4.

## 2. THE COMMUNITY CLIMATE SYSTEM MODEL, VERSION 3

The CCSM3 is a fully coupled climate model composed of atmosphere, ocean, sea ice, and land surface components developed by the National Center for Atmospheric Research (NCAR) [*Collins et al.*, 2006a]. The model was released in May 2004 and was a participating model of the Fourth IPCC Assessment [*Meehl et al.*, 2007]. The model has incorporated a number of updates from the previous version that was described by *Kiehl and Gent* [2004] and includes an improved representation of cloud process and the mixed ocean layer [*Holland et al.*, 2006b]. The atmospheric component is referred to as the Community Atmosphere Model, or CAM3, and is based on an Eulerian spectral dynamical core [*Collins et al.*, 2006b]. Standard simulations use a configuration with 26 hybrid sigma levels in the vertical and horizontal resolution with triangular truncation of 85 wave numbers. This is equivalent to an equatorial grid spacing of about 1.4°. The orography used in CAM3 is shown in

Figure 1 for the Arctic region. While topographic features are greatly smoothed such as in the North American Rockies, the Alps, and in particular the Ural Mountains, there is a good representation of the Putoran Plateau in central Russia and mountain ranges in Siberia. The ocean component is based on the Parallel Ocean Program (POP), originally developed at Los Alamos National Laboratory [*Smith and Gent*, 2002]. The POP model is configured for 40 levels in the vertical and a horizontal grid spacing of approximately $1° × 1°$ but with a grid such that the North Pole is located within the land surface of Greenland in order to avoid the difficulty of converging meridians within the ocean domain. In the Arctic, the Bering Strait and the Canadian Arctic Archipelago, with the exception of the Nares Strait between Greenland and Ellesmere Island, are open. The Community Sea Ice Model (CSIM) uses the same grid as the POP model. The CSIM is documented by *Briegleb et al.* [2004] and uses a dynamic-thermodynamic scheme that includes a subgrid-scale ice thickness distribution [*Bitz et al.*, 2001; *Lipscomb*, 2001]. The ice rheology is based on the elastic-viscous plastic method of *Hunke and Dukowicz* [1997] with a parameterization of subgrid-scale ridging and rafting following *Rothrock* [1975] and *Thorndike et al.* [1975]. Thermodynamics are from *Bitz and Lipscomb* [1999], which has multiple vertical layers and accounts for the thermodynamic influences of brine pockets within the ice cover.

The performance of the CCSM3 in northern high latitudes has been evaluated in several recent studies. *Holland et al.* [2006b] noted the Northern Hemisphere sea ice spatial distribution and seasonal cycle are very reasonable in comparison to the passive microwave observational record (Plate 1), although the wintertime ice cover is too excessive in the Labrador Sea and in the Sea of Okhotsk. These problems have since been corrected using an improved ocean viscosity parameterization [*Jochum et al.*, 2008]. *Stroeve et al.* [2007] found that the observed 20th century downward trend in late summer sea ice cover of $-7.8 ± 0.6\%$ per decade was most closely approximated by an ensemble member of the CCSM3 ($-5.4 ± 0.4\%$ per decade) of the IPCC simulations examined. The simulated ice thickness is also found to be in close agreement with observation, with average values greater than 4 m adjacent to the Canadian Arctic Archipelago and values of about 3 m in the central Arctic [*Holland et al.*, 2006c]. This is shown in Plate 2 for winter months, along with averaged ice drift vectors. The average ice thickness distribution compares favorably with observational compilations such as those of *Bourke and Garrett* [1987]. *Gerdes and Köberle* [2007] also found that the ice thickness distribution compared favorably to hindcast simulations of the Arctic Ocean Model Intercomparison Project. For the control simulation, the ice velocity vectors shown in Plate 2 are roughly in agreement with observation with a well-defined Beaufort Gyre circulation centered on the Pacific side of the Arctic and a transpolar drift pattern with major export through the Fram Strait. The difficulty in achieving a realistic sea ice cover such as that found in CCSM3 is underscored by comparisons with other climate models. *Zhang and Walsh* [2006] note the large diversity in the treatment of sea ice in contemporary models. Zhang and Walsh found large uncertainties in simulations of the seasonal cycle of Arctic sea ice. This result is echoed in the Fourth IPCC Assessment report, which indicated that improvement in the representation of sea ice over previous assessments is not obvious and that "... in many models the regional distribution of sea ice is poorly simulated..." [*Randall et al.*, 2007, p. 616].

While the 20th century sea ice distribution and seasonal cycle in CCSM3 simulations are exceptional, other aspects of the Arctic climate system simulated by the CCSM3 have been found to differ significantly from observation. These differences have been noted in multimodel assessments of the mean atmospheric state and the surface energy budget [e.g., *Chapman and Walsh*, 2007; *Sorteberg et al.*, 2007]. In particular, *Chapman and Walsh* [2007] found the CCSM3 annual mean sea level pressure distribution over the central Arctic to be 5 to 10 hPa less than observed values and a notable outlier in comparison to other models. Evaluations performed for this study confirm these results and indicate that the 20th century wintertime climatology of the CCSM3 does not contain a closed Beaufort High sea level pressure as is generally found in analyzed fields [*DeWeaver and Bitz*, 2006; see *Chapman and Walsh*, 2007, Figure 7]. *DeWeaver and Bitz* [2006] found that the vertical structure of the Beaufort High in the CCSM3 is baroclinic as opposed to the barotropic configuration found in reanalyses and that the strength of the simulated high is slightly stronger at a lower model horizontal resolution.

Two other aspects of the CCSM3 model are briefly noted here. *Gorodetskaya et al.* [2008] evaluated the simulated Arctic cloud properties in comparison to in situ observations from the Surface Heat Budget of the Arctic Ocean (SHEBA) experiment [*Uttal et al.*, 2002]. In contrast to two other climate models examined in the study, the CCSM3 was found to contain a high liquid water content, which exceeded observed values. This produced an optically thick cloud cover which decreased the amount of shortwave radiation absorbed by the Arctic Ocean during summer months. However, this was found to be more than offset by an increased downwelling longwave flux throughout the year [*Gorodetskaya and Tremblay*, this volume]. Averaged surface fluxes using the IPCC control simulation forced with constant 1990 trace gas values are compared with SHEBA station values

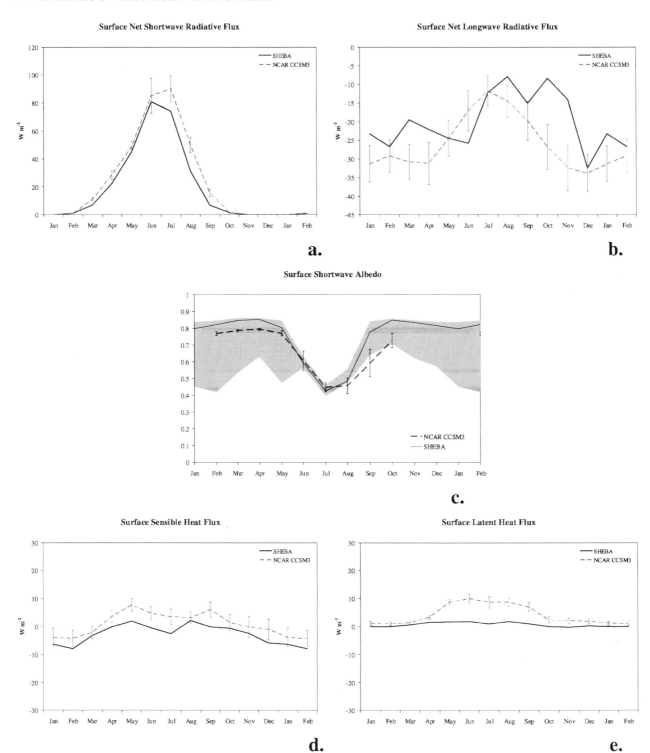

**Figure 2.** Monthly averaged surface energy budget components from observed SHEBA station data and corresponding values from the NCAR CCSM3 1990 control simulation. Model values are from the grid box colocated with the monthly mean position of the SHEBA station. Error bars correspond to the standard deviation over simulation years 250–599. Shown are the (a) net (downwelling minus upwelling) shortwave and (b) longwave radiative fluxes in W m$^{-2}$, (c) albedo, and (d) sensible and (e) latent heat fluxes in W m$^{-2}$. Shaded area of the albedo plot indicates the range of SHEBA observations.

[*Persson et al.*, 2002] in Figure 2. The error bars shown in Figure 2 indicate the standard deviation over simulation years 250–599. Although the simulated net shortwave radiative flux exceeds observations in July and August by 15 to 18 W m$^{-2}$ (Figure 2a), this is a significantly smaller overestimate than for other climate models examined by *Gorodetskaya et al.* [2008] for the SHEBA location (not shown). Additionally, the simulated surface albedo is within the range of observations used in SHEBA. The magnitude of the simulated net longwave flux shown in Figure 2b is generally greater than observation. Simulated sensible and latent heat fluxes are small components of the total energy budget in agreement with observation. However, the pronounced annual cycle in the latent heat flux is not well reproduced (Figures 2d and 2e).

Finally, *Tremblay et al.* [2007] examined simulated Arctic Ocean temperature and salinity profiles in comparison to four other climate models. The CCSM3 was found to have a good representation of surface stratification. This is important for simulating the effects of Atlantic warm water intrusions into the Arctic, which are typically insulated from the surface ice pack by a sharply defined halocline. However, the cold halocline layer in the model is absent (as in every other model evaluated), and the Atlantic waters were found to be 1°C too warm and somewhat saltier.

In summary, Arctic sea ice characteristics and 20th century trends are well represented in the CCSM3; however, this is achieved with anomalous cyclonic circulation in the atmosphere and a larger poleward Atlantic oceanic heat transport than observed. These discrepancies highlight the difficulties in simulating the Arctic climate system.

## 3. RESULTS

Figure 3 (top) shows the time series of the September Northern Hemisphere sea ice extent from the CCSM3 control simulation for model years 250 through 599 using constant 1990 solar, trace gas, and aerosol forcing. For reference, the record maximum and minimum values of September sea ice extent from the HadISST1 climatology [*Rayner et al.*, 2003] from the period 1979–2006 are shown along with the record September 2007 minimum extent. There is considerable variability in the time series, and occasionally the 2007 minimum extent value is reached. In Figure 3 (bottom), it may be seen that the interannual variability in September sea ice extent is larger than for March, the month of maximum ice extent. The standard deviation is $0.32 \times 10^6$ km$^2$ for the March time series and $0.54 \times 10^6$ km$^2$ for the September time series. This is in agreement with the available satellite record, which shows greater variability and trends in the September sea ice extent [*Serreze et al.*, 2007]. There is a correlation between 20-year

running means of winter and summer sea ice extent, suggesting the presence of low-frequency climate variability that affects the ice cover in both seasons. The correlation between the two filtered time series is 0.28 over the full period. If a period of low March ice cover during the years 350–400 is excluded, however, this correlation is then 0.61. This period of low March ice cover (year 350–400) is associated with anomalous concentrations in the Sea of Okhotsk.

From visual inspection of Figure 3, it may be seen that there are extended periods of both positive and negative trends; it is noted that there are extended periods of the simulation where the magnitude and trend of the September minimum operates within the bounds of 1979–2006 observed sea ice extent. For example, September ice extent decreases for the years 300 to 340 at a rate of 6% per decade. This is similar to the observed trend of 8% per decade over the 1979–2006 satellite era [*Stroeve et al.*, 2007]. Of particular interest are three marked events occurring in model years 451, 490, and 556–558. Each of these events is approximately 3 standard deviations less than the simulated long-term September mean extent, and they roughly match the observed dramatic minimum which occurred in September 2007, with the first two events being slightly less than this value. These events largely occur in the absence of significant ice anomalies during the preceding winter months. For example, the third event occurs over three consecutive September months in which ice anomalies are near zero or are positive in the intervening months of November through April of each year, as referenced to the 350-year simulation period examined. The monthly anomalies for each event occur in July through October, with September incurring the largest anomaly. The spatial distribution of these three events is shown in Plate 3. For reference, the simulated long-term average September extent is shown as a red line.

From Plate 3, it may be seen that the ice loss for the first two events occurs primarily on the Pacific side of the basin. For model year 451, the losses are particularly evident in the Chukchi, Laptev, and East Siberian seas. For model year 490, September ice cover is notably absent in the Beaufort, Chukchi, East Siberian, and Laptev seas, while the third event shows losses primarily adjacent to the Siberian coastline and in the central Arctic. The spatial distribution of these losses are reminiscent of dramatic reductions in the summertime ice cover from the recent observational record of 2002–2007 [*Ogi and Wallace*, 2007]. These simulated events are now examined in more detail.

### 3.1. Simulation Year 451 Event

As described previously, this event is characterized by substantial ice cover losses throughout the Arctic Basin but

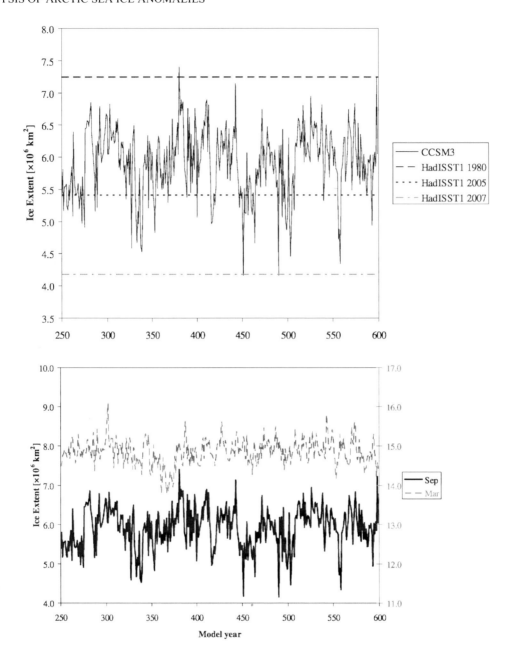

**Figure 3.** (top) Time series of September monthly sea ice extent (concentration greater than 20%) for CCSM3 control simulation, in $10^6$ km². The observed minimum (2005) and maximum (1980) September sea ice extent from the 1979–2006 period is indicated along with the record minimum extent from 2007. (bottom) Time series of September monthly sea ice extent compared with March. Note the right vertical axis corresponding to the time series for March.

with particular emphasis in the Chukchi, East Siberian, and Laptev seas (Plate 3a). Figure 4 shows time series of simulated climate variables that are associated with this event as well as the time series of the September sea ice extent anomaly. The purpose of Figure 4 is to present an overview of the model variables that have been examined, which necessarily requires a prioritization of the results but which shows the

evolution of the simulation over the time prior to and after the period of interest. The variables shown characterize the Arctic sea ice cover, the wintertime export, and the monthly atmospheric circulation.

The time series of the first principal component of Northern Hemisphere sea level pressure variability, otherwise known as the Arctic Oscillation (AO) index [*Thompson and*

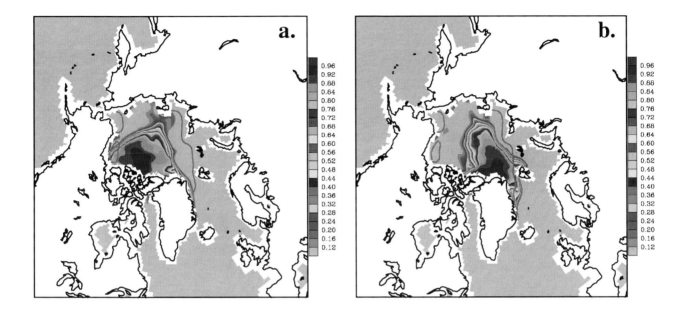

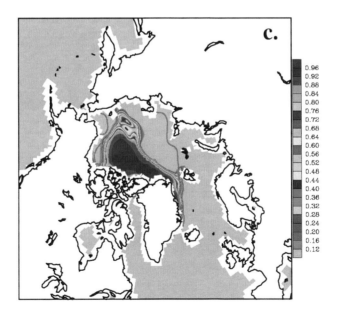

**Plate 3.** CCSM3 1990 control simulation September sea ice concentration for simulation years (a) 451, (b) 490, and (c) 558. The red line indicates the 20% contour of years 250–599 average September sea ice concentration.

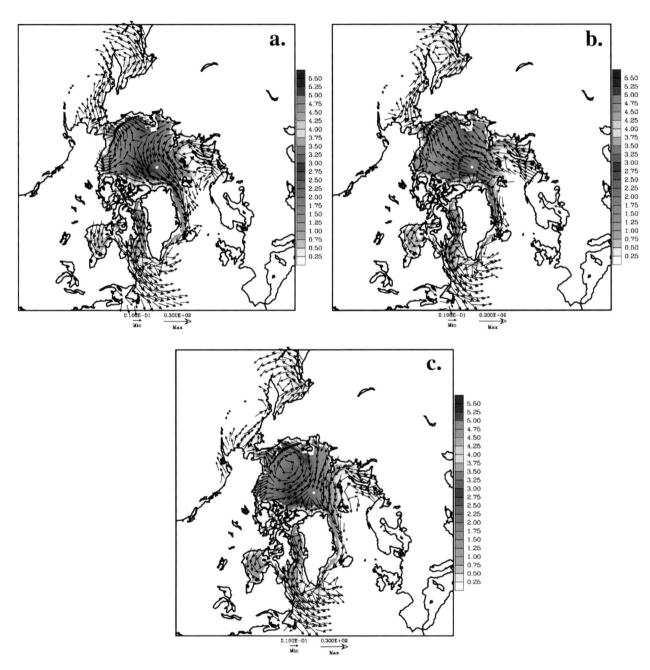

**Plate 4.** Simulated ice thickness (shaded) and ice velocity for the CCSM3 control simulation averaged for the months of November through April for model years (a) 451, (b) 452, and (c) 453. Ice thickness is shaded every 0.25 m. The maximum ice velocity vector shown corresponds to 0.3 m s$^{-1}$.

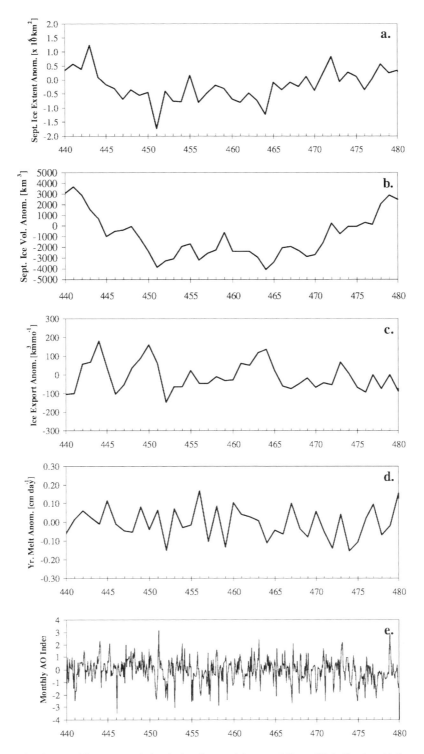

**Figure 4.** Time series from CCSM3 control simulation for model years 440 to 480 indicating (a) September Northern Hemisphere sea ice extent expressed as an anomaly of the averaging period 250–599 in $10^6$ km², (b) September Northern Hemisphere sea ice volume anomaly in km³, (c) Fram Strait southward ice export anomaly averaged for November through April in km³ month⁻¹, (d) January through December sea ice melt anomaly in cm d⁻¹, and (e) the first principal component of the monthly Northern Hemisphere sea level pressure anomaly as computed over the period 250–599 and representing 23.6% of the variance.

*Wallace*, 1998], is shown in Figure 4e. The time series of the AO index is closely related to the NAO [*Deser*, 2000] and describes 24% of the variance in the monthly Northern Hemisphere extratropical sea level pressure field over the 350-year period computed. From Figures 4a and 4e, it is seen that the event for simulation year 451 coincides with a brief period of extreme positive values in the AO index in February and March of that year. The AO index for February of simulation year 451 is the largest monthly value in over a 200-year period of the model simulation. This corresponds to an increase in wintertime ice velocities in the Lincoln Sea and through Fram Strait (see Plate 4a). For reference, Plate 4a may be compared with the average ice velocity field shown in Plate 2. Plate 4a also indicates anomalous export through the Barents Sea. However, the total ice volume for the Arctic during this event is already less than average values. For example, regions of ice thickness greater than 2.5 m cover a much smaller region of the central Arctic during simulation year 451 than in the averaged depiction. This may also be seen in the time series of ice volume shown in Figure 4b. In this time series, the total ice volume anomaly for the Arctic decreased from a long-term maximum at the beginning of model year 441 to a negative anomaly of approximately 3700 $km^3$ in simulation year 451. It may then be inferred that the Arctic ice volume had been preconditioned to thinner ice in the years prior to 451 and that the extraordinary minimum ice extent of $4.2 \times 10^6$ $km^2$ for that September is the result of both dynamic and thermodynamic processes. Using the long-term average annual extent of sea ice, the corresponding ice thickness anomaly is equivalent to about 0.5 m (see Figure 4b).

In Figure 4c, the time series of wintertime ice export through Fram Strait is shown. The export is computed on the CCSM3 native ocean grid. For monthly values (not shown), ice export through the Fram Strait corresponds closely with the AO index, and very large export values of greater than 800 $km^3$ month$^{-1}$ correspond to the large values of the February and March AO index for the year 451. However, on a seasonal average, the Fram Strait ice export was not exceptional for the year 451, as seen in Figure 4c. Figure 4c shows that the ice volume preconditioning occurred in two periods of anomalous export and melting. In the export time series, these 3- to 5-year periods occurred in simulation years centered on years 444 and 450. It is concluded that the ice extent anomaly of 451 is the result of both thermodynamic and dynamic processes occurring in a preconditioned system.

In Figure 4a, it is seen that the ice extent recovers considerably in the following summer. Negative ice extent anomalies for the years 452–454 range from $0.4 \times 10^6$ to $0.8 \times 10^6$ $km^2$, and by year 455 the ice extent anomaly is positive. This contrasts with the ice volume anomaly, which averages greater

than (negative) 2600 $km^3$ over the period 451–464. Subsequent recovery over simulation years 465–472 is associated with decreases in winter ice export and sea ice melt (Figures 4c and 4d). Sea ice melt is the sum of top, lateral, and basal terms and is averaged over the central Arctic Ocean poleward of 80°N in the Atlantic Sector (90°W to 105°E) and poleward of 70°N elsewhere. The melt curve in Figure 4d shows a great deal of year-to-year variability but typically lower values during this period of recovery in ice volume.

In terms of ice motion, the recovery period immediately after the ice extent anomaly is dramatic. Plate 4b indicates a highly anomalous velocity pattern with surface ice everywhere drifting from the Fram Strait to the Bering Strait in the presence of an expanded Siberian High. Simulation year 453 sees a return to a more normal circulation pattern with the return of a dominant Beaufort Gyre in the Canada Basin (Plate 4c). The contrast in velocity patterns between sea ice extent depletion and recovery years emphasizes the role of the Beaufort Gyre in maintaining the cover during the summer months. Oceanic heat transports (not shown) computed for the major passages in the North Atlantic do not show significant variability during this event.

Arctic summertime conditions during simulation year 451 are dominated by the sea level pressure anomaly patterns in the prior winter and concurrent summer. Pan-Arctic averages of near-surface air temperature and precipitation for the June–September months of year 451 are found to be near climatology values. In Figure 5, the winter and summer fields of sea level pressure anomaly are contoured, and hatched areas indicate statistical significance. The patterns emphasize the role of the wintertime circulation, with low-pressure anomalies in excess of 6 hPa over a large region of the central Arctic Ocean. In summer, high pressure is located over the Canadian Arctic Archipelago; however, the Arctic Ocean is again shaded by negative sea level pressure anomalies. The event may be seen as more analogous to the September 2002 event as described by *Serreze et al.* [2003], in which persistent low pressure dispersed the ice pack, as opposed to more recent events where high pressure over the Arctic Ocean (and increased downwelling shortwave radiation absorbed at the surface) played a more dominant role. As the summer air temperatures are found to be near climatology and there is no significant melt signal, it would seem plausible that the event was dynamically induced during a period of low sea ice concentration.

### 3.2. Simulation Year 490 Event

As described previously, the spatial pattern of the September sea ice minimum event for simulation year 490 is shown in Plate 3b. Similar to the year 451 event, year 490 sea ice

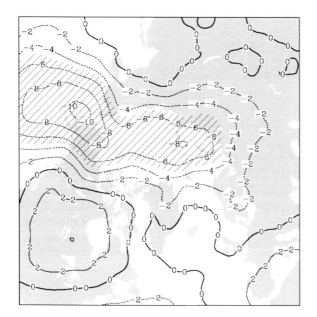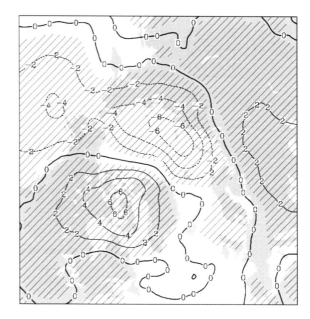

**Figure 5.** Sea level pressure anomaly from the CCSM control simulation for model year 451, averaged for (left) November through April and (right) May through September. The contour interval is 2 hPa. Hatched areas denote significance at the 95th percentile.

anomalies for September are largely focused on the Pacific side of the Arctic, with significant losses in the Beaufort and Chukchi seas as well as in all of the Eurasian marginal seas. Figure 6 shows the time series of several simulated Arctic climate variables along with the ice extent anomalies for the years before and after year 490. As seen in Figure 6a, the ice extent anomaly abruptly drops from near climatology values to a local minimum of $0.18 \times 10^6$ km² for simulation year 490. In the following September, the ice extent is slightly larger than the long-term average, and there are subsequent positive and negative anomalies over the following 20 years.

This variability in the ice extent time series is in substantial contrast with the time series of ice volume. As seen in Figure 6b, the extent minimum occurs during a long-term downward trend in ice volume which begins in year 481 and declines by approximately 7400 km³ over the following 18 years. The ice volume minimum in the year 499 of 8188 km³ (which is plotted in Figure 6b as an anomaly of –4916 km³) is the lowest value in the 350-year time series that is considered here. The decrease in ice volume is characterized by above-average summer ice melt throughout the period and particularly during the years 496–499. Positive AO index values over the time period (not shown) and enhanced ice export through Fram Strait (Figure 6c) play a lesser role in the decrease of ice volume over the period 481 to 499. As

with the previous minimum ice extent event, the time series of ocean heat transport is unremarkable.

In the presence of weak atmospheric wind forcing, a near-normal ice export, and oceanic heat transport but increased melting, this event may be characterized as a thermodynamic event which occurred during a period of low ice volume. Shown in Figure 6e is the summertime net surface radiative flux that has been averaged over the central Arctic region as defined earlier for ice melt. This curve peaks in year 490 and is approximately 5 W m⁻² greater than found in the prior decade. A great deal of this peak in the radiative flux is associated with the shortwave component. The net surface shortwave flux exceeds the climatological value by more than 5 W m⁻² in May of that year and by more than 20 W m⁻² in June.

The effects of the increased radiative flux on the sea ice cover are illustrated in Figure 7, which shows the annual cycle of sea ice melt compared with long-term average of the simulation. For reference, adjacent years in the time series are also plotted. For example, it may be seen in Figure 6d that ice melt was anomalously large in simulation year 485. This is reflected in Figure 7; however, the anomaly is centered with the climatological annual cycle. In contrast, the melting of sea ice in year 490 began much earlier and was anomalously large in March (which may not be noticeable in Figure 7). By complementary association, the presence

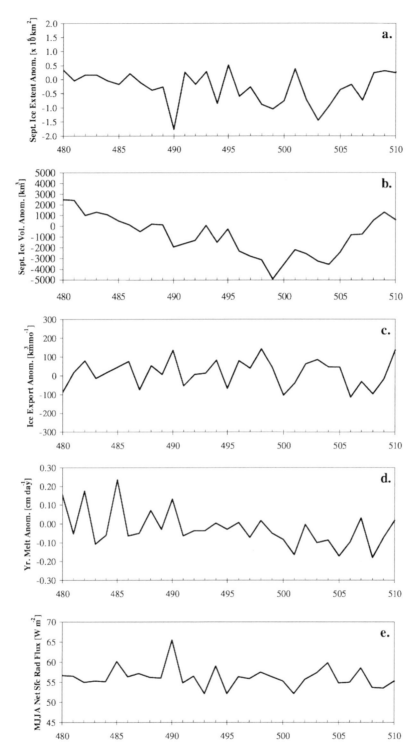

**Figure 6.** Time series from CCSM3 control simulation for model years 480 to 510 indicating (a) September Northern Hemisphere sea ice extent expressed as an anomaly of the averaging period 250–599 in 10⁶ km², (b) September Northern Hemisphere sea ice volume anomaly in km³, (c) Fram Strait southward ice export anomaly for November through April in km³ month⁻¹, (d) January through December central Arctic sea ice melt anomaly in cm d⁻¹, and (e) the July through September net surface radiative flux in W m⁻² averaged for the region from 70°N to 90°N.

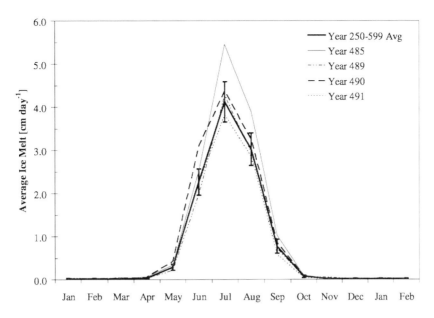

**Figure 7.** Simulated average rate of sea ice melt over the central Arctic basin for CCSM3 control simulation years 485, 489, 490, and 491, and for the long-term average, in cm d⁻¹. Error bars denote the standard deviation over the period 250–599.

of early melt implies that sea ice growth in the Arctic ended earlier in the preceding winter. These conditions of an early spring, which occurred in the presence of declining sea ice volume, appear to have prompted the remarkable September sea ice extent minimum in the simulation.

A dramatic picture of conditions in simulation year 490 is seen in Figure 8, which shows the summertime near-surface air temperature anomalies. While air temperatures are likely restricted by the presence of sea ice and cold water temperatures in the central Arctic, values greater than 3°C

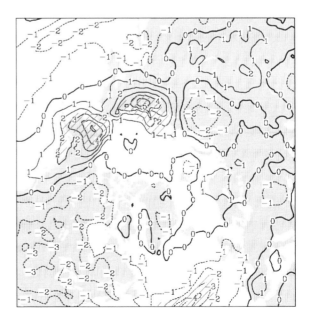

**Figure 8.** Stevenson screen reference height air temperature anomaly from the CCSM control simulation for model years (left) 489 and (right) 490, averaged for the months of May through September. The contour interval is 1°C. Hatched areas denote significance at the 95th percentile.

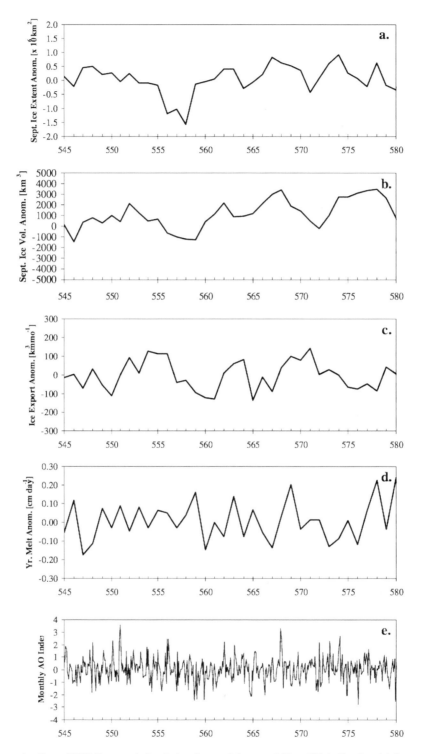

**Figure 9.** Time series from CCSM3 control simulation for model years 545 to 580 indicating (a) September Northern Hemisphere sea ice extent expressed as an anomaly of the averaging period 250–599 in $10^6$ km², (b) September Northern Hemisphere sea ice volume anomaly in km³, (c) Fram Strait southward ice export anomaly for November through April in km³ month⁻¹, (d) January through December sea ice melt anomaly in cm d⁻¹, and (e) the first principal component of the monthly Northern Hemisphere sea level pressure anomaly as computed over the period 250–599.

are found in Alaska and Siberian regions, indicating conditions favorable to large-scale melting of the ice pack. Year 489 is also shown in Figure 8 for comparison, which does not show significant temperature anomalies near the Arctic Basin. The rapid recovery of the sea ice extent, combined with the continuing decline in sea ice volume is indicative of large-scale redistribution of sea ice with a decrease in mean ice thickness. Ten years after the minimum ice cover event (year 500) the sea ice volume began to increase again until full recovery in year 510. This ice volume recovery is concurrent with anomalously low summer melt and low ice export, particularly in years 500 and 506–509. In summary, unusually warm conditions and an enhanced net shortwave surface flux in the late winter and early spring resulted in an ice extent anomaly, which occurred in the presence of a long-term downward trend in ice volume. While the extent immediately recovered, the volume anomaly recovered some 20 years later, as a result of decreased export and melt.

### 3.3. Simulation Years 556–558 Event

As seen in Plate 3c, the spatial pattern of the September ice minimum for the third event beginning in simulation year 556 is predominantly focused on the Atlantic side of the basin, with large losses in the Eurasian marginal seas and in the central Arctic. The ice cover in the Chukchi and Beaufort seas in contrast is not significantly affected, and there is ice present along the northeast Greenland coast. Time series

of Arctic climate variables for this event are shown in Figure 9. This third extraordinary sea ice minimum event in the simulation differs from the previous two in that the decease is sustained over a period of 3 years, suggesting the presence of either a sustained forcing mechanism or the extended absence of an essential recovery mechanism. In the model year following the 3-year minimum ice cover event, sea ice extent immediately returns to near-normal values and tends toward positive anomalies over the following 2-decade period. As seen in Figure 9, the third year of the event denotes the largest anomaly, corresponding to simulation year 558, in which the sea ice cover has decreased by a further $0.37 \times 10^6$ km$^2$. This event also differs from the previous two in that there is a lack of any significant ice volume anomaly prior to the event. As shown in Figure 9b, ice volume anomalies are negative throughout the 3-year period by an average of 950 km$^3$, which is approximately half the standard deviation of the 350-year time series that is considered here. Local maxima in the time series of winter ice export and summer melting anomalies are apparent in the time series leading up to the event. In particular, a sustained anomalous export over the five winters preceding the first year of the minimum ice coverage event is evident in Figure 9c, while an ice melt anomaly of 0.10 cm d$^{-1}$ occurs during simulation year 556 (Figure 9d). While changes in export and melt may serve to initiate the ice extent anomaly, the absence of any significant loss in ice volume suggests a redistribution of the ice from atmospheric forcing.

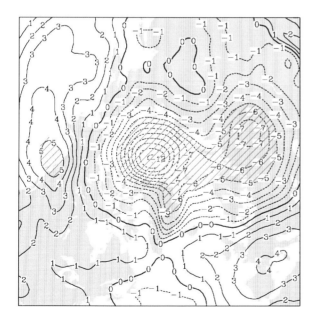
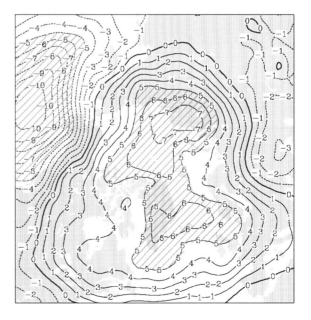

**Figure 10.** Sea level pressure anomaly from the CCSM control simulation averaged for November through April for (left) model year 556 and after the event in (right) year 559. The contour interval is 1 hPa. Hatched areas denote significance at the 95th percentile.

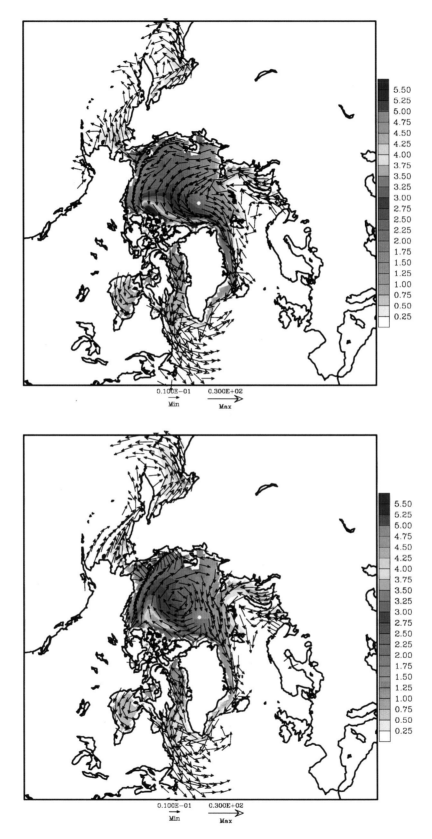

**Plate 5.** Simulated ice thickness (shaded) and ice velocity for the CCSM3 control simulation averaged for the months of November through April for model years (top) 556 and (bottom) 559. Ice thickness is shaded every 0.25 m. The maximum ice velocity vector shown corresponds to 0.3 m s⁻¹.

While the time series of the monthly AO index values shown in Figure 9e does not indicate a consistent anomaly, this does not preclude the presence of strong cyclonic circulation within the Arctic. In fact, strong low pressure throughout the central Arctic is found for the simulation years 556–558 and is illustrated for year 556 in Figure 10. Sustained negative wintertime sea level pressure anomalies of greater than 12 hPa are shown for the central Arctic, and significant negative anomalies even extend to the middle latitudes near the Ural Mountains in eastern Russia. The effects on the sea ice coverage may be seen in the ice thickness contours and ice velocity vectors shown in Plate 5. For the year 556, a notable absence of the Beaufort Gyre in the central Arctic may be seen, similar to the year 451 event as shown in Plate 4a. Sea ice under these conditions does not recirculate within the central Arctic; rather, it is seen to be compressed along the Canadian Arctic Archipelago or to exit directly through Fram Strait. Plate 5 indicates anomalously thick ice of greater than 5 m along the Canadian Arctic Archipelago, suggesting a compaction of the sea ice into those locations.

In Figure 9c, it is seen that the recovery period is marked by a period of low sea ice export, which extends for several years to year 561. In Plate 5 (right), the pattern of the Beaufort Gyre is found to be reestablished after the anomaly, and thicker sea ice may be seen to be extending into the central Arctic Basin.

## 4. DISCUSSION

In this study, three events of extraordinary Northern Hemisphere minimum sea ice cover in an extended control simulation of a coupled climate model are investigated. The events are episodic and have differing forcing characteristics. A pervasive characteristic of these events is the absence of the climatological Beaufort Gyre, which serves to recirculate multiyear sea ice within the Arctic Basin and restricts export. The first two events are associated with ice volume anomalies that were established by anomalous export and melting in the decade prior to each event. In each case, the extent anomaly recovered to normal conditions very rapidly, while the ice volume anomaly continued for an extended period of time. In the case of the second event, a minimum volume anomaly was established 10 years after the minimum sea ice extent event, while complete recovery of Northern Hemisphere sea ice volume to normal conditions was an additional 10 years later. In the third event, anomalously low pressure in the central Arctic over a 3-year period produced a redistribution of the sea ice cover, resulting in an ice extent anomaly but not a sea ice volume anomaly. The ice extent subsequently returned to normal conditions in the presence of anticyclonic circulation forced by high atmospheric pressure.

Although the three events and their recoveries are heterogeneous, they highlight mechanisms that have been identified with sea ice loss in the contemporary Arctic. These include the influence of the prevailing wintertime atmospheric circulation in producing enhanced ice export and the preconditioning of the Arctic through volume losses to produce large extent anomalies at a later time [e.g., *Rigor et al.*, 2002]. The second event is important in demonstrating enhanced losses through surface radiative fluxes during the melt season [e.g., *Francis et al.*, 2005]. Finally, the third event denotes a coupled model simulation of the scenario proposed by *Holloway and Sou* [2002] in showing an extent anomaly in the absence of a volume anomaly. Such a condition requires a sustained, concurrent atmospheric forcing to produce the extent anomaly, and recovery is found to occur immediately after the forcing is removed.

Over the extended period of the simulation, several of the climate variables examined here show strong correlations with the modeled September sea ice cover; however, these variables are not directly involved in the remarkable minimum ice cover events that have been examined. Shown in Figure 11 is the time series of September sea ice extent in comparison with the seasonal AO index and with the North Atlantic oceanic poleward heat transport. Atmospheric circulation is found in this study to be a key initiator of the first and third events. In the first event, the anomalous AO index occurred in February and March of the year of the anomaly, while the seasonal average of the index was unremarkable. In the case of third event, low-pressure anomalies extended well beyond the high latitudes, thus obfuscating the canonical AO index. It may be seen in Figure 11 that there is a large amount of high-frequency variability in the seasonal AO index but that there is a correlation with sea ice extent at low frequencies. A cross-spectral analysis indicates a broad spectrum of coherence for periods from 29 through 58 years that is significant at the 99.9% confidence level. Similarly, the oceanic heat transport shows a correlation with sea ice extent at low frequencies. The ocean heat flux and September sea ice extent show significant coherence centered at a period of 25 years. As may be seen in the Figure 11, however, the global ocean does not appear to play a direct role in influencing the anomalous events examined here. This is in marked contrast with simulations using 21st century forcing scenarios in which the ocean plays a more direct role [*Holland et al.*, 2006a].

An important question arising from this study is the interpretation of the presence of such events and their rapid recovery in a control simulation, assuming an absolute validity of the model in reproducing the Arctic climate (a tenuous assumption). One may speculate that such events are evidently part of the internal, natural climate variability and that evidence

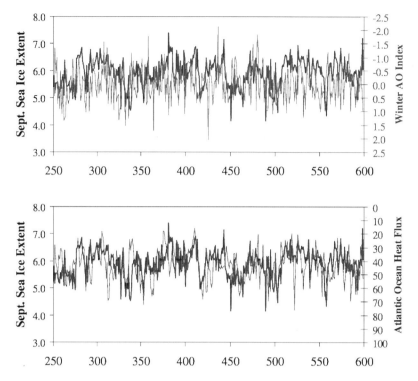

**Figure 11.** (top) Time series of simulated Northern Hemisphere September sea ice extent in $10^6$ km$^2$ and the AO index averaged over winter months (NDJFMA). Note the reversed scale for the AO index. (bottom) The same sea ice time series compared with the North Atlantic Ocean heat flux. This is given as the sum of heat fluxes through the Davis and Fram straits, between Svalbard and Franz Josef Land, and between Franz Josef and Nova Zemlya. Note the reversed scale for the heat flux in PW.

might exist for such events over the past millennium. Such evidence is difficult to obtain. The HadISST1 data set extends prior to the satellite era to 1900, which corresponds to the beginning of Russian historical records of sea ice extent [*Polyakov et al.*, 2003]. *Lemke et al.* [2007] review these available data sources. Prior to the 20th century, a limited number of proxy information are available. Recent sedimentary coring of the Lomonosov Ridge has provided evidence for the onset of a perennial Arctic sea ice cover at about 13 Ma [*Krylov et al.*, 2008]. Variations in this cover have been examined in more recent records. Using ice cores, ocean cores, and mammalian and archeological histories, *Fisher et al.* [2006] notes the evidence for a minimum ice cover occurring in the early Holocene at 10,500–9000 years before present. Proxy characterizations of ice cover are not as quantitative as satellite observations; however, the evidence indicates that Davis Strait and Bering Sea bowhead whales were able to intermingle during this time. The most readily accessible proxy information is from transported driftwood, which circulates within the Arctic in the presence of an ice cover [e.g., *Dyke et al.*, 1997; *Bennike*, 2004; *Johansen*,

2001; *Tremblay et al.*, 1997]. Studies by *Dyke et al.* [1997] suggest a change in driftwood patterns occurring 4000 years before present. The studies of Dyke et al. and Tremblay et al. emphasize the patterns of drift in the Arctic, and suggest an enhanced Beaufort Gyre over the last 4000 years, and a more variable gyre pattern earlier in the record. As this study has indicated, the strength of the Beaufort Gyre has important consequences for Arctic ice extent. These records have a temporal resolution of about 10 years and thus would not be capable of recording episodic, single-year anomalies of the kind examined in this study. Moreover, it is important to expressly distinguish "natural" variability from the internal variability using 1990 anthropogenic forcings, which are clearly not preindustrial. As part of this study, an extended 1860 control simulation of the Geophysical Fluid Dynamics Laboratory (GFDL) of NOAA CM2 [*Delworth et al.*, 2006] was examined and found to contain a single ice anomaly event over a comparable simulation period to the time examined here. The GFDL model has significantly different sea ice characteristics than those of the CCSM3. Nevertheless, it is likely proper to view the time series of extended control simulations as probability distributions of the sea ice coverage that are possible for the given forcing. This is consistent with the results of *Holland et al.* [this volume], who found that the frequency of sea ice extent variance changes with the mean state. The identified minimum events indicate the susceptibility of the Arctic to seasonally ice-free conditions under postindustrial anthropogenic forcing. Notwithstanding known climate model deficiencies, the control simulation in this context suggests that Arctic sea ice extent anomalies of the magnitude observed in the September 2007 minimum are plausible under 1990 forcing conditions. It is reasonable then to conclude that more severe anomalies are achievable with more acute forcing conditions that are in excess of those of 1990.

*Acknowledgments.* The authors thank Robert Newton and two anonymous reviews for their help in the manuscript revision. Output fields of the CCSM3 model were obtained from the National Center for Atmospheric Research (NCAR). NCAR is sponsored by the National Science Foundation. HadISST1 monthly sea ice fields were obtained from the UK Met Office Hadley Centre. Surface energy flux data from the SHEBA field study were obtained from the Polar Science Center at the Applied Physics Laboratory, University of Washington. This research was sponsored by the National Science Foundation under grant ARC-05-20496 and an NSERC Discovery Grant awarded to the second author. This is contribution 7196 of the Lamont-Doherty Earth Observatory of Columbia University.

# REFERENCES

Bennike, O. (2004), Holocene sea-ice variations in Greenland: Onshore evidence, *Holocene*, *14*, 607–613.

Bitz, C. M., and W. H. Lipscomb (1999), An energy-conserving thermodynamic model of sea ice, *J. Geophys. Res.*, *104*, 15,669–15,677.

Bitz, C. M., M. M. Holland, M. Eby, and A. J. Weaver (2001), Simulating the ice-thickness distribution in a coupled climate model, *J. Geophys. Res.*, *106*, 2441–2463.

Bourke, R. H., and R. P. Garrett (1987), Sea ice thickness distribution in the Arctic Ocean, *Cold Regions Sci. Technol.*, *13*, 259–280.

Boyd, T. J., M. Steele, R. D. Muench, and J. T. Gunn (2002), Partial recovery of the Arctic Ocean halocline, *Geophys. Res. Lett.*, *29*(14), 1657, doi:10.1029/2001GL014047.

Briegleb, B. P., C. M. Bitz, E. C. Hunke, W. H. Lipscomb, M. M. Holland, J. L. Schramm, and R. E. Moritz (2004), Scientific description of the sea ice component of the Community Climate System Model, version three, *NCAR Tech. Note NCAR/TN-463+STR*, 70 pp., Natl. Cent. for Atmos. Res., Boulder, Colo.

Carmack, E. C., R. W. Macdonald, R. G. Perkin, F. A. McLaughlin, and R. J. Pearson (1995), Evidence for warming of Atlantic water in the southern Canadian Basin of the Arctic Ocean: Results from the Larsen-93 expedition, *Geophys. Res. Lett.*, *22*, 1061–1064.

Chapman, W. L., and J. E. Walsh (2007), Simulations of Arctic temperature and pressure by global coupled models, *J. Clim.*, *20*, 609–632.

Collins, W. D., et al. (2006a), The Community Climate System Model version 3 (CCSM3), *J. Clim.*, *19*, 2122–2143.

Collins, W. D., P. J. Rasch, B. A. Boville, J. J. Hack, J. R. McCaa, D. L. Williamson, B. P. Briegleb, C. M. Bitz, S.-J. Lin, and M. Zhang (2006b), The formulation and atmospheric simulation of the Community Atmosphere Model version 3 (CAM3), *J. Clim.*, *19*, 2144–2161.

Delworth, T. L., et al. (2006), GFDL's CM2 global coupled climate models. Part I: Formulation and simulation characteristics, *J. Clim.*, *19*, 643–674.

Deser, C. (2000), On the teleconnectivity of the "Arctic Oscillation," *Geophys. Res. Lett.*, *27*, 779–782.

DeWeaver, E., and C. M. Bitz (2006), Atmospheric circulation and its effect on Arctic sea ice in CCSM3 simulations at medium and high resolution, *J. Clim.*, *19*, 2415–2436.

Dickson, R. R., T. J. Osborn, J. W. Hurrell, J. Meincke, J. Blindheim, B. Adlandsvik, T. Vinje, G. Alekseev, and W. Maslowski (2000), The Arctic Ocean response to the North Atlantic Oscillation, *J. Clim.*, *13*, 2671–2696.

Dyke, A. S., J. England, E. Reimnitz, and H. Jetté (1997), Changes in driftwood delivery to the Canadian Arctic Archipelago: The hypothesis of postglacial oscillations of the transpolar drift, *Arctic*, *50*, 1–16.

Fisher, D., A. Dyke, R. Koerner, J. Bourgeois, C. Kinnard, C. Zdanowicz, A. de Vernal, C. Hillaire-Marcel, J. Savelle, A. Rochon (2006), Natural variability of Arctic sea ice over the Holocene, *Eos Trans. AGU*, *87*(28), 273.

Francis, J. A., E. Hunter, J. R. Key, and X. Wang (2005), Clues to variability in Arctic minimum sea ice extent, *Geophys. Res. Lett.*, *32*, L21501, doi:10.1029/2005GL024376.

Gerdes, R., and C. Köberle (2007), Comparison of Arctic sea ice thickness variability in IPCC Climate of the 20th century experiments and in ocean-sea ice hindcasts, *J. Geophys. Res.*, *112*, C04S13, doi:10.1029/2006JC003616.

Gorodetskaya, I., and L.-B. Tremblay (2008), Arctic cloud properties and radiative forcing from observations and their role in sea ice predicted by the NCAR CCSM3 model during the 21st century, this volume.

Gorodetskaya, I. V., L.-B. Tremblay, B. Liepert, M. A. Cane, and R. I. Cullather (2008), The influence of cloud and surface properties on the Arctic Ocean shortwave radiation budget in coupled models, *J. Clim.*, *21*, 866–882.

Grotefendt, K., K. Logemann, D. Quadfasel, and S. Ronski (1998), Is the Arctic Ocean warming?, *J. Geophys. Res.*, *103*, 27,679–27,687.

Holland, M. M., C. M. Bitz, and B. Tremblay (2006a), Future abrupt reductions in the summer Arctic sea ice, *Geophys. Res. Lett.*, *33*, L23503, doi:10.1029/2006GL028024.

Holland, M. M., J. Finnis, and M. C. Serreze (2006b), Simulated Arctic Ocean freshwater budgets in the twentieth and twenty-first centuries, *J. Clim.*, *19*, 6221–6242.

Holland, M. M., C. M. Bitz, E. C. Hunke, W. H. Lipscomb, and J. L. Schramm (2006c), Influence of the sea ice thickness distribution on polar climate in CCSM3, *J. Clim.*, *19*, 2398–2414.

Holland, M., C. M. Bitz, B. Tremblay, and D. A. Bailey (2008), The role of natural versus forced change in future rapid summer Arctic ice loss, this volume.

Holloway, G., and T. Sou (2002), Has Arctic sea ice rapidly thinned?, *J. Clim.*, *15*, 1691–1701.

Hunke, E. C., and J. K. Dukowicz (1997), An elastic-viscous-plastic model for sea ice dynamics, *J. Phys. Oceanogr.*, *27*, 1849–1867.

Hurrell, J. W. (1995), Decadal trends in the North Atlantic Oscillation: Regional temperatures and precipitation, *Science*, *269*, 676–679.

Jochum, M., G. Danabasoglu, M. Holland, Y.-O. Kwon, and W. G. Large (2008), Ocean viscosity and climate, *J. Geophys. Res.*, *113*, C06017, doi:10.1029/2007JC004515.

Johansen, S. (2001), A dendrochronological analysis of driftwood in Northern Dvina delta and on northern Novaya Zemlya, *J. Geophys. Res.*, *106*, 19,929–19,938.

Jones, E. P. (2001), Circulation in the Arctic Ocean, *Polar Res.*, *20*, 139–146.

Kiehl, J. T., and P. R. Gent (2004), The Community Climate System Model, version 2, *J. Clim.*, *17*, 3666–3682.

Krylov, A. A., I. A. Andreeva, C. Vogt, J. Backman, V. V. Krupskaya, G. E. Grikurov, K. Moran, and H. Shoji (2008), A shift in heavy and clay mineral provenance indicates a middle Miocene onset of a perennial sea ice cover in the Arctic Ocean, *Paleoceanography*, *23*, PA1S06, doi:10.1029/2007PA001497.

Kwok, R. (2000), Recent changes in Arctic Ocean sea ice motion associated with the North Atlantic Oscillation, *Geophys. Res. Lett.*, *27*, 775–778.

Lemke, P., et al. (2007), Observations: Changes in snow, ice and frozen ground, in *Climate Change 2007: The Physical Science Basis: Contribution of Working Group I to the Fourth Assessment Report of the Intergovernmental Panel on Climate Change*, edited by S. Solomon et al., pp. 337–383, Cambridge Univ. Press, Cambridge, U. K.

Lipscomb, W. H. (2001), Remapping the thickness distribution in sea ice models, *J. Geophys. Res.*, *106*, 13,989–14,000.

Meehl, G. A., et al. (2007), Global climate projections, in *Climate Change 2007: The Physical Science Basis: Contribution of Working Group I to the Fourth Assessment Report of the Intergovernmental Panel on Climate Change*, edited by S. Solomon et al., pp. 747–845, Cambridge Univ. Press, Cambridge, U. K.

Ogi, M., and J. M. Wallace (2007), Summer minimum Arctic sea ice extent and the associated summer atmospheric circulation, *Geophys. Res. Lett.*, *34*, L12705, doi:10.1029/2007GL029897.

Parkinson, C. L. (2006), Earth's cryosphere: Current state and recent changes, *Annu. Rev. Environ. Resour.*, *31*, 33–60.

Persson, P. O. G., C. W. Fairall, E. L. Andreas, P. S. Guest, and D. K. Perovich (2002), Measurements near the Atmospheric Surface Flux Group tower at SHEBA: Near-surface conditions and surface energy budget, *J. Geophys. Res.*, *107*(C10), 8045, doi:10.1029/2000JC000705.

Polyakov, I. V., G. V. Alekseev, R. V. Bekryaev, U. S. Bhatt, R. Colony, M. A. Johnson, V. P. Karklin, D. Walsh, and A. V. Yulin (2003), Long-term ice variability in Arctic marginal seas, *J. Clim.*, *16*, 2078–2085.

Quadfasel, D., A. Sy, D. Wells, and A. Turnik (1991), Warming in the Arctic, *Nature*, *350*, 385.

Randall, D. A., et al. (2007), Climate models and their evaluation, in *Climate Change 2007: The Physical Science Basis: Contribution of Working Group I to the Fourth Assessment Report of the Intergovernmental Panel on Climate Change*, edited by S. Solomon et al., pp. 589–662, Cambridge Univ. Press, Cambridge, U. K.

Rayner, N. A., D. E. Parker, E. B. Horton, C. K. Folland, L. V. Alexander, D. P. Rowell, E. C. Kent, and A. Kaplan (2003), Global analyses of sea surface temperature, sea ice, and night marine air temperature since the late nineteenth century, *J. Geophys. Res.*, *108*(D14), 4407, doi:10.1029/2002JD002670.

Rigor, I. G., J. M. Wallace, and R. L. Colony (2002), Response of sea ice to the Arctic Oscillation, *J. Clim.*, *15*, 2648–2663.

Rothrock, D. A. (1975), The energetics of the plastic deformation of pack ice by ridging, *J. Geophys. Res.*, *80*, 4514–4519.

Serreze, M. C., J. E. Walsh, F. S. Chapin III, T. Osterkamp, M. Dyurgerov, V. Romanovsky, W. C. Oechel, J. Morison, T. Zhang, and R. G. Barry (2000), Observational evidence of recent change in the northern high-latitude environment, *Clim. Change*, *46*, 159–207.

Serreze, M. C., J. A. Maslanik, T. A. Scambos, F. Fetterer, J. Stroeve, K. Knowles, C. Fowler, S. Drobot, R. G. Barry, and T. M. Haran (2003), A record minimum Arctic sea ice extent and area in 2002, *Geophys. Res. Lett.*, *30*(3), 1110, doi:10.1029/2002GL016406.

Serreze, M. C., M. M. Holland, and J. Stroeve (2007), Perspectives on the Arctic's shrinking sea-ice cover, *Science*, *315*, 1533–1536.

Smith, R., and P. Gent (2002), Reference manual for the Parallel Ocean Program (POP) ocean component of the Community Climate System Model (CCSM2.0 and 3.0), *Rep. LAUR-02–2484*, Los Alamos Natl. Lab., Los Alamos, N. M.

Sorteberg, A., V. Kattsov, J. E. Walsh, and T. Pavlova (2007), The Arctic surface energy budget as simulated with the IPCC AR4 AOGCMs, *Clim. Dyn.*, *29*, 131–156.

Steele, M., and T. Boyd (1998), Retreat of the cold halocline layer in the Arctic Ocean, *J. Geophys. Res.*, *103*, 10,419–10,435.

Stroeve, J. C., M. C. Serreze, F. Fetterer, T. Arbetter, W. Meier, J. Maslanik, and K. Knowles (2005), Tracking the Arctic's shrinking ice cover: Another extreme September minimum in 2004, *Geophys. Res. Lett.*, *32*, L04501, doi:10.1029/2004GL021810.

Stroeve, J., M. M. Holland, W. Meier, T. Scambos, and M. Serreze (2007), Arctic sea ice decline: Faster than forecast, *Geophys. Res. Lett.*, *34*, L09501, doi:10.1029/2007GL029703.

Stroeve, J., M. Serreze, S. Drobot, S. Gearheard, M. Holland, J. Maslanik, W. Meier, and T. Scambos (2008), Arctic sea ice extent plummets in 2007, *Eos Trans. AGU*, *89*(2), 13.

Thompson, D. W. J., and J. M. Wallace (1998), The Arctic Oscillation signature in the wintertime geopotential height and temperature fields, *Geophys. Res. Lett.*, *25*, 1297–1300.

Thorndike, A. S., D. A. Rothrock, G. A. Maykut, and R. Colony (1975), Thickness distribution of sea ice, *J. Geophys. Res.*, *80*, 4501–4513.

Tremblay, L.-B., L. A. Mysak, and A. S. Dyke (1997), Evidence from driftwood records for century-to-millennial scale variations of the high latitude atmospheric circulation during the Holocene, *Geophys. Res. Lett.*, *24*, 2027–2030.

Tremblay, L. B., M. M. Holland, I. V. Gorodetskaya, and G. A. Schmidt (2007), An ice-free Arctic? Opportunities for computational science, *Comp. Sci. Eng.*, *9*, 65–74.

Uttal, T., et al. (2002), Surface heat budget of the Arctic Ocean, *Bull. Am. Meteorol. Soc.*, *83*, 255–275.

Walsh, D., I. Polyakov, L. Timokhov, and E. Carmack (2007), Thermohaline structure and variability in the eastern Nansen Basin as seen from historical data, *J. Mar. Res.*, *65*, 685–714.

Wang, X., and J. R. Key (2005), Arctic surface, cloud, and radiation properties based on the AVHRR Polar Pathfinder dataset. Part II: Recent trends, *J. Clim.*, *18*, 2575–2593.

Yang, J., J. Comiso, D. Walsh, R. Krishfield, and S. Honjo (2004), Storm-driven mixing and potential impact on the Arctic Ocean, *J. Geophys. Res.*, *109*, C04008, doi:10.1029/2001JC001248.

Zhang, X., and J. E. Walsh (2006), Toward a seasonally ice-covered Arctic Ocean: Scenarios from the IPCC AR4 model simulations, *J. Clim.*, *19*, 1730–1747.

R. I. Cullather, Lamont-Doherty Earth Observatory of Columbia University, P.O. Box 1000, 61 Route 9 W, Palisades, NY 10964-1000, USA. (cullat@ldeo.columbia.edu)

L.-B. Tremblay, Department of Atmospheric and Oceanic Sciences, McGill University, 805 Sherbrooke Street West, Montreal, QC H2A 2K6, Canada.

# A Bayesian Network Modeling Approach to Forecasting the 21st Century Worldwide Status of Polar Bears

Steven C. Amstrup

*Alaska Science Center, U.S. Geological Survey, Anchorage, Alaska, USA*

Bruce G. Marcot

*Pacific Northwest Research Station, USDA Forest Service, Portland, Oregon, USA*

David C. Douglas

*Alaska Science Center, U.S. Geological Survey, Juneau, Alaska, USA*

To inform the U.S. Fish and Wildlife Service decision, whether or not to list polar bears as threatened under the Endangered Species Act (ESA), we projected the status of the world's polar bears (*Ursus maritimus*) for decades centered on future years 2025, 2050, 2075, and 2095. We defined four ecoregions based on current and projected sea ice conditions: seasonal ice, Canadian Archipelago, polar basin divergent, and polar basin convergent ecoregions. We incorporated general circulation model projections of future sea ice into a Bayesian network (BN) model structured around the factors considered in ESA decisions. This first-generation BN model combined empirical data, interpretations of data, and professional judgments of one polar bear expert into a probabilistic framework that identifies causal links between environmental stressors and polar bear responses. We provide guidance regarding steps necessary to refine the model, including adding inputs from other experts. The BN model projected extirpation of polar bears from the seasonal ice and polar basin divergent ecoregions, where ≈2/3 of the world's polar bears currently occur, by mid century. Projections were less dire in other ecoregions. Decline in ice habitat was the overriding factor driving the model outcomes. Although this is a first-generation model, the dependence of polar bears on sea ice is universally accepted, and the observed sea ice decline is faster than models suggest. Therefore, incorporating judgments of multiple experts in a final model is not expected to fundamentally alter the outlook for polar bears described here.

Arctic Sea Ice Decline: Observations, Projections, Mechanisms, and Implications
Geophysical Monograph Series 180
This paper is not subject to U.S. copyright. Published in 2008 by the American Geophysical Union.
10.1029/180GM14

## 1. INTRODUCTION

Polar bears depend upon sea ice for access to their prey and for other aspects of their life history [*Stirling and Øritsland*, 1995; *Stirling and Lunn*, 1997; *Amstrup*, 2003]. Observed declines in sea ice availability have been associated with reduced body condition, reproduction, survival, and population size for polar bears in parts of their range [*Stirling et al.*, 1999; *Obbard et al.*, 2006; *Stirling and Parkinson*, 2006; *Regehr et al.*, 2007b]. Observed [*Comiso*, 2006] and projected [*Holland et al.*, 2006] sea ice declines have led to the hypothesis that the future welfare of polar bears may be diminished worldwide and to the proposal by the U.S. Fish and Wildlife Service (FWS) to list the polar bear as a threatened species under the Endangered Species Act [*U.S. Fish and Wildlife Service*, 2007].

Classification as "threatened" requires determination that a species is likely to become "endangered" within the "foreseeable future" throughout all or a significant portion of its range. An "endangered" species is any species that is in danger of extinction throughout all or a significant portion of its range. For polar bears, the "foreseeable future" was defined as 45 years from now [*U.S. Fish and Wildlife Service*, 2007]. Here we describe a method for combining available information on polar bear life history and ecology with projections of the future state of Arctic sea ice to project the future worldwide status of polar bears. We present our forecast in a "compared to now" setting where projections for the decade of 2045–2054 are compared to the "present" period of 1996–2006. For added perspective, we looked to the nearer term as well as beyond the defined foreseeable future by comparing projections for the periods 2020–2029, 2070–2079, and 2090–2099 to the present. Also, we looked back to the period of 1985–1995. Hence we examined six time periods in total.

Our view of the present and past was based on sea ice conditions derived from satellite data. Our future forecasts were based on information derived from general circulation model (GCM) projections of the extent and spatiotemporal distribution of sea ice, our understanding of how polar bears have responded to ongoing changes in sea ice, and projections of how polar bears are likely to respond to future changes. This paper synthesizes information in nine Administrative Reports prepared by the U.S. Geological Survey and delivered to the FWS in 2007 (http://www.usgs.gov/newsroom/special/polar_bears/) plus other recent literature.

Polar bears occur throughout portions of the Northern Hemisphere where the sea is ice covered for all or much of the year. Polar bears are thought to have branched off of brown bear (*Ursus arctos*) stocks as long ago as 250,000 years, but they appear in the fossil record no earlier than 120,000 years ago [*Talbot and Shields*, 1996; *Hufthamer*, 2001; *Ingolfsson and Wiig*, 2007]. Since moving offshore, behavioral and physical adaptations have allowed polar bears to increasingly specialize at hunting seals from the surface of the ice [*Stirling*, 1974; *Smith*, 1980; *Stirling and Øritsland*, 1995].

Over much of their range, polar bears are nutritionally dependent on the ringed seal (*Phoca hispida*). Polar bears occasionally catch belugas (*Delphinapterus leucas*), narwhals (*Monodon monocerus*), walrus (*Odobenus rosmarus*), and harbor seals (*P. vitulina*) [*Smith*, 1985; *Calvert and Stirling*, 1990; *Smith and Sjare*, 1990; *Stirling and Øritsland*, 1995; *Derocher et al.*, 2002]. Walruses can be seasonally important in some parts of the polar bear range [*Parovshchikov*, 1964; *Ovsyanikov*, 1996]. Bearded seals (*Erignathus barbatus*) can be a large part of their diet where they are common and are probably the second most common prey of polar bears [*Derocher et al.*, 2002]. The most common prey of polar bears, however, is the ringed seal [*Smith and Stirling*, 1975; *Smith*, 1980]. The relationship between ringed seals and polar bears is so close that the abundance of ringed seals in some areas appears to regulate the density of polar bears, while polar bear predation, in turn, regulates density and reproductive success of ringed seals in other areas [*Hammill and Smith*, 1991; *Stirling and Øritsland*, 1995]. Across much of the polar bear range, their dependence on ringed seals is close enough that the abundances of ringed seals have been estimated by knowing the abundances of polar bears [*Stirling and Øritsland*, 1995; *Kingsley*, 1998]. Although polar bears occasionally catch seals on land or in open water [*Furnell and Oolooyuk*, 1980], they consistently catch seals and other marine mammals only at the air-ice-water interface.

Like all bears, polar bears are opportunistic and will take a broad variety of foods when available. When stranded on land for long periods polar bears will consume coastal marine and terrestrial plants and other terrestrial foods [*Derocher et al.*, 1993]. Polar bears have been observed hunting caribou [*Derocher et al.*, 2000; *Brook and Richardson*, 2002], and they rarely have been observed fishing [*Townsend*, 1911; *Dyck and Romber*, 2007]. They will eat eggs, catch flightless (molting) birds, take human refuse, and consume a variety of plant materials [*Russell*, 1975; *Lunn and Stirling*, 1985; *Derocher et al.*, 1993; *Smith and Hill*, 1996; *Stempniewicz*, 1993, 2006]. Although individual bears may gain short-term energetic rewards from alternate foods, available data suggest that polar bears gain little benefit at the population level from these sources [*Ramsay and Hobson*, 1991]. Maintenance of polar bear populations appears dependent upon marine prey, largely ringed seals.

Although polar bears occur in most ice-covered regions of the Northern Hemisphere [*Stefansson*, 1921], they are not evenly dispersed. They are observed most frequently in shallow water areas nearshore and in other areas, called polynyas, where currents and upwellings keep the winter ice cover from freezing solid. These shore leads and polynyas create a zone of active unconsolidated sea ice that is small in geographic area but contributes ~50% of the total productivity in Arctic waters [*Sakshaug*, 2004]. Polar bears have been shown to focus their annual activity areas over these regions [*Stirling et al.*, 1981; *Amstrup and DeMaster*, 1988; *Stirling*, 1990; *Stirling and Øritsland*, 1995; *Stirling and Lunn*, 1997; *Amstrup et al.*, 2000, 2004, 2005]. Ice over waters less than 300 m deep is the most preferred habitat of polar bears throughout the polar basin [*Durner et al.*, 2008].

Polar bears inhabit regions with very different sea ice characteristics. The southern reaches of their range includes areas where sea ice is seasonal. There, polar bears are forced onto land where they are food deprived for extended periods each year. Other polar bears live in the harshest and most northerly climes of the world where the ocean is ice covered year-round. Still others occupy the pelagic regions of the polar basin where there are strong seasonal changes in the character and especially distribution of the ice. The common denominator is that all polar bears regardless of where they live make seasonal movements to maximize their foraging time on sea ice that is suitable for hunting [*Amstrup*, 2003].

## 2. METHODS

### 2.1. Overview

We used a Bayesian network (BN) model [*Marcot et al.*, 2006] to forecast future population status of polar bears in each of four distinct ecoregions. The BN model incorporated projections of sea ice change as well as anticipated likelihoods of changes in several other potential population stressors. In the following sections, we provide detailed descriptions of the four polar bear ecoregions. We describe the process we used to make projections of the amount and distribution of future sea ice habitat. Finally, we provide details of the BN population stressor model we used to project the future status of polar bears.

### 2.2. Polar Bear Ecoregions

Polar bears are distributed throughout regions of the Arctic and subarctic where the sea is ice covered for large portions of the year. Telemetry studies have demonstrated spatial segregation among groups or stocks of polar bears in different regions of their circumpolar range [*Schweinsburg and Lee*, 1982; *Amstrup* 1986, 2000; *Garner et al.*, 1990, 1994; *Messier et al.*, 1992; *Amstrup and Gardner*, 1994; *Ferguson et al.*, 1999]. As a result of patterns in spatial segregation suggested by telemetry data, survey and reconnaissance, marking and tagging, and traditional knowledge, the Polar Bear Specialist Group (PBSG) of the International Union for the Conservation of Nature recognizes 19 partially discrete polar bear groups [*Aars et al.*, 2006]. Although there is

considerable overlap in areas occupied by members of these groups [*Amstrup et al.*, 2004, 2005], they are thought to be ecologically meaningful [*Aars et al.*, 2006] and are managed as subpopulations (Plate 1).

We recognized that many of the 19 subpopulations share more similarities than differences and pooled them into four ecological regions (Plate 1). We defined "ecoregions" on the basis of observed temporal and spatial patterns of ice melt, freeze, and advection, observations of how polar bears respond to those patterns, and how general circulation models (GCMs) forecast future ice patterns in each ecoregion.

The seasonal ice ecoregion (SIE) includes the two subpopulations of bears which occur in Hudson Bay, as well as the bears of Foxe Basin, Baffin Bay, and Davis Strait. The sum of the members of these five subpopulations is thought to include about 7500 polar bears [*Aars et al.*, 2006]. All five share the characteristic that the sea ice, on which the polar bears hunt, melts entirely in summer and bears are forced ashore for extended periods of time during which they are food deprived.

The archipelago ecoregion (AE) includes the channels between the Canadian Arctic Islands. This ecoregion includes approximately 5000 polar bears representing six subpopulations recognized by the PBSG [*Aars et al.*, 2006]. These subpopulations are Kane Basin, Norwegian Bay, Viscount-Melville Sound, Lancaster Sound, M'Clintock Channel, and the Gulf of Boothia. Much of this region is characterized by heavy annual and multiyear (perennial) ice that historically has filled the interisland channels year-round. Polar bears remain on the sea ice, therefore, throughout the year.

In the polar basin as in the AE, polar bears mainly stay on the sea ice year-round. In our analyses, we split the polar basin into two ecoregions. This split was based upon the different patterns of sea ice formation and advection [*Rigor et al.*, 2002; *Rigor and Wallace* 2004; *Maslanik et al.*, 2007; *Meier et al.*, 2007; *Ogi and Wallace*, 2007]. The polar basin divergent ecoregion (PBDE) is characterized by extensive formation of annual sea ice that is typically advected toward the central polar basin, against the Canadian Arctic Islands and Greenland, or out of the polar basin through Fram Strait. The PBDE lies between ~127°W and 10°E and includes the southern Beaufort, Chukchi, East Siberian-Laptev, Kara, and Barents sea subpopulations. There are no population estimates for the Kara Sea region. Assuming that 1000 bears live in the Kara Sea, this ecoregion could be home to approximately 8500 polar bears [*Aars et al.*, 2006].

The polar basin convergent ecoregion (PBCE) is the remainder of the polar basin including the east Greenland Sea, the continental shelf areas adjacent to northern Greenland and the Queen Elizabeth Islands, and the northern Beaufort Sea (Plate 1). There are thought to be approximately

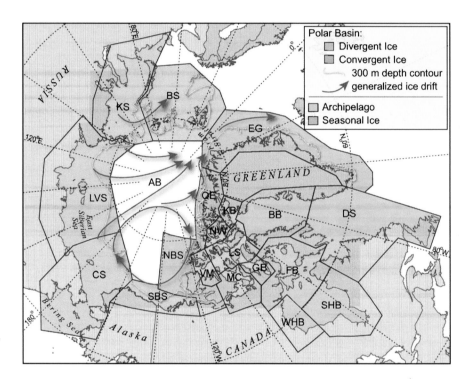

**Plate 1.** Map of four polar bear ecoregions defined by grouping recognized subpopulations which share seasonal patterns of ice motion and distribution. The polar basin divergent ecoregion (PBDE) (purple) includes Southern Beaufort Sea (SBS), Chukchi Sea (CS), Laptev Sea (LVS), Kara Sea (KS), and the Barents Sea (BS). The polar basin convergent ecoregion (PBCE) (blue) includes East Greenland (EG), Queen Elizabeth (QE), and Northern Beaufort Sea (NBS). The seasonal ice ecoregion (SIE) (green) includes southern Hudson Bay (SHB), western Hudson Bay (WHB), Foxe Basin (FB), Davis Strait (DS), and Baffin Bay (BB). The archipelago ecoregion (AE) (yellow) includes Gulf of Boothia (GB), M'Clintock Channel (MC), Lancaster Sound (LS), Viscount-Melville Sound (VM), Norwegian Bay (NW), and Kane Basin (KB).

1200 polar bears in the Northern Beaufort Sea subpopulation [*Aars et al.*, 2006], but numbers of bears in the rest of this ecoregion are poorly known. There are no estimates for the east Greenland subpopulation, but we assumed there currently may be up to 1000 bears there. We modified the PBSG recognized subpopulation boundaries of this ecoregion by redefining a Queen Elizabeth Islands subpopulation (QE). QE had formerly included the continental shelf region and interisland channels between Prince Patrick Island and the northeast corner of Ellesmere Island [*Aars et al.*, 2006]. We extended its boundary to northwest Greenland. This area is characterized by heavy multiyear ice, except for a recurring lead system that runs along the Queen Elizabeth Islands from the northeastern Beaufort Sea to northern Greenland [*Stirling*, 1980]. Over 200 polar bears could be resident here, and some bears from other regions have been recorded moving through the area [*Durner and Amstrup*, 1995; *Lunn et al.*, 1995]. Like the Northern Beaufort Sea subpopulation, QE occurs in a region of the polar basin that recruits ice as it is advected from the PBDE [*Comiso*, 2002; *Rigor and Wallace*, 2004; *Belchansky et al.*, 2005; *Holland et al.*, 2006; *Durner et al.*, 2008; *Ogi and Wallace*, 2007; *Serreze et al.*, 2007]. Assuming these rough estimates are close, up to 2400 bears might presently occupy the PBCE.

We did not incorporate the central Arctic Basin into our analyses. This area was defined to contain a separate subpopulation by the PBSG in 2001 [*Lunn et al.*, 2002] to recognize bears that may reside outside the territorial jurisdictions of the polar nations. The Arctic Basin region is characterized by very deep water which is known to be unproductive [*Pomeroy*, 1997]. Available data are conclusive that polar bears prefer sea ice over shallow water (<300 m deep) [*Amstrup et al.*, 2000, 2004; *Durner et al.*, 2008], and it is thought that this preference reflects increased hunting opportunities over more productive waters. Tracking studies indicate that few if any bears are year-round residents of the central Arctic Basin. For all of these reasons, we did not include the Arctic Basin in our analyses.

### 2.3. Sea Ice Habitat Variables

Our BN model incorporated changes in area and spatiotemporal distribution of sea ice habitat along with other "stressors" that might help predict the future of polar bears. We used monthly averaged ice concentration estimates derived from passive microwave satellite imagery for the observational period 1979–2006 [*Cavalieri et al.*, 1996]. Sea ice data for the future were derived from monthly sea ice concentration projections of 10 GCMs. The GCMs we used were included in the Intergovernmental Panel of Climate Change (IPCC) Fourth Assessment Report (AR4) (Table 1).

These included hindcast ice estimates from the 20th Century Experiment (20C3M) and projection estimates for the 21st century forced with the "business as usual" Special Report on Emissions Scenarios (SRES) A1B emissions scenario [*Nakićenović et al.*, 2000]. We obtained GCM ice projection outputs of nine models from the World Climate Research Programme's Coupled Model Intercomparison Project phase 3 (CMIP3) multimodel data set [*Meehl et al.*, 2007a]. We obtained projections from the 10th model (Community Climate System Model, version 3 (CCSM3)) directly from the National Center for Atmospheric Research in its native CCSM grid format (D. Bailey and M. Holland, NCAR, personal communication, 2007). We obtained and analyzed one run (run 1) for each GCM, except CCSM3 for which we obtained eight runs. In our analyses we included the mean of the eight CCSM3 runs as a single member of our 10-model ensemble.

We selected the 10 GCMs from a larger group of 20 based on their ability to simulate (20C3M) the mean Northern Hemisphere ice extent for September 1953–1995 to within 20% of the observed September mean (Had1SST [*Rayner et al.*, 2003]). This selection method emulated that used by *Stroeve et al.* [2007], except we used a 50% ice concentration threshold [*DeWeaver*, 2007] to define ice extent (as opposed to 15%). We chose a 50% threshold because other studies have shown that polar bears prefer medium to high sea ice concentrations [*Arthur et al.*, 1996; *Ferguson et al.*, 2000; *Durner et al.*, 2006, 2008].

Sea ice grids among the 10 GCMs we analyzed had various model-specific spatial resolutions ranging from ~1 × 1 to 3 × 4 degrees of latitude × longitude. To facilitate integration with our analyses of observational data, we resampled the GCM grids to match the gridded 25 km resolution passive microwave sea ice concentration maps from the National Snow and Ice Data Center. Each native GCM grid of sea ice concentration was converted to an Arc/Info (version 9.2; ESRI, Redlands, California, United States) point coverage and projected to polar stereographic coordinates (central meridian 45°W, true scale 70°N). A triangular irregular network (TIN) (Arc/Info) was created from the point coverage using ice concentration as the z value, and a 25 km pixel resolution grid was generated by sampling the TIN surface. Effectively, this procedure oversampled the original GCM resolution using linear interpolation.

### 2.3.1. Total annual habitat area.
For input to our models, we defined two area-based metrics of habitat availability to polar bears. The first was an expression of the yearly extent of "total available ice habitat," and the second, which was available in the polar basin only, was an expression of "total optimal habitat." We derived "total available ice habitat" from both observed and projected Arctic-wide sea ice con-

**Table 1.** Sea Ice Simulations and Projections Produced by Ten General Circulation Models[a]

| GCM Model ID | Country | Grid Resolution ( Latitude × Longitude) | Number of Runs |
|---|---|---|---|
| ncar_ccsm3_0 | USA | 1.0 × 1.0 | 8 |
| cccma_cgcm3_1 | Canada | 3.8 × 3.8 | 1 |
| cnrm_cm3 | France | 1.0 × 2.0 | 1 |
| gfdl_cm2_0 | USA | 0.9 × 1.0 | 1 |
| giss_aom | USA | 3.0 × 4.0 | 1 |
| ukmo_hadgem1 | UK | 0.8 × 1.0 | 1 |
| ipsl_cm4 | France | 1.0 × 2.0 | 1 |
| miroc3_2_medres | Japan | 1.0 × 1.4 | 1 |
| miub_echo_g | Germany/Korea | 1.5 × 2.8 | 1 |
| mpi_echam5 | Germany | 1.0 × 1.0 | 1 |

[a]GCMs were developed for the Intergovernmental Panel on Climate Change Fourth Assessment Report (IPCC AR4) [*Meehl et al.*, 2007b] to define ice covariates for polar bear RSF models and to project future sea ice distributions used in our BN model. Note that we used the ensemble mean of the 8 available runs to represent CCSM3 outputs. Sea ice estimates for the period of observational records were derived from the 20th Century Experiment (20C3M). All 21st century projections were forced with the "business as usual" SRES-A1B emissions scenario [*Nakićenović et al.*, 2000].

centration maps as the annual 12-month sum of sea ice extent over the continental shelves (<300 m depth) in each ecoregion. Ice extent was defined as the aerial cover (square kilometers) of all pixels with ≥50% ice concentration. Since deep water is uncommon in the AE and SIE, we considered those entire ecoregions to effectively reside over the continental shelf, meaning total ice habitat equated to the total annual amount of ice cover summed over all 12 months.

We quantified optimal polar bear habitat using the resource selection functions (RSFs) of *Durner et al.* [2008]. RSFs are quantitative expressions of the habitats animals choose to utilize, relative to the habitats that are available to them [*Manly et al.*, 2002]. Estimates of preferred habitat were derived only in the polar basin because only there were sufficient radio-tracking data available to build RSF models. The satellite imagery captured dynamics of the available sea ice habitats, while the satellite telemetry indicated the choices bears made. *Durner et al.* [2008] developed the RSFs with 1985–1995 location data from satellite radio-tagged female polar bears (n=12,171 locations from 333 bears), monthly passive microwave ice concentration maps [*Cavalieri et al.*, 1996], and digital bathymetry and coastline maps. Discrete-choice modeling distinguished between the available and chosen habitats based on six environmental covariates: ocean depth, distance to land, ice concentration, and distances to the 15%, 50%, and 75% ice edges. *Durner et al.* [2008] used 1985–1995 tracking data to establish a baseline of preferred polar bear habitat selection criteria, because during this early period of our study, year-round polar bear movements were less restricted and hence more likely

to represent preferences than during the more recent years of reduced sea ice extent.

Optimal polar bear habitat was defined to be any mapped pixel with an RSF value in the upper 20% of the seasonally averaged (1985–1995) RSF scores [*Durner et al.*, 2008]. This approach created a foundation that allowed us to examine whether future ice projections indicated increases, decreases, or stability in the area, summed over all 12 months, of optimal polar bear habitat, relative to our earliest decade of empirical observations. Like "total ice habitat," optimal habitat had the disadvantage of not being able to resolve seasonal changes.

We note that expressing change on the basis of annual square kilometer months tends to minimize the potential effects of large seasonal swings in habitat availability. Whereas the yearly average sea ice extent has declined at a rate of 3.6% per decade during 1979–2006, the mean September sea ice extent has declined at a rate of 8.4% per decade [*Meier et al.*, 2007]. Further, because all GCMs project extensive winter sea ice through the 21st century in most ecoregions [*Durner et al.*, 2008], the severity of summer periods of food deprivation may be hidden by extensive sea ice in winter when data are pooled annually. Although polar bears are well adapted to a feast and famine diet [*Watts and Hansen*, 1987], there apparently are limits to their ability to sustain long periods of food deprivation [*Regehr et al.*, 2007b]. We recognized our measures of change in square kilometer months were largely insensitive to these seasonal effects. Two other sea ice variables included in our model, the distance and duration of ice retreat from the continental shelf, do, however, reflect projected seasonal fluctuations (see below).

*2.3.2. Seasonal habitat availability.* Recognizing the potential importance of the seasonal separation of sea ice cover from preferred continental shelf foraging areas, and duration of such separation, we determined the number of ice-free months over the continental shelf and the average shelf-to-ice distance in both the observed and GCM-projected ice concentration maps. An ice-free month occurred in an ecoregion when <50% of the shelf area was covered by sea ice of ≥50% concentration. Shelf ice distance was the mean distance from every shelf pixel in a polar basin ecoregion to the nearest ice-covered pixel (>50% concentration) during the month of minimum ice extent. This described how far polar bears occupying sea ice habitats would be from their preferred continental shelf foraging areas. The average shelf-to-ice distance was not calculated for the SIE and AE because we considered those ecoregions to be composed entirely of shelf waters.

## 2.4. Bayesian Network Population Stressor Model

A Bayesian network is a graphical model that represents a set of variables (nodes) linked by probabilities [*Neopolitan*, 2003; *McCann et al.*, 2006]. Nodes can represent correlates or causal variables that affect some outcome of interest, and links define which specific variables directly affect which other specific variables. BNs can combine expert knowledge and empirical data into the same modeling structure. Crafting a BN augments understanding of relationships and sensitivities among the elements of a causal web and provides insights into the workings of the system that otherwise would not have been evident. BNs have become an accepted and popular modeling tool in many fields [*Pourret et al.*, 2008] including ecological and environmental sciences [e.g., *Aalders*, 2008; *Uusitalo*, 2007]. Each node in a BN model typically has two or more mutually exclusive states, the probabilities of which sum to one. Prior probabilities are distributed as discontinuous Dirichlet functions in the form of $D(x) = \lim_{m \to \infty} \lim_{n \to \infty} \cos^{2n}(m!\pi x)$, which is a multivariate, n state generalization of the two-state Beta distribution with state probabilities being continuous within [0,1]. BN nodes can represent categorical, ordinal, or continuous variable states or constant (scalar) values and typically have an associated probability table that describes either prior (unconditional) probabilities of each state for input nodes or conditional probabilities of each state for nodes that directly depend on other nodes (see *Marcot et al.* [2006] for a description of the underlying statistics). States $S$ of output nodes contain posterior probabilities that are calculated conditional upon nodes $H$ that directly affect them, using Bayes theorem, as $P(S|H) = [P(H|S)P(S)/P(H)]$ (see *Jensen* [2001] and *Marcot* [2006] for further explanation of the statistical basis of

BNs). BNs are "solved" by specifying the values of input nodes and having the model calculate posterior probabilities of the output node(s) through "Bayesian learning" [*Jensen*, 2001]. BNs are useful for modeling systems where empirical data are lacking, but variable interactions and their uncertainties can be depicted based on expert judgment [*Das*, 2000]. They are also particularly useful in efforts to synthesize large amounts of divergent quantitative and qualitative information to answer "what if" kinds of questions.

Developing a BN model entails depicting the "causal web" of interacting variables [*Marcot et al.*, 2006] in an influence diagram, assigning states to each node, and assigning probabilities to each node that define the conditions under which each state could occur. We used the modeling shell Netica® (Norsys, Inc.) and followed guidelines for developing BN models developed by *Jensen* [2001], *Cain* [2001], and *Marcot et al.* [2006].

Our BN stressor model was based on the knowledge of one polar bear expert (S. Amstrup), who established the model structure and probability tables according to expected influences among variables. B. Marcot served as a "knowledge engineer" and provided guidance to help structure the expert's knowledge into an appropriate BN format. Amstrup compiled an initial list of ecological correlates which were organized into an influence diagram (Plate 2). With discussion and questioning, Marcot guided Amstrup through several stages to a final model structure.

The BN model structure was divided into three kinds of nodes: (1) input nodes were anthropogenic stressors or environmental variables, states of input nodes were parameterized with unconditional probabilities; (2) summary nodes, sometimes called latent variables [e.g., *Bollen*, 1989], collect and summarize effects of multiple input nodes, states of these were parameterized with conditional probability tables; and (3) output nodes that represented numerical, distribution, and overall population responses to the suite of inputs. Probabilities of the various states of output nodes are derived through Bayesian learning. We developed the model structure in an iterative fashion adding variables for which we could hypothesize important roles. Published as well as unpublished information on how polar bears respond to changes in sea ice allowed us to parameterize the model to ensure it responded to particular input conditions in ways that paralleled responses of polar bear populations that have been observed or for which there are strong prevailing hypotheses among polar bear biologists worldwide.

To assure our outcomes were relevant to the question whether to list polar bears as a threatened species, we designed the summary nodes in the BN model to include four of the five major listing factors used to determine a species' status according to the Endangered Species Act [*U.S. Fish and Wildlife Service*, 2007]. We included summary nodes for factor A, habitat threats; factor B, overutilization; factor C, disease and predation; and factor E, other natural or manmade factors. We did not include factor D, inadequacy of existing regulatory mechanisms, because our model focused on ecosystem effects; however, regulatory aspects could be seamlessly added at a future time.

*2.5. Parameterizing the Bayesian Network Model*

We averaged the sea ice parameters for each GCM over decadal periods to generate metrics that were less sensitive to the intrinsic variability of GCM projections that occurs at annual timescales. The BN model was applied to each of the four ecoregions at six decadal time periods: 1985–1995, 1996–2006, 2020–2029, 2045–2054, 2070–2079, and 2090–2099. For convenience, we hereinafter refer to these six time periods, in relation to the present, as years –10, 0, 25, 50, 75, and 95. Analyses included observed habitat conditions from the satellite passive microwave data for years –10 and 0 and future habitat conditions projected by GCM ice projections for future years. To capture the full range of uncertainty in GCM outputs, we solved the BN model using sea ice parameters from the (1) GCM multimodel (ensemble) means, (2) GCM that projected the minimum ice extent, and (3) GCM that projected the maximum ice extent, for each ecoregion in each time period. Inputs other than sea ice features included various categories of anthropogenic stressors [*Barrett*, 1981] such as harvest, pollution, oil and gas development, shipping, and direct bear-human interactions. Inputs also included other environmental factors that could affect polar bear populations such as availability of primary and alternate prey and foraging areas and occurrence of parasites, disease, and predation [*Ramsay and Stirling*, 1984]. Whereas the ice habitat factors were entered into the BN model as ranges of values (e.g., ice retreat of 0–200 or 200–800 km beyond current measures), other potential stressors were included as ordinal or qualitative categories (Tables D1a and D1b).

Because we were interested in forecasting changes from current conditions, states of each node were expressed categorically as "compared to now." That is, an outcome state could represent a condition similar to present, better than present, or worse than present. Here, now or year "0" means the 1996–2006 period when referring to observations and 2000–2009 when referring to sea ice model projections. Before the BN model was run, we specified the states for each input node that seemed most plausible (Tables D1a and D1b).

States of environmental correlates were established under each combination of time step, ecoregion, and GCM model outputs. We ensured that input conditions matched the current understanding of polar bear ecology and parameterized the conditional probability tables to assure that node structures were specified in accordance with available polar bear data or expert understanding of data. We checked the validity of the model parameterization by testing whether the BN model responded to particular input conditions in ways that paralleled responses of polar bear populations to conditions that have been observed.

When the model is run, it calculates posterior probabilities of outcomes by applying standard Bayesian learning to the values assigned to each input variable. The relative influence of each input node, in terms of inherent model sensitivity structure, is determined by the values assigned in the conditional probability tables that underlie each summary or output node in the network. One input variable can be given greater influence than another if the result of a change in the first variable is thought to have a greater influence on the outcome states of the summary or output node than the second, and if the conditional probabilities are assigned accordingly. For example, it may be thought that the temporal absence of sea ice from the continental shelf is more important to the availability of foraging habitat than is the distance to which the ice retreats while it is absent. If data or projections suggest both measurements change in parallel, then temporal absence would have the greater final influence. If, however, data or projections show there is a greater change in distance than in time of absence, then distance may have the greatest contribution to posterior (outcome) probabilities even though its weight in the conditional probability table might be lower than temporal absence.

We used three different methods to arrive at final model structure: (1) sensitivity analyses of subparts of the model, (2) solving the model backward by specifying outcome states and evaluating if the most likely input states that were returned were plausible according to what we know about polar bears now, and (3) running the model (and subparts) forward to ascertain if the summary and output nodes responded as expected given the states of the input nodes. Our goals were to ensure that input conditions matched the current understanding of polar bear ecology and that the model responded to particular input conditions in ways that paralleled observed responses of polar bear populations.

As fully specified, the BN model consisted of 38 nodes, 44 links, and 1667 conditional probability values specified by the modelers (Plate 3 and Appendices A and B). The model was solved for each combination of four ecoregions, six time periods, and three future GCM scenarios (ensemble mean, maximum, and minimum).

The input data to run each combination were specified by summarizing the respective GCM-derived habitat variables

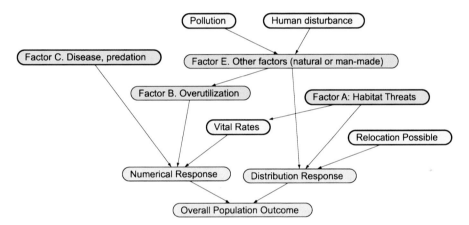

**Plate 2.** Basic influence diagram for the Bayesian network polar bear population stressor model showing the role of four listing factor categories (orange) used by U.S. Fish and Wildlife Service.

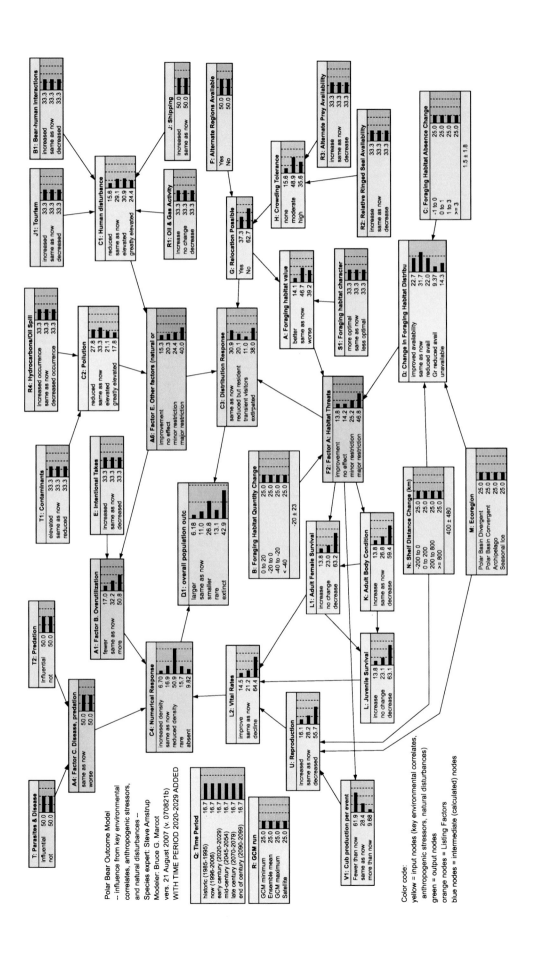

Polar Bear Outcome Model
-- influence from key environmental
correlates, anthropogenic stressors,
and natural disturbances --

Species expert: Steve Amstrup
Modeler: Bruce G. Marcot
vers. 21 August 2007 (v. 070821b)
WITH TIME PERIOD 2020-2029 ADDED

Color code:
yellow = input nodes (key environmental correlates,
anthropogenic stressors, natural disturbances)

green = output nodes

orange nodes = Listing Factors

blue nodes = intermediate (calculated) nodes

**Plate 3**

and the professional judgment of polar bear expert S. Amstrup (Tables D1a and D1b). Because BN models combine expert judgment and interpretation with quantitative and qualitative empirical information, inputs from multiple experts are sometimes used to structure and parameterize a "final" model. Because the model presented was parameterized with the judgment of only one polar bear expert, it should be viewed as a first-generation version. Accordingly, it will be refined through formally developed processes (see section 4) at a future time.

### 2.6. Output States of the Bayesian Network Model

The final outcomes of BN model runs were statements of relative probabilities that the population in each ecoregion would be larger than now, same as now, smaller than now, rare, or extinct. Responses of polar bears to projected habitat changes and other potential stressors could affect polar bear distribution or polar bear numbers independently in some cases, or they could affect both distribution and numbers simultaneously. Principal results of the BN model are levels of relative probabilities for the potential states at output nodes. In the polar bear BN population stressor model, outcomes of greatest interest were (1) those related to listing factors used by the FWS, (2) the distribution responses, (3) numerical responses, and (4) the overall population response.

We defined our output nodes (shown in Plate 3) in such a way that their possible states could be assessed empirically through future field observations. Potential states at the three principal output nodes are described below.

#### 2.6.1. Node C4: Numerical response. This node represents the anticipated numerical response of polar bears in an ecoregion based upon the sum total of the identified factors which are likely to have affected numbers of polar bears in each ecoregion:

- increased density, polar bear density detectable as significantly greater than that at year 0, where density can be expressed in terms of number of polar bears per unit area of optimal habitat (thus expressing "ecological density") or of total (optimal plus suboptimal) habitat (thus expressing "crude density");
- same as now, equivalent to the density at year 0;
- reduced density, polar bear density less than that at year 0 but greater than one half of the density at year 0;
- rare, polar bear density less than half of that at year 0;
- absent, polar bears are not demonstrably present.

#### 2.6.2. Node C3: Distribution response. Distribution refers here to the functional response of polar bears (namely, movement and spatial redistribution of bears) to changing conditions:

- same as now, polar bear distribution equivalent to that at year 0;
- reduced but resident, bears would still occur in the area but their spatial distribution would be more limited than at year 0;
- transient visitors, changing conditions would seasonally limit distribution of polar bears;
- extirpated, complete or effective year-round dearth of polar bears in the area.

#### 2.6.3. Node D1: Overall population outcome. Overall population outcome refers to the collective influence of both numerical response (node C4) and distribution response (node C3). It incorporates the full suite of effects from all anthropogenic stressors, natural disturbances, and environmental conditions on the expected occurrence and levels of polar bear populations in the ecoregion. Overall population outcome states were defined as follows:

- larger, polar bear populations have a numerical response greater than at present (year 0) and a distribution response at least the same as at present;
- same as now, polar bear populations numerically and distributionally indistinguishable from present;
- smaller, polar bears at a reduced density and distributed the same as at present or density same as at present but occur as transient visitors;
- rare, polar bears are numerically difficult to detect and have a distribution response same as at present, or occur as small numbers of transient visitors;
- extinct, polar bears are numerically absent or distributionally extirpated.

Here, the "extinct" state refers to conditions of (1) complete absence of the species (N=0) from an ecoregion; or (2) numbers and distributions below a "quasi-extinction" level, that refers to a nonzero population level at or below which the population is near extinction [*Ginzburg et al.*, 1982; *Otway et al.*, 2004]; or (3) functional extinction, that refers to being so scarce as to be near extinction and contributing negligibly to ecosystem processes [*Sekercioglu et al.*, 2004; *McConkey and Drake*, 2006].

---

**Plate 3.** (Opposite) Full Bayesian network population stressor model developed to forecast polar bear population outcomes in the 21st century. Values shown in the bottom of nodes B, N, and C represent expected values +/- 1 standard deviation which are automatically calculated and displayed by the Netica® modeling shell for continuous nodes with defined state values, based on Gaussian distributions.

Outcomes from the BN model are expressions of probability that each outcome state will occur (e.g., X% extinct, Y% rare, and Z% smaller). It is important here to understand that these probability values are provided without error bars and should not, in themselves, be interpreted as absolute measures of the certainty of any particular outcome. Rather, probabilities of outcome states of the model should be viewed in terms of their general direction and overall magnitudes. When predictions result in high probability of one outcome state and low or zero probabilities of all other states, there is low overall uncertainty of predicted results. When projected probabilities of various states are more equally distributed or when two or more states have large probability, there is greater uncertainty in the outcome. In these cases, careful consideration should be given to large probabilities representing particular states even if those probabilities are not the largest.

*2.7. Sensitivity of the Bayesian Network Model*

Knowledge of polar bears, their dependence on sea ice, the ways in which sea ice changes have been observed to affect polar bears, and professional judgment regarding how ecological and human factors may differ if sea ice changes occur as projected were used to populate the conditional probability tables in the BN model. Because our model incorporated the professional judgment of only one polar bear expert, it is reasonable to ask how robust the results might be to input probabilities which could vary among other experts. It also is appropriate to ask whether it is likely that future sea ice change, to which model outcomes are very sensitive, could fall into ranges that would result in qualitatively different outcomes than our BN model projects. Finally, it is appropriate to ask the extent to which model outcomes may be altered by active management of the states of nodes which represent variables humans could control.

We addressed questions about the ability of changes in human activities to alter the BN output states by fixing inputs humans could control and examining differences in the overall outcomes. We evaluated the extent to which sea ice projections would have to differ to make qualitative differences in outcomes by holding all non-ice variables at uniform priors and allowing ice variables only to vary at future time steps. Comparing those results to the range of ice conditions projected by our GCMs provided a sense of just how much the realized future ice conditions would have to vary from those projected to make a difference in population outcomes. Finally, although we cannot second guess how other polar bear experts may recommend parameterizing and structuring a BN model, comparison of model runs with preset values provides some sense of how much differently the model would have to be parameterized to pro-

vide qualitatively different outcome patterns than those we obtained.

We ran overall sensitivity analyses to determine the degree to which each input and summary variable influenced the outcome variables. For discrete and categorical variables, sensitivity was calculated in the modeling shell Netica® as the degree of entropy reduction (reduction in the disorder or variation) at one node relative to the information represented in other nodes of the model. That is, the sensitivity tests indicate how much of the variation in the node in question is explained by each of the other nodes considered. The degree of entropy reduction, $I$, is the expected reduction in mutual information of an output variable $Q$, with $q$ states, due to a finding of an input variable $F$, with $f$ states. For discrete variables, $I$ is measured in terms of information bits and is calculated as

$$I = H(Q) - H(Q|F) = \sum_q \sum_f \frac{P(q,f)\log_2[P(q,f)]}{P(q)P(f)},$$

where $H(Q)$ is the entropy of $Q$ before new findings are applied to input node $F$ and $H(Q|F)$ is the entropy of $Q$ after new findings are applied to $F$. In Netica®, entropy reduction is also termed mutual information.

For continuous variables, sensitivity is calculated as variance reduction $VR$, which is the expected reduction in variation, $V(Q)$, of the expected real value of the output variable $Q$ due to the value of input variable $F$, and is calculated as

$$VR = V(Q) - V(Q|F),$$

where

$$V(Q) = \sum_q P(q)[X_q - E(Q)]^2,$$

$$V(Q|F) = \sum_q P(q|f)[X_q - E(Q|f)]^2,$$

$$E(Q) = \sum_q P(q)X_q,$$

and where $X_q$ is the numeric real value corresponding to state $q$, $E(Q)$ is the expected real value of $Q$ before new findings are applied, $E(Q|F)$ is the expected real value of $Q$ after findings $f$ are applied to $F$, and $V(Q)$ is the variance in the real value of $Q$ before any new findings [*Marcot et al.*, 2006].

### 3. RESULTS

*3.1. Bayesian Network Model Outcomes*

The most probable BN model outcome, for both the SIE and PBDE, was "extinct" (Table 2 and Plate 4). In all but

**Table 2.** Results of the Bayesian Network Population Stressor Model, Showing the Most Probable Outcome State, and Probabilities, in Percentiles, of Each State for Overall Population Outcome (Node D1) for Four Polar Bear Ecoregions[a]

| Time Period | Basis | Node D1: Overall Population Outcome | | | | | |
|---|---|---|---|---|---|---|---|
| | | Most Probable Outcome | Larger (%) | Same as Now (%) | Smaller (%) | Rare (%) | Extinct (%) |
| *Seasonal Ice Ecoregion* | | | | | | | |
| Year −10 | Satellite data | larger | 93.92 | 5.75 | 0.30 | 0.02 | 0.00 |
| Year 0 | Satellite data | same_as_now | 21.85 | 43.72 | 18.98 | 8.37 | 7.07 |
| Year 25 | GCM minimum | extinct | 0.37 | 3.87 | 25.41 | 22.59 | 47.76 |
| Year 50 | GCM minimum | extinct | 0.05 | 0.61 | 9.79 | 12.36 | 77.19 |
| Year 75 | GCM minimum | extinct | 0.00 | 0.09 | 3.48 | 8.28 | 88.15 |
| Year 95 | GCM minimum | extinct | 0.00 | 0.09 | 3.48 | 8.28 | 88.15 |
| Year 25 | Ensemble mean | extinct | 0.37 | 3.87 | 25.41 | 22.59 | 47.76 |
| Year 50 | Ensemble mean | extinct | 0.05 | 0.61 | 9.79 | 12.36 | 77.19 |
| Year 75 | Ensemble mean | extinct | 0.00 | 0.09 | 3.48 | 8.28 | 88.15 |
| Year 95 | Ensemble mean | extinct | 0.00 | 0.09 | 3.48 | 8.28 | 88.15 |
| Year 25 | GCM maximum | extinct | 0.37 | 3.87 | 25.41 | 22.59 | 47.76 |
| Year 50 | GCM maximum | extinct | 0.24 | 2.20 | 24.37 | 19.35 | 53.85 |
| Year 75 | GCM maximum | extinct | 0.01 | 0.18 | 5.17 | 9.52 | 85.11 |
| Year 95 | GCM maximum | extinct | 0.01 | 0.18 | 5.17 | 9.52 | 85.11 |
| *Archipelago Ecoregion* | | | | | | | |
| Year −10 | Satellite data | same_as_now | 22.51 | 34.73 | 31.48 | 8.72 | 2.56 |
| Year 0 | Satellite data | larger | 69.48 | 29.26 | 1.06 | 0.19 | 0.00 |
| Year 25 | GCM minimum | same_as_now | 14.23 | 36.36 | 32.10 | 6.58 | 10.73 |
| Year 50 | GCM minimum | smaller | 4.57 | 12.93 | 51.34 | 20.60 | 10.56 |
| Year 75 | GCM minimum | extinct | 0.89 | 3.16 | 32.07 | 19.34 | 44.54 |
| Year 95 | GCM minimum | extinct | 1.38 | 4.65 | 33.38 | 19.51 | 41.07 |
| Year 25 | Ensemble mean | same_as_now | 14.23 | 36.36 | 32.10 | 6.58 | 10.73 |
| Year 50 | Ensemble mean | smaller | 4.57 | 12.93 | 51.34 | 20.60 | 10.56 |
| Year 75 | Ensemble mean | extinct | 1.05 | 3.34 | 32.25 | 26.07 | 37.30 |
| Year 95 | Ensemble mean | extinct | 1.38 | 4.65 | 33.38 | 19.51 | 41.07 |
| Year 25 | GCM maximum | same_as_now | 14.23 | 36.36 | 32.10 | 6.58 | 10.73 |
| Year 50 | GCM maximum | smaller | 5.83 | 15.93 | 52.35 | 18.01 | 7.88 |
| Year 75 | GCM maximum | smaller | 4.42 | 12.40 | 49.36 | 22.96 | 10.85 |
| Year 95 | GCM maximum | extinct | 1.38 | 4.65 | 33.38 | 19.51 | 41.07 |
| *Polar Basin Divergent Ecoregion* | | | | | | | |
| Year −10 | Satellite data | larger | 99.78 | 0.22 | 0.00 | 0.00 | 0.00 |
| Year 0 | Satellite data | same_as_now | 24.16 | 56.60 | 13.36 | 4.73 | 1.14 |
| Year 25 | GCM minimum | extinct | 0.00 | 0.97 | 18.98 | 23.00 | 57.05 |
| Year 50 | GCM minimum | extinct | 0.00 | 0.00 | 2.86 | 10.58 | 86.55 |
| Year 75 | GCM minimum | extinct | 0.00 | 0.00 | 3.07 | 10.91 | 86.02 |
| Year 95 | GCM minimum | extinct | 0.00 | 0.00 | 3.88 | 12.23 | 83.89 |
| Year 25 | Ensemble mean | extinct | 0.21 | 2.37 | 27.43 | 24.69 | 45.30 |
| Year 50 | Ensemble mean | extinct | 0.00 | 0.18 | 6.16 | 13.34 | 80.33 |
| Year 75 | Ensemble mean | extinct | 0.00 | 0.00 | 2.86 | 10.58 | 86.55 |
| Year 95 | Ensemble mean | extinct | 0.00 | 0.00 | 3.88 | 12.23 | 83.89 |
| Year 25 | GCM maximum | extinct | 0.25 | 2.43 | 27.88 | 27.58 | 41.86 |
| Year 50 | GCM maximum | extinct | 0.00 | 0.18 | 6.16 | 13.34 | 80.33 |
| Year 75 | GCM maximum | extinct | 0.00 | 0.07 | 4.46 | 12.00 | 83.47 |
| Year 95 | GCM maximum | extinct | 0.00 | 0.09 | 5.73 | 13.84 | 80.33 |

**Table 2.** (continued)

| Time Period | Basis | Node D1: Overall Population Outcome | | | | | |
|---|---|---|---|---|---|---|---|
| | | Most Probable Outcome | Larger (%) | Same as Now (%) | Smaller (%) | Rare (%) | Extinct (%) |
| | | *Polar Basin Convergent Ecoregion* | | | | | |
| Year −10 | Satellite data | larger | 98.39 | 1.61 | 0.00 | 0.00 | 0.00 |
| Year 0 | Satellite data | larger | 71.69 | 27.49 | 0.63 | 0.19 | 0.00 |
| Year 25 | GCM minimum | larger | 64.86 | 28.47 | 6.09 | 0.49 | 0.09 |
| Year 50 | GCM minimum | extinct | 0.26 | 2.30 | 27.98 | 31.59 | 37.87 |
| Year 75 | GCM minimum | extinct | 0.00 | 0.39 | 9.68 | 13.24 | 76.70 |
| Year 95 | GCM minimum | extinct | 0.00 | 0.39 | 9.68 | 13.24 | 76.70 |
| Year 25 | Ensemble mean | same_as_now | 18.23 | 41.81 | 26.37 | 5.16 | 8.43 |
| Year 50 | Ensemble mean | extinct | 0.48 | 2.72 | 29.27 | 32.46 | 35.06 |
| Year 75 | Ensemble mean | extinct | 0.00 | 0.27 | 8.40 | 15.10 | 76.23 |
| Year 95 | Ensemble mean | extinct | 0.02 | 0.44 | 9.49 | 12.75 | 77.30 |
| Year 25 | GCM maximum | same_as_now | 18.23 | 41.81 | 26.37 | 5.16 | 8.43 |
| Year 50 | GCM maximum | extinct | 0.14 | 1.24 | 21.15 | 30.71 | 46.77 |
| Year 75 | GCM maximum | extinct | 0.02 | 0.46 | 12.64 | 24.46 | 62.41 |
| Year 95 | GCM maximum | extinct | 0.02 | 0.44 | 10.51 | 16.52 | 72.52 |

[a]See Plate 3.

the earliest time periods, we forecasted low probabilities for all other outcome states in these two ecoregions. The low probability afforded to outcome states other than extinct suggested a clear trend in these ecoregions toward probable extirpation by mid century. Forecasts were less severe in other ecoregions. At year 50, probability of the "extinct" state was only 8–10% in the AE. At all time steps in the AE, and at year 50 in the PBCE, considerable probability fell into outcome states other than extinct (Table 2 and Plate 4). The distribution of probabilities for the states of overall population outcome suggests polar bears could persist in all ecoregions through the early part of the century, through mid century in the PBCE and through the end of the century in the AE (Table 2 and Plate 4).

Future conditions affected node C3, polar bear distribution, more than they affected node C4, polar bear numbers. "Extirpated" was the most probable outcome at mid century for node C3 in the PBDE and SIE. The most probable outcome in these ecoregions for node C4, however, was reduced density [see *Amstrup et al.*, 2007]. This probably reflects the high relative certainty that areas where ice is absent for too long will not support many bears, and the relative uncertainty regarding how population dynamics features may change while the sea ice is retreating. Modeled future polar bear distributions were driven by the FWS listing factor "Habitat Threats" (node F2, Table 3), as well as by node D, habitat distribution. Distribution and availability of habitat, especially in the

SIE and PBDE (Table 3) appear to be the most salient threats to polar bears. We also assumed that deteriorating sea ice would be accompanied by worsening conditions listed under FWS listing factors C, disease and predation, etc., and E, other natural or man-made factors (nodes A4 and A6) (Table 4). We included year 25 in our projections to help provide context for mid century projections and beyond and to help understand the transition from current to future conditions. It is important to emphasize, however, that polar bears have a long life span. Many individuals alive now could still be alive during the decade of 2020–2029. Hence, projecting changes between now and then incorporates the uncertainty of tradeoffs between functional and numerical responses, as well as the greater uncertainties in sea ice status in the nearer term.

### 3.2. Sensitivity Structure of the Bayesian Network Model

We conducted 10 tests on the BN population stressor model to determine its sensitivity structure (Appendix C). The BN model was well balanced in that sensitivity of overall population outcome (node D1, sensitivity test 1) was not dominated by a single or small group of input variables. Considering that "ecoregion" and "availability of alternate regions" are in essence habitat variables, 6 of the top 7 variables explaining overall outcome were sea ice related and together explained 87% of the variation in overall population outcome (node D1, Appendix C and Plate 5).

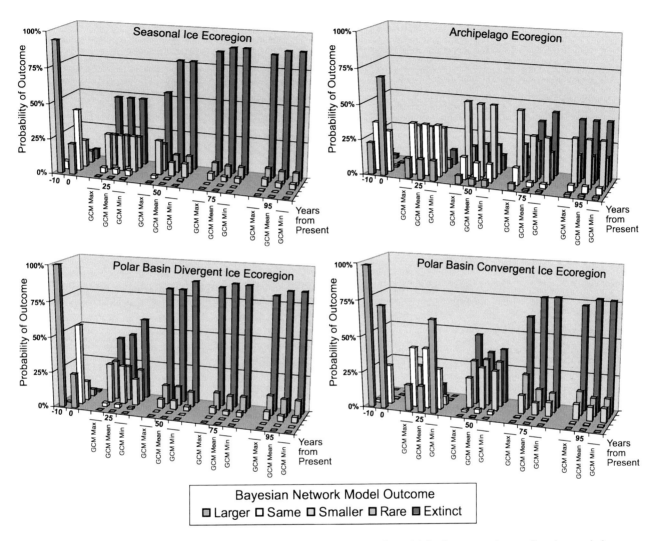

**Plate 4.** Projected polar bear population outcomes of Bayesian network model for four ecoregions at five time periods relative to present. Present and prior decade (years 0 and −10) sea ice conditions were from observed record. Future ice conditions were based on the ensemble mean of 10 GCMs and the two GCMs that forecasted maximum and minimum ice extent in each ecoregion at each time period. Note that strength of dominant outcomes (tallest bars) is inversely proportional to heights of competing outcomes. Outcome definitions are as follows: larger, more abundant than present (year 0) plus distribution at least the same as at present; same, numerical and distribution responses similar to present; smaller, reduced in numbers and distribution; rare, numerically rare but occupying similar distribution or reduced numerically but spatially represented as transient visitors; and extinct, numerically absent or distributionally extirpated.

**Table 3.** Results of the Bayesian Network Population Stressor Model, Showing the Most Probable Outcome States, and Probabilities of Each State, in Percentiles, for Habitat Threats and Direct Mortalities Summary Variables (Nodes F2 and A1) for Four Polar Bear Ecoregions[a]

| Time Period | Basis | Node F2: Factor A—Habitat Threats | | | | | Node A1: Factor B—Overutilization | | | |
| --- | --- | --- | --- | --- | --- | --- | --- | --- | --- | --- |
| | | Most Probable Outcome | Improvement (%) | No Effect (%) | Minor Restriction (%) | Major Restriction (%) | Most Probable Outcome (%) | Fewer (%) | Same as Now (%) | More (%) |
| *Seasonal Ice Ecoregion* | | | | | | | | | | |
| Year −10 | Satellite data | improvement | 94.60 | 5.00 | 0.40 | 0.00 | fewer | 100.00 | 0.00 | 0.00 |
| Year 0 | Satellite data | no_effect | 26.41 | 36.84 | 23.02 | 13.72 | same_as_now | 0.00 | 100.00 | 0.00 |
| Year 25 | GCM minimum | major_restriction | 0.60 | 10.52 | 43.40 | 45.48 | same_as_now | 0.00 | 62.60 | 37.40 |
| Year 50 | GCM minimum | major_restriction | 0.08 | 2.00 | 16.64 | 81.28 | same_as_now | 0.00 | 62.60 | 37.40 |
| Year 75 | GCM minimum | major_restriction | 0.00 | 0.00 | 4.72 | 95.28 | same_as_now | 0.00 | 60.00 | 40.00 |
| Year 95 | GCM minimum | major_restriction | 0.00 | 0.00 | 4.72 | 95.28 | same_as_now | 0.00 | 60.00 | 40.00 |
| Year 25 | Ensemble mean | major_restriction | 0.60 | 10.52 | 43.40 | 45.48 | same_as_now | 0.00 | 62.60 | 37.40 |
| Year 50 | Ensemble mean | major_restriction | 0.08 | 2.00 | 16.64 | 81.28 | same_as_now | 0.00 | 62.60 | 37.40 |
| Year 75 | Ensemble mean | major_restriction | 0.00 | 0.00 | 4.72 | 95.28 | same_as_now | 0.00 | 60.00 | 40.00 |
| Year 95 | Ensemble mean | major_restriction | 0.00 | 0.00 | 4.72 | 95.28 | same_as_now | 0.00 | 60.00 | 40.00 |
| Year 25 | GCM maximum | major_restriction | 0.60 | 10.52 | 43.40 | 45.48 | same_as_now | 0.00 | 62.60 | 37.40 |
| Year 50 | GCM maximum | major_restriction | 0.40 | 9.68 | 43.60 | 46.32 | same_as_now | 0.00 | 62.60 | 37.40 |
| Year 75 | GCM maximum | major_restriction | 0.00 | 0.08 | 9.60 | 90.32 | same_as_now | 0.00 | 60.00 | 40.00 |
| Year 95 | GCM maximum | major_restriction | 0.00 | 0.08 | 9.60 | 90.32 | same_as_now | 0.00 | 60.00 | 40.00 |
| *Archipelago Ecoregion* | | | | | | | | | | |
| Year −10 | Satellite data | no_effect | 39.00 | 44.60 | 16.40 | 0.00 | same_as_now | 4.80 | 53.00 | 42.20 |
| Year 0 | Satellite data | improvement | 88.56 | 10.43 | 1.01 | 0.00 | same_as_now | 0.00 | 100.00 | 0.00 |
| Year 25 | GCM minimum | no_effect | 20.80 | 38.40 | 32.80 | 8.00 | same_as_now | 0.00 | 75.14 | 24.86 |
| Year 50 | GCM minimum | no_effect | 32.48 | 41.28 | 22.30 | 3.94 | more | 0.00 | 0.00 | 100.00 |
| Year 75 | GCM minimum | minor_restriction | 4.08 | 24.32 | 40.32 | 31.28 | more | 0.00 | 30.00 | 70.00 |
| Year 95 | GCM minimum | minor_restriction | 4.08 | 24.32 | 40.32 | 31.28 | same_as_now | 0.00 | 60.00 | 40.00 |
| Year 25 | Ensemble mean | no_effect | 20.80 | 38.40 | 32.80 | 8.00 | same_as_now | 0.00 | 75.14 | 24.86 |
| Year 50 | Ensemble mean | no_effect | 32.48 | 41.28 | 22.30 | 3.94 | more | 0.00 | 0.00 | 100.00 |
| Year 75 | Ensemble mean | minor_restriction | 4.96 | 25.44 | 39.84 | 29.76 | more | 0.00 | 30.00 | 70.00 |
| Year 95 | Ensemble mean | minor_restriction | 4.08 | 24.32 | 40.32 | 31.28 | same_as_now | 0.00 | 60.00 | 40.00 |
| Year 25 | GCM maximum | no_effect | 20.80 | 38.40 | 32.80 | 8.00 | same_as_now | 0.00 | 75.14 | 24.86 |
| Year 50 | GCM maximum | improvement | 41.92 | 38.40 | 17.06 | 2.62 | more | 0.00 | 0.00 | 100.00 |
| Year 75 | GCM maximum | no_effect | 32.48 | 41.28 | 22.30 | 3.94 | more | 0.00 | 0.00 | 100.00 |
| Year 95 | GCM maximum | minor_restriction | 4.08 | 24.32 | 40.32 | 31.28 | same_as_now | 0.00 | 60.00 | 40.00 |
| *Polar Basin Divergent Ecoregion* | | | | | | | | | | |
| Year −10 | Satellite data | improvement | 99.68 | 0.32 | 0.00 | 0.00 | fewer | 100.00 | 0.00 | 0.00 |
| Year 0 | Satellite data | no_effect | 30.20 | 47.24 | 20.54 | 2.02 | same_as_now | 0.00 | 100.00 | 0.00 |
| Year 25 | GCM minimum | major_restriction | 0.00 | 2.16 | 45.44 | 52.40 | same_as_now | 0.00 | 60.00 | 40.00 |
| Year 50 | GCM minimum | major_restriction | 0.00 | 0.00 | 0.00 | 100.00 | same_as_now | 0.00 | 60.00 | 40.00 |
| Year 75 | GCM minimum | major_restriction | 0.00 | 0.00 | 0.00 | 100.00 | same_as_now | 0.00 | 60.60 | 39.40 |
| Year 95 | GCM minimum | major_restriction | 0.00 | 0.00 | 0.00 | 100.00 | same_as_now | 0.00 | 63.00 | 37.00 |
| Year 25 | Ensemble mean | minor_restriction | 0.72 | 12.72 | 48.96 | 37.60 | same_as_now | 0.00 | 60.00 | 40.00 |
| Year 50 | Ensemble mean | major_restriction | 0.00 | 0.36 | 9.80 | 89.84 | same_as_now | 0.00 | 60.00 | 40.00 |
| Year 75 | Ensemble mean | major_restriction | 0.00 | 0.00 | 0.00 | 100.00 | same_as_now | 0.00 | 60.00 | 40.00 |
| Year 95 | Ensemble mean | major_restriction | 0.00 | 0.00 | 0.00 | 100.00 | same_as_now | 0.00 | 63.00 | 37.00 |
| Year 25 | GCM maximum | minor_restriction | 0.89 | 13.45 | 48.62 | 37.04 | same_as_now | 0.00 | 60.00 | 40.00 |
| Year 50 | GCM maximum | major_restriction | 0.00 | 0.36 | 9.80 | 89.84 | same_as_now | 0.00 | 60.00 | 40.00 |
| Year 75 | GCM maximum | major_restriction | 0.00 | 0.00 | 5.08 | 94.92 | same_as_now | 0.00 | 60.00 | 40.00 |
| Year 95 | GCM maximum | major_restriction | 0.00 | 0.00 | 5.08 | 94.92 | same_as_now | 0.00 | 63.60 | 36.40 |

**Table 3.** (continued)

| Time Period | Basis | Node F2: Factor A—Habitat Threats | | | | | Node A1: Factor B—Overutilization | | | |
|---|---|---|---|---|---|---|---|---|---|---|
| | | Most Probable Outcome | Improvement (%) | No Effect (%) | Minor Restriction (%) | Major Restriction (%) | Most Probable Outcome (%) | Fewer (%) | Same as Now (%) | More (%) |
| | | | | | *Polar Basin Convergent Ecoregion* | | | | | |
| Year −10 | Satellite data | improvement | 97.48 | 2.52 | 0.00 | 0.00 | fewer | 100.00 | 0.00 | 0.00 |
| Year 0 | Satellite data | improvement | 88.56 | 10.43 | 1.01 | 0.00 | same_as_now | 0.00 | 100.00 | 0.00 |
| Year 25 | GCM minimum | improvement | 87.20 | 11.20 | 1.60 | 0.00 | same_as_now | 0.00 | 89.00 | 11.00 |
| Year 50 | GCM minimum | minor_restriction | 1.10 | 14.38 | 48.19 | 36.32 | same_as_now | 0.00 | 60.00 | 40.00 |
| Year 75 | GCM minimum | major_restriction | 0.00 | 0.00 | 23.60 | 76.40 | same_as_now | 0.00 | 60.00 | 40.00 |
| Year 95 | GCM minimum | major_restriction | 0.00 | 0.00 | 23.60 | 76.40 | same_as_now | 0.00 | 60.00 | 40.00 |
| Year 25 | Ensemble mean | no_effect | 20.80 | 38.40 | 32.80 | 8.00 | same_as_now | 0.00 | 89.00 | 11.00 |
| Year 50 | Ensemble mean | minor_restriction | 1.25 | 15.49 | 49.10 | 34.16 | same_as_now | 0.00 | 60.00 | 40.00 |
| Year 75 | Ensemble mean | major_restriction | 0.00 | 0.00 | 17.65 | 82.35 | same_as_now | 0.00 | 60.00 | 40.00 |
| Year 95 | Ensemble mean | major_restriction | 0.00 | 0.24 | 22.16 | 77.60 | same_as_now | 0.00 | 60.00 | 40.00 |
| Year 25 | GCM maximum | no_effect | 20.80 | 38.40 | 32.80 | 8.00 | same_as_now | 0.00 | 89.00 | 11.00 |
| Year 50 | GCM maximum | major_restriction | 0.29 | 4.22 | 45.49 | 50.00 | same_as_now | 0.00 | 60.00 | 40.00 |
| Year 75 | GCM maximum | major_restriction | 0.00 | 0.58 | 25.18 | 74.24 | same_as_now | 0.00 | 60.00 | 40.00 |
| Year 95 | GCM maximum | major_restriction | 0.00 | 0.35 | 23.13 | 76.52 | same_as_now | 0.00 | 60.00 | 40.00 |

[a]See Plate 3.

**Table 4.** Results of the Bayesian Network Population Stressor Model, Showing the Most Probable Outcome States, and Probabilities of Each State, for Disease/Predation and Other Disturbance Factors Variables (Nodes A4 and A6) for Four Polar Bear Ecoregions[a]

| Time Period | Basis | Node A4: Factor C—Disease and Predation | | | Node A6: Factor E—Other Factors (Natural or Man-Made) | | | | |
|---|---|---|---|---|---|---|---|---|---|
| | | Most Probable Outcome | Same as Now (%) | Worse (%) | Most Probable Outcome | Improvement (%) | No Effect (%) | Minor Restriction (%) | Major Restriction (%) |
| | | | | | *Seasonal Ice Ecoregion* | | | | |
| Year −10 | Satellite data | same_as_now | 100.00 | 0.00 | improvement | 84.80 | 15.20 | 0.00 | 0.00 |
| Year 0 | Satellite data | same_as_now | 100.00 | 0.00 | no_effect | 0.00 | 100.00 | 0.00 | 0.00 |
| Year 25 | GCM minimum | worse | 30.00 | 70.00 | major_restriction | 0.00 | 0.00 | 13.00 | 87.00 |
| Year 50 | GCM minimum | worse | 0.00 | 100.00 | major_restriction | 0.00 | 0.00 | 13.00 | 87.00 |
| Year 75 | GCM minimum | worse | 0.00 | 100.00 | major_restriction | 0.00 | 0.00 | 0.00 | 100.00 |
| Year 95 | GCM minimum | worse | 0.00 | 100.00 | major_restriction | 0.00 | 0.00 | 0.00 | 100.00 |
| Year 25 | Ensemble mean | worse | 30.00 | 70.00 | major_restriction | 0.00 | 0.00 | 13.00 | 87.00 |
| Year 50 | Ensemble mean | worse | 0.00 | 100.00 | major_restriction | 0.00 | 0.00 | 13.00 | 87.00 |
| Year 75 | Ensemble mean | worse | 0.00 | 100.00 | major_restriction | 0.00 | 0.00 | 0.00 | 100.00 |
| Year 95 | Ensemble mean | worse | 0.00 | 100.00 | major_restriction | 0.00 | 0.00 | 0.00 | 100.00 |
| Year 25 | GCM maximum | worse | 30.00 | 70.00 | major_restriction | 0.00 | 0.00 | 13.00 | 87.00 |
| Year 50 | GCM maximum | worse | 0.00 | 100.00 | major_restriction | 0.00 | 0.00 | 13.00 | 87.00 |
| Year 75 | GCM maximum | worse | 0.00 | 100.00 | major_restriction | 0.00 | 0.00 | 0.00 | 100.00 |
| Year 95 | GCM maximum | Worse | 0.00 | 100.00 | major_restriction | 0.00 | 0.00 | 0.00 | 100.00 |
| | | | | | *Archipelago Ecoregion* | | | | |
| Year −10 | Satellite data | same_as_now | 100.00 | 0.00 | major_restriction | 4.80 | 20.00 | 34.80 | 40.40 |
| Year 0 | Satellite data | same_as_now | 100.00 | 0.00 | no_effect | 0.00 | 100.00 | 0.00 | 0.00 |
| Year 25 | GCM minimum | same_as_now | 100.00 | 0.00 | no_effect | 0.00 | 46.40 | 42.20 | 11.40 |
| Year 50 | GCM minimum | worse | 30.00 | 70.00 | major_restriction | 0.00 | 0.00 | 28.00 | 72.00 |

**Table 4.** (continued)

| Time Period | Basis | Node A4: Factor C—Disease and Predation | | | Node A6: Factor E—Other Factors (Natural or Man-Made) | | | | |
|---|---|---|---|---|---|---|---|---|---|
| | | Most Probable Outcome | Same as Now (%) | Worse (%) | Most Probable Outcome | Improvement (%) | No Effect (%) | Minor Restriction (%) | Major Restriction (%) |
| Year 75 | GCM minimum | worse | 0.00 | 100.00 | major_restriction | 0.00 | 0.00 | 0.00 | 100.00 |
| Year 95 | GCM minimum | worse | 0.00 | 100.00 | major_restriction | 0.00 | 0.00 | 0.00 | 100.00 |
| Year 25 | Ensemble mean | same_as_now | 100.00 | 0.00 | no_effect | 0.00 | 46.40 | 42.20 | 11.40 |
| Year 50 | Ensemble mean | worse | 30.00 | 70.00 | major_restriction | 0.00 | 0.00 | 28.00 | 72.00 |
| Year 75 | Ensemble mean | worse | 0.00 | 100.00 | major_restriction | 0.00 | 0.00 | 0.00 | 100.00 |
| Year 95 | Ensemble mean | worse | 0.00 | 100.00 | major_restriction | 0.00 | 0.00 | 0.00 | 100.00 |
| Year 25 | GCM maximum | same_as_now | 100.00 | 0.00 | no_effect | 0.00 | 46.40 | 42.20 | 11.40 |
| Year 50 | GCM maximum | worse | 30.00 | 70.00 | major_restriction | 0.00 | 0.00 | 28.00 | 72.00 |
| Year 75 | GCM maximum | worse | 30.00 | 70.00 | major_restriction | 0.00 | 0.00 | 0.00 | 100.00 |
| Year 95 | GCM maximum | worse | 0.00 | 100.00 | major_restriction | 0.00 | 0.00 | 0.00 | 100.00 |
| | | | | *Polar Basin Divergent Ecoregion* | | | | | |
| Year −10 | Satellite data | same_as_now | 100.00 | 0.00 | improvement | 100.00 | 0.00 | 0.00 | 0.00 |
| Year 0 | Satellite data | same_as_now | 100.00 | 0.00 | no_effect | 0.00 | 100.00 | 0.00 | 0.00 |
| Year 25 | GCM minimum | worse | 0.00 | 100.00 | major_restriction | 0.00 | 0.00 | 0.00 | 100.00 |
| Year 50 | GCM minimum | worse | 0.00 | 100.00 | major_restriction | 0.00 | 0.00 | 0.00 | 100.00 |
| Year 75 | GCM minimum | worse | 0.00 | 100.00 | major_restriction | 0.00 | 0.00 | 3.00 | 97.00 |
| Year 95 | GCM minimum | worse | 0.00 | 100.00 | major_restriction | 0.00 | 0.00 | 15.00 | 85.00 |
| Year 25 | Ensemble mean | worse | 0.00 | 100.00 | major_restriction | 0.00 | 0.00 | 0.00 | 100.00 |
| Year 50 | Ensemble mean | worse | 0.00 | 100.00 | major_restriction | 0.00 | 0.00 | 0.00 | 100.00 |
| Year 75 | Ensemble mean | worse | 0.00 | 100.00 | major_restriction | 0.00 | 0.00 | 0.00 | 100.00 |
| Year 95 | Ensemble mean | worse | 0.00 | 100.00 | major_restriction | 0.00 | 0.00 | 15.00 | 85.00 |
| Year 25 | GCM maximum | worse | 0.00 | 100.00 | major_restriction | 0.00 | 0.00 | 0.00 | 100.00 |
| Year 50 | GCM maximum | worse | 0.00 | 100.00 | major_restriction | 0.00 | 0.00 | 0.00 | 100.00 |
| Year 75 | GCM maximum | worse | 0.00 | 100.00 | major_restriction | 0.00 | 0.00 | 0.00 | 100.00 |
| Year 95 | GCM maximum | worse | 0.00 | 100.00 | major_restriction | 0.00 | 0.00 | 18.00 | 82.00 |
| | | | | *Polar Basin Convergent Ecoregion* | | | | | |
| Year −10 | Satellite data | same_as_now | 100.00 | 0.00 | improvement | 100.00 | 0.00 | 0.00 | 0.00 |
| Year 0 | Satellite data | same_as_now | 100.00 | 0.00 | no_effect | 0.00 | 100.00 | 0.00 | 0.00 |
| Year 25 | GCM minimum | same_as_now | 100.00 | 0.00 | no_effect | 0.00 | 80.00 | 10.00 | 10.00 |
| Year 50 | GCM minimum | worse | 0.00 | 100.00 | major_restriction | 0.00 | 0.00 | 0.00 | 100.00 |
| Year 75 | GCM minimum | worse | 0.00 | 100.00 | major_restriction | 0.00 | 0.00 | 0.00 | 100.00 |
| Year 95 | GCM minimum | worse | 0.00 | 100.00 | major_restriction | 0.00 | 0.00 | 0.00 | 100.00 |
| Year 25 | Ensemble mean | same_as_now | 100.00 | 0.00 | no_effect | 0.00 | 80.00 | 10.00 | 10.00 |
| Year 50 | Ensemble mean | worse | 0.00 | 100.00 | major_restriction | 0.00 | 0.00 | 0.00 | 100.00 |
| Year 75 | Ensemble mean | worse | 0.00 | 100.00 | major_restriction | 0.00 | 0.00 | 0.00 | 100.00 |
| Year 95 | Ensemble mean | worse | 0.00 | 100.00 | major_restriction | 0.00 | 0.00 | 0.00 | 100.00 |
| Year 25 | GCM maximum | same_as_now | 100.00 | 0.00 | no_effect | 0.00 | 80.00 | 10.00 | 10.00 |
| Year 50 | GCM maximum | worse | 0.00 | 100.00 | major_restriction | 0.00 | 0.00 | 0.00 | 100.00 |
| Year 75 | GCM maximum | worse | 0.00 | 100.00 | major_restriction | 0.00 | 0.00 | 0.00 | 100.00 |
| Year 95 | GCM maximum | worse | 0.00 | 100.00 | major_restriction | 0.00 | 0.00 | 0.00 | 100.00 |

[a]See Plate 3.

Relative to the FWS listing factors, overall population outcome was most influenced by stressors related to factor A (habitat threats). Influences from factor B (overutilization), factor E (other natural or man-made factors), and factor C (disease and predation) provided progressively less influence (Appendix C, sensitivity test 2).

Recognizing the sensitivity of model outcomes to changes in sea ice, we reran the BN population stressor model under two sets of fixed conditions to determine whether management of human activities on the ground might be able to alter sea ice–driven outcomes. In these "influence runs" we set the states for all nodes over which humans might be able to exert control (e.g., harvest, contaminants, oil, and gas development) first to "same as now," and then to "improved conditions." After doing so, projected probabilities of extinction were lower at every time step (Plate 6). At and beyond mid century, extinction was still the most probable outcome in the PBDE and SIE. However, extinction did not become the most probable outcome in the PBDE and SIE until mid century. And in the SIE, with model runs based on GCMs retaining the maximum sea ice, extinction, as the most probable outcome, was avoided until year 75. Recall that extinction was the most probable outcome in these ecoregions at year 25 in the original model runs. In contrast, results of these influence runs suggested that on the ground management of human activities could improve the fate of polar bears in the AE and PBCE through the latter part of the century (Plate 6). In summary, for our 50-year foreseeable future, it appeared that management of localized human activities could benefit polar bears in the PBCE and especially in the AE but was likely to have little qualitative effect on the future of polar bears in the PBDE and SIE if sea ice continues to decline as projected.

To examine how much different, than projected, future sea ice would need to be to cause a qualitative change in our overall outcomes, we composed another influence run in which we set the values for all non-ice inputs to uniform prior probabilities. That is, we assumed complete uncertainty with regard to future food availability, oil and gas activity, contaminants, disease, etc. Then, we ran the model to determine how changes in the sea ice states alone, specified by our ensemble of GCMs, would affect our outcomes. This exercise illustrated that in order to obtain any qualitative change in the probability of extinction in any of the ecoregions, sea ice projections would need to leave more sea ice, at all future time steps, than even the maximum-ice GCM projection we used (Plate 6).

## 4. DISCUSSION

### 4.1. Uncertainty

Analyses in this paper contain four main categories of uncertainty: (1) uncertainty in our understandings of the biological, ecological, and climatological systems; (2) uncertainty in the representation of those understandings in models and statistical descriptions; (3) uncertainty in predictions of species abundance and distribution, and (4) uncertainty in model credibility, acceptability, and appropri-

ateness of model structure. All of these can influence model predictions. Uncertainty in our understanding of complex ecosystems is virtually inevitable. We have, however, dealt with this as well as possible by incorporating a broad sweep of available information regarding polar bears and their environment. How to best represent our understanding of the system in models, the second source of uncertainty, can be structured in various ways. Here, we captured and represented expert understanding of polar bear habitats and populations in a manner that can be reviewed, tested, verified, calibrated, and amended as appropriate. We have attempted to open the "black box" so to speak and to fully expose all formulas and probabilities. We also used sensitivity testing to understand the dynamics of BN model predictions [*Johnson and Gillingham*, 2004] (Appendix C). After BN models of this type are modified through peer review or revised by incorporating the knowledge from more than one expert into the model parameterization, any variation in resulting models can represent the divergence (or convergence) of expertise and judgment among multiple specialists.

Also included in the second category of uncertainty are those associated with statistical estimation of parameters, including measurement and random errors. The sea ice parameters we used in our polar bear models were derived from GCM outputs that possess their own wide margins of uncertainty [*DeWeaver*, 2007]. Hence, the magnitude and distribution of errors associated with our sea ice parameters were unknown. To compensate for these unknowns, we accommodated a broad range of sea ice uncertainties by analyzing the 10-member ensemble GCM mean, as well as the minimum and maximum GCM ice forecasts. In the case of polar bear population estimates, many are known so poorly that the best we have are educated guesses. Pooling subpopulations where numbers are merely guesses with those where precise estimates are available, to gain a range-wide perspective, prevents meaningful calculation and incorporation of specific error terms. We recognize that difficulty, but because our projections are expressed in the context of a comparison to present conditions, we largely avoid the issue. That is, whatever the population size is now, the future size is expressed relative to that and all errors are carried forward.

The third category, uncertainty in predictions of species abundance and distribution can be subject to errors because of spatial autocorrelation, dispersal and movement of organisms, and biotic and environmental interactions [*Guisan et al.*, 2006]. We addressed these error sources by deriving estimates of ice habitat area separately for each ecoregion from the GCM models because sea ice formation, melt, and advection occur differently in each ecoregion. The BN population stressor model accounted explicitly for potential movement of polar bears (e.g., availability of alternative regions) and for biotic and environmental interactions (as expressed in the conditional probability tables; see Appendix B). The spread of probabilities among the BN outcome states, reflect the combinations of uncertainties in states across all other variables, as reflected in each of their conditional probability tables (Appendix B). This spread carries important information for the decision maker who needs to weigh alternative outcomes in a risk assessment (see below).

Finally, uncertainty in model predictions entails addressing model credibility, acceptability, and appropriateness of the model structure. We made every effort to ensure that the model structure was appropriate and credible and that the inputs (Tables D1a and D1b) and conditional probability tables (Appendix B) were parameterized according to best available knowledge of polar bears and their environment. We explored the logic and structure of our BN model through sensitivity analyses, running the model backward from particular states to ensure it returned the appropriate starting point, and performing particular "what if" experiments (e.g., by fixing values in some nodes and watching how values at other nodes respond). We are as confident as we can be at this point of development that our BN model is performing in a plausible manner and providing outcomes that can be useful in qualitatively forecasting the potential future status of polar bears.

Although this manuscript and the model it describes have been peer reviewed by additional polar bear experts, the model structure and parameterizations were based upon the judgments of only one expert. Therefore, additional criteria of model validation must be addressed through subsequent peer review of the model parameters and structure [*Marcot et al.*, 1983; *Marcot*, 1990, 2006; *Marcot et al.*, 2006]. This requirement means the model presented here should be viewed as a first-generation alpha level model [*Marcot et al.*, 2006]. The next development steps have been described in detail by *Marcot et al.* [2006] and include peer review of the alpha model by other subject matter experts and consideration of their judgments regarding model parameterization; reconciliation of the peer reviews by the initial expert; updating the model to a beta level that incorporates the reviews; and testing the beta model for accuracy with existing data (e.g., determining if it matches historic or current known conditions). Additional updating of the model can include incorporation of new data or analyses if available. Throughout this process, sensitivity testing is used to verify model performance and structure. This framework has been used successfully for developing a number of BN models of rare species of plants and animals [*Marcot et al.*, 2001, 2006; *Raphael et al.*, 2001; *Marcot*, 2006]. Model variants that may have emerged in this process would represent the range of expert judgments and experiences (possibly verified with new data), and this range could be important information for decision making.

Because these additional steps in development have not yet been completed, it is important to view probabilities of outcome states of our first-generation model in terms of their general direction and overall magnitudes rather than focusing on the exact numerical probabilities of the outcomes. When predictions result in high probability of one population outcome state and low or zero probabilities of all other states, there is low overall uncertainty of predicted results. When projected probabilities of various states are more equally distributed, however, careful consideration should be given to large probabilities representing particular outcomes even if those probabilities are not the largest. Consistency of pattern among scenarios (e.g., different GCM runs) also is important to note. If the most probable outcome has a much higher probability than all of the other states and if the pattern across time frames and GCM models is consistent, confidence in that outcome pattern is high. If, on the other hand, probabilities are more uniformly spread among different states and if the pattern varies among scenarios, importance of the numerically most probable outcome should be tempered in view of the competing outcomes. This approach takes advantage of the information available from the model while recognizing that it is still in development. It also conforms to the concept of viewing the model as a tool describing relative probabilistic relationships among major levels of population response under multiple stressors.

### 4.2. Bayesian Network Model Outcomes

In the BN model, for each scenario run, the spread of population outcome probabilities (or at least nonzero possibilities) represented how individual uncertainties propagate and compound across multiple stressors. Beyond year 50, "extinct" was the most probable overall outcome state for all polar bear ecoregions, except the AE (Plate 4 and Table 2). For the decade of 2020–2029, outcomes were intermediate between the present (year 0) and the foreseeable future (year 50) time frames. We projected that polar bear numbers in the AE and PBCE could remain the same as now through the earlier decade, becoming smaller by mid century. In the SIE and PBDE, polar bears appeared to be headed toward extinction soon. However, probabilities they may persist in the PBDE and SIE were much higher at year 25 than at mid century (Plate 4). Although our BN model suggests polar bears are most likely to be absent from the PBDE and SIE by mid century, there is much uncertainty regarding when, between now and then, they might disappear from these ecoregions.

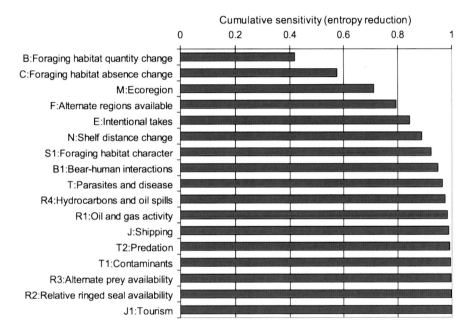

**Plate 5.** Cumulative sensitivity of overall population outcome (node D1, Plate 3) to all input variables (yellow boxes, Plate 3) in the Bayesian network population stressor model. The 17 input variables on the vertical axis are listed, top to bottom, in decreasing order of their individual influence on overall population outcome (see Appendix C, sensitivity test 1). The horizontal axis represents the cumulative proportion of total entropy reduction (mutual information) from the input variables. For example, the first two variables, foraging habitat quantity change and foraging habitat absence change, together explain 58% of the variation or uncertainty in the overall population outcome.

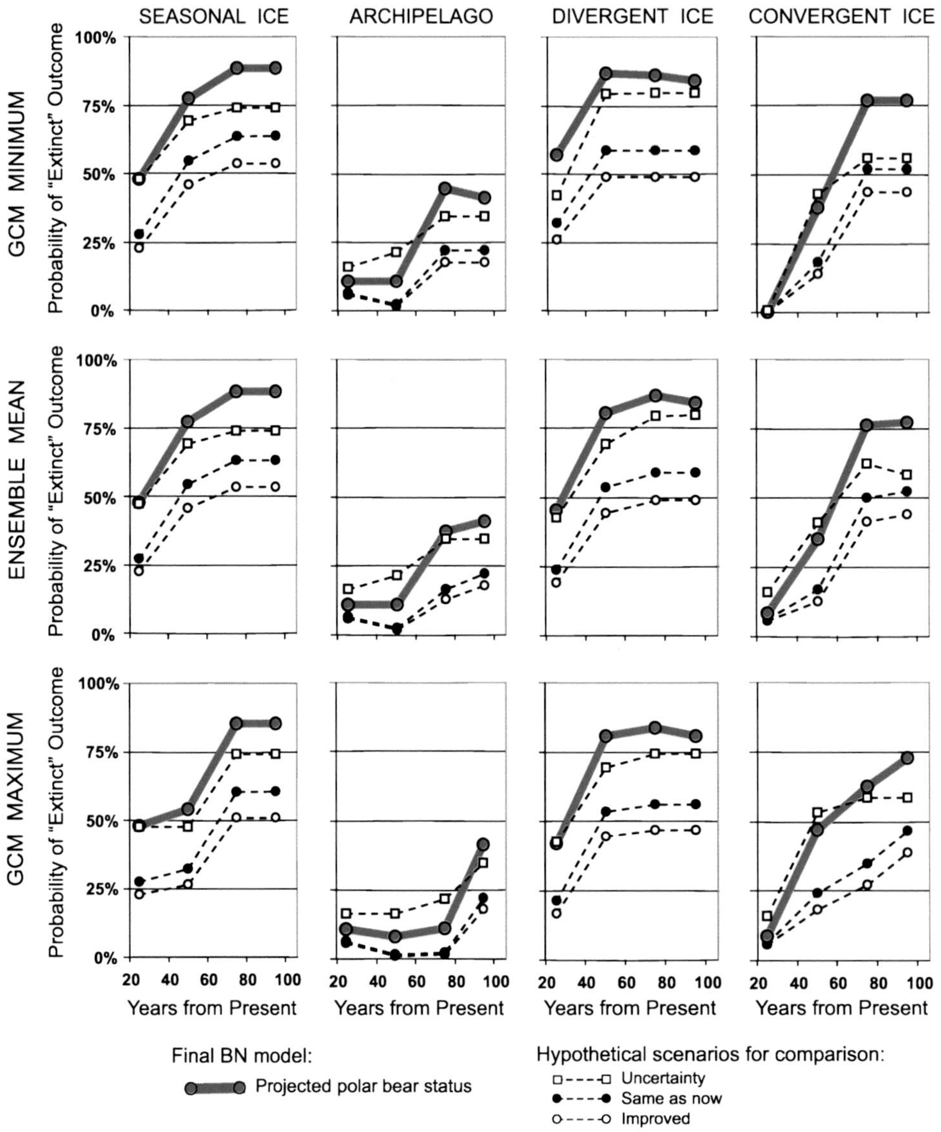

**Plate 6.** Probability of "extinct" outcome projected by Bayesian network (BN) model of worldwide polar bear populations. Shown are probabilities provided by the mean of a 10-member ensemble of general circulation models and the individual models which leave the maximum and minimum amount of sea ice at each time stop. Red line illustrates extinction probabilities from Table 2. Open circles illustrate results from setting all possible on the ground human influences to more favorable for bears than they are now. Solid circles illustrate holding all on the ground human influences as they are now. Squares illustrate results when all inputs, except ice nodes N, B and C, are held to uniform probabilities (i.e., total uncertainty). Only in the AE and PBCE does it appear that manipulating on the ground human activities can substantively influence overall outcomes at mid century and beyond.

Because polar bears are tied to the sea ice for obtaining food, major changes in the quantity and distribution of sea ice must result in similar changes in polar bear distribution. Therefore, the distributional effects of projected changes are most apparent. Whereas it is fairly certain that polar bears will not remain in areas where habitat absence is too prolonged to make seasonal use practical, it is less certain how many bears from areas of former habitat may be sustained in areas with remaining habitat. It is not surprising, therefore, that overall outcomes projected for polar bears appeared to be driven more by distributional effects than numerical effects. This is largely due to the parameterization in the model. Some input variables such as hunting or direct bear/human interactions might be expected to most immediately affect bear numbers rather than distribution. History has shown, however, that these things can be managed effectively to maintain sustainable populations when habitats are adequate. Our model incorporated the manageability of these human effects in the conditional probability tables. In contrast, polar bears cannot be maintained where their habitats are absent, and GCM projections suggest existing habitat areas will be progressively declining. Regardless of whether some concentration of numbers is possible in areas with remaining habitat, and there is great uncertainty regarding the relevance of this to the future, polar bears are not likely to survive in any numbers in areas where their current ice habitats no longer exist.

The most probable outcomes for factor A (Habitat Threats) of the proposal to list polar bears as a threatened species were "major restriction" (Table 3). Numerical responses of polar bears to future circumstances were forecast to be more modest than changes in distribution. In all regions, reduced density was the most probable outcome for numerical response. One way to interpret that outcome may be that where habitat remains, polar bears will remain even if in reduced numbers. This is consistent with our BN model results suggesting that polar bears may persist in the AE through the end of the 21st century. Declines in distribution and number are likely to be faster and more profound in the PBDE and the SIE than elsewhere. Sea ice availability in both the PBDE and SIE already is declining rapidly in these ecoregions [*Meier et al.*, 2007; *Stirling and Parkinson*, 2006]. The loss of sea ice habitats in the PBDE is projected to continue, and possibly to accelerate [*Holland et al.*, 2006; *Durner et al.*, 2008; *Stroeve et al.*, 2007].

Plate 7 illustrates how distribution changes driven by changes in the sea ice appeared to be the major factor leading to our dire predictions of the future for polar bears. For projection purposes, we binned the number of additional months during which the sea ice was projected to be absent from the continental shelf (node C) into four categories

which included the range from 1 month less than current (−1) to ≥3 months longer than current. Similarly, we binned the maximum distance the ice edge could move away from the shelf (node N) into four categories including the range from 200 km less than current (−200) to ≥800 km additional distance. It is clear from node D (see Appendix B) that we parameterized the model such that more distant ice retreat and longer ice absence meant reduced availability of critical foraging habitats as documented by *Durner et al.* [2008]. Such reduced availability has been shown to have negative impacts on polar bears [*Regehr et al.*, 2007a]. In the PBDE as an example, the general circulation models that we used to project future ice conditions, indicated values for nodes C and N will range from 1.8 to 2.2 additional months of ice absence and 234 km to 1359 km additional ice distance by mid century. Similarly, foraging habitat quantity is projected to decline between 16 and 32% by mid century. As Plate 7 illustrates even the smaller of these values for temporal and spatial retreat of sea ice place factor A node F2 (see Appendix B) into the category of major habitat restriction. That, in turn, pushes the distribution response toward extirpated which pushes the most probable overall population outcome into the "extinct" category. The outcome percentages in this example differ from the overall outcomes presented in Table 2 because results shown in this example occurred without changing any other inputs included in the full model. Hence, this result provides an example of how the projected changes in sea ice alone influenced the dire projections of our BN model. Outcomes in the PBDE are even more dire when the GCMs that lose the most ice are used or when we look farther into the future. In contrast, as Table 2 and Plate 4 illustrate, outcomes are less alarming in the PBCE and AE because of the more modest changes projected for sea ice in those regions. Sensitivity analyses described below confirm this role of sea ice in driving the expected future for polar bears.

*4.3. Sensitivity Analyses*

Sensitivity analyses offer an opportunity to interpret model outcomes at every level. The overall population outcome was most sensitive to change in habitat quantity (node B) and temporal habitat availability (node C). The other major habitat variable, change in distance between ice and the continental shelf (node N) was the 6th most influential factor on the overall population outcome, despite its being relevant only to the polar basin ecoregions. Our BN model recognized that sea ice characteristics, and how polar bears respond to them, differed among the four ecoregions. In the SIE, for example, all members of the subpopulation are forced ashore when the ice melts entirely in summer. In the

PBDE, by comparison, some bears retreat to shore, while most follow the sea ice as it retreats far offshore in summer. The fact that ecoregion and the availability of alternate ecoregions together explained 22% of the variation in overall population outcome was further evidence of the importance of sea ice habitat and its regional differences.

Another habitat variable, "foraging habitat character" (node S1), was ranked 7th among variables having influence on the overall population outcome. This qualitative variable relating to sea ice character was included to allow for the fact that in addition to changes in quantity and distribution of sea ice, subtle changes in the composition of sea ice could affect polar bears. For example, longer open water periods and warmer winters have resulted in thinner ice in the polar basin region [*Lindsay and Zhang*, 2005; *Holland et al.*, 2006; *Belchansky et al.*, 2008]. *Fischbach et al.* [2007] concluded that increased prevalence of thinner and less stable ice in autumn has resulted in reduced sea ice denning among polar bears of the southern Beaufort Sea.

Observations during polar bear field work suggest that the thinning of the sea ice also has resulted in increased roughness and rafting among ice floes. Compared to the thicker ice that dominated the polar basin decades ago, thinner ice is more easily deformed, even late in the winter. Whether or not thinner ice is satisfactory for seals, the extensive areas of jagged pressure ridges that can result when ice is more easily deformed may not be well suited to polar bear foraging. These changes appear to reduce foraging effectiveness of polar bears, and it is suspected the changes in ice conditions may have contributed to recent cannibalism and other unusual foraging behaviors [*Stirling et al.*, 2008]. Also, thinner, rougher ice, interspersed with more open water, may be an impediment to the travels of young cubs. Physical difficulties in navigating this "new" ice environment could explain recent observed increases in mortality of first-year cubs [*Rode et al.*, 2007]. The fact that six of the seven variables most influential on overall outcome were sea ice related and explained 87% of the variation in that outcome corroborates the well established link between polar bears and sea ice.

The 5th ranked potential stressor to which overall population outcome was sensitive was intentional takes. Historically, the direct killing of polar bears by humans for subsistence or sport has been the biggest challenge to polar bear welfare [*Amstrup*, 2003]. Our model suggests that harvest of polar bears may remain an important factor in the population dynamics of polar bears in the AE and PBCE, as sea ice retreats.

It is important to remember that there is great uncertainty in the exact way the potential stressors we modeled may change in the future. Also, the degree of uncertainty differs among the variables we included in our model. There is rela-

tively great certainty that the spatiotemporal distribution of sea ice will decline through the coming century. There is less certainty in just how much it will decline by a specified decade. The short, intermediate, and long-term effects of that decline on food availability for polar bears are largely unknown [*Bluhm and Gradinger*, 2008]. Here, we assumed that declining sea ice means declining food availability for polar bears, with the decline in food mirroring that of the sea ice. Spatiotemporal reductions in sea ice cover, however, could fundamentally alter the structure and function of the Arctic ecosystem. Such changes could result in different timing and level of productivity. It seems clear that continued declines in sea ice ultimately will mean reduced year-round food availability and declines in polar bear numbers and distribution. We cannot rule out, however, that increases in productivity could result in transitory increases in food availability for bears. Such changes would not alter the ultimate predictions made here, polar bears are clearly tied to the sea ice for access to their food. Such changes could, however, alter the temporal sensitivity of our outcomes to values at input nodes.

### 4.4. Strength of Evidence

Our BN population stressor model projects that sea ice and sea ice related factors will be the dominant driving force affecting future distributions and numbers of polar bears through the 21st century. Despite caveats regarding the early stage of development of our BN model, there are several reasons to believe that the directions and general magnitudes of its outcomes are reasonable.

First, they are consistent with hypothesized effects of global warming on polar bears [*Derocher et al.*, 2004] and with recent observations of how decreasing spatiotemporal distribution of sea ice has affected polar bears in some portions of their range [*Stirling and Derocher*, 1993; *Stirling and Parkinson*, 2006; *Stirling et al.*, 1999, 2007, 2008; *Ainley et al.*, 2003; *Ferguson et al.*, 2005; *Amstrup et al.*, 2006; *Hunter et al.*, 2007; *Regehr et al.*, 2007a, 2007b; *Rode et al.*, 2007]. The high sensitivity of our overall model outcomes to sea ice habitat changes is consistent with the recent PBSG decision, based mainly on projected changes in sea ice, that polar bears should be reclassified as vulnerable [*Aars et al.*, 2006].

Second, results of influence runs to assess the ability of on the ground activities to alter outcomes were not qualitatively different, during most time periods, from previous runs for the PBDE and SIE (Plate 6). Maintaining current conditions (other than sea ice) in the PBDE and SIE or improving conditions on the ground appeared to have some ability to reduce the risk of extinction being the most probable outcome during the first couple decades of this century. This effect

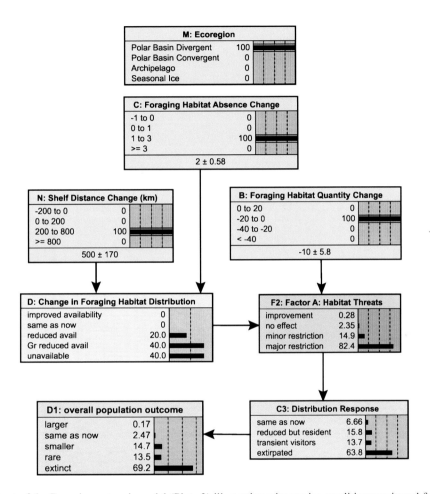

**Plate 7.** A subset of the Bayesian network model (Plate 3) illustrating why sea ice conditions projected for year 50 and beyond have such dire consequences for polar bears of the PBDE. Values of nodes N, C, and B are specified by the GCM outputs (see Tables D1a and D1b). Shown here are the input categories from the sea ice models projecting the smallest retreats of sea ice by mid century. Nodes D, F2, C3, and D1 are calculated by the Bayesian network model according to conditional probabilities specified in Appendix B. For illustration of the influence of sea ice on outcomes, we show future inputs only for the subset of nodes dealing with sea ice values here; therefore, outcomes shown here differ from those in Table 2. As illustrated in Table 2, results are more negative when other GCM outputs are used and when time frames farther into the future are examined.

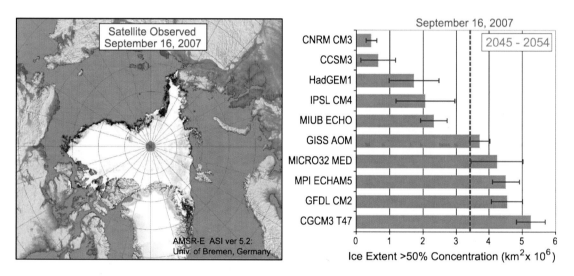

**Plate 8.** Area of sea ice extent (>50% ice concentration) on 16 September 2007, compared to 10 Intergovernmental Panel of Climate Change Fourth Assessment Report GCM mid century projections of ice extent for September 2045–2054 (mean ±1 standard deviation, n = 10 years). Ice extent for 16 September 2007 was calculated using near-real-time ice concentration estimates derived with the NASA Team algorithm and distributed by the National Snow and Ice Data Center (http://nsidc.org). Note that five of the models we used in our analyses project more perennial sea ice at mid century than was observed in 2007. This suggests our projections for the future status of polar bears may be conservative.

largely disappears by mid century, however, and management of localized human activities appears to have little qualitative effect on the future of polar bears in the PBDE and SIE if sea ice continues to decline as projected. The fact that sea ice has been declining more rapidly than even our minimum ice GCMs project (see below) suggests even possible transitory benefits of on the ground interventions may be illusory in these ecoregions. In contrast, our BN model suggested that managing human controlled stressors could qualitatively lower the probabilities of extinction in the PBCE and AE. Such management has played an important role in the past recovery of polar bears [Amstrup, 2003] and apparently could continue to be important in regions where sea ice habitat remains.

Third, influence runs in which we specified complete uncertainty in all inputs except sea ice illustrated that in order to obtain any qualitative change in the probability of extinction in any of the ecoregions, sea ice availability in future time periods must be greater than even the maximum-ice GCM projections we used (Plate 6). This eventuality may be unlikely in light of the fact that most GCMs simulate more ice than has actually been observed during 1979 to present [Durner et al., 2008; Stroeve et al., 2007]. It also seems unlikely in light of the most recent observations that September sea ice extent (15% concentration) in 2007 was $4.13 \times 10^6$ km$^2$. This is lower than the previous record set in 2005 by nearly $1.2 \times 10^6$ km$^2$ (National Snow and Ice Data Center, 1 October 2007 press release, available at http://nsidc.org/news/index.html). When ice extent based on a 50% concentration threshold, which is probably near the lower limit of ice cover useful to polar bears [Durner et al., 2008], is calculated, the area in 2007 was down to approximately $3.5 \times 10^6$ km$^2$. Five of the GCM models in our 10-member ensemble forecast more September sea ice in 2050 than was observed in 2007 (Plate 8). Perhaps even more telling than the overall "faster-than-forecasted" ice decline, is the observation that much of the AE, the region where we forecasted polar bears to remain until late in the century, was ice free in September 2007 (Plate 8). Hence, the probability that more sea ice than projected will be available during the rest of this century seems low.

Finally, a polar bear BN population stressor model would have to be structured and parameterized very differently to project qualitatively different outcomes than we have here. Yet it seems unlikely that other polar bear experts would do that. Evidence for the polar bear's reliance on sea ice is replete. Although they are opportunistic and will take terrestrial foods, including human refuse when available, and may benefit from such activity [Lunn and Stirling, 1985; Derocher et al., 1993], polar bears are largely dependent on the productivity of the marine environment. Refuse, for example, is of limited availability throughout the polar bear range, and could at best benefit relatively few individuals. Also, polar bears are inefficient in preying on terrestrial animals [Brook and Richardson, 2002; Stempniewicz, 2006]. Perhaps most importantly, polar bears have evolved a strategy designed to take advantage of the high fat content of marine mammals [Best, 1984]. Available terrestrial foods are, with few exceptions, not rich enough or cannot be gathered efficiently enough to support large bodied bears [Welch et al., 1997; Rode et al., 2001; Robbins et al., 2004]. Because polar bears are the largest of the bears, it is unlikely that terrestrial arctic habitats which are depauperate from the standpoint of bear food could support them in anything like current numbers. Empirical evidence of this is provided by the fact that Arctic grizzly bears are the smallest grizzly bears found anywhere and they occur at the lowest densities [Miller et al., 1997]. Habitats adjacent to present polar bear ranges just do not seem likely to support large numbers of the much larger-bodied polar bears. Although polar bears in Hudson Bay, which are forced onto land all summer, are known to consume a wide variety of foods; they gain little energetic benefit from those foods [Ramsay and Hobson, 1991]. The only foods, other than their ringed and bearded seal staples, that are known to be energetically important to polar bears in some regions are other marine mammals [Iverson et al., 2006]. Polar bears, it appears, are obligately dependent on the surface of the sea ice for capture of the prey necessary to maintain their populations.

In short, although other polar bear experts might structure a model differently and populate the conditional probability tables differently than we have here, it seems unlikely that those differences would be great enough to make a qualitative difference in the outcomes projected by our BN population stressor model for mid century and beyond.

## 5. CONCLUSIONS

We used a first-generation BN population stressor model to forecast future populations of polar bears worldwide. Outcomes of this model suggested that declines in the spatiotemporal distribution of sea ice habitat along with other potential stressors will severely impact future polar bear populations. Polar bears in the PBDE and SIE, home to approximately two thirds of the current world population will likely disappear by mid century. Management of localized human activities is unlikely to mitigate those mid century losses. Polar bears in and around the AE appear likely to persist late into the 21st century, especially if on-the-ground human activities, particularly human harvests, are carefully managed.

A declining habitat base, corresponding with FWS listing factor A (habitat threats), is the overriding factor in projections of declining numbers and distribution of polar bears. Other factors which correspond with FWS listing factors B, C, and E, and which could result in additional population stress on polar bears, are likely to exacerbate effects of habitat loss. To qualitatively alter outcomes projected by our models, it appears future sea ice would have to be far more extensive than is projected by even the more conservative of the general circulation models we used. Because recently observed declines in sea ice extent continue to outpace most GCM projections [*Stroeve et al.*, 2007], more extensive sea ice in the future seems unlikely unless greenhouse gas emissions are reduced. This study and the others on which it was based establish that the future security of polar bears over much of their present range is threatened in an ecological context. In May 2008, Secretary of Interior Dirk Kempthorne, upon review of this and other available information, decided this ecological threat required policy actions and declared polar bears a threatened species under the definition of the Endangered Species Act [*U.S. Fish and Wildlife Service*, 2008].

## APPENDIX A: DOCUMENTATION OF THE BAYESIAN NETWORK POLAR BEAR POPULATION STRESSOR MODEL

Appendix A documents the structure of the Bayesian network (BN) population stressor model. We used the BN modeling shell Netica® (Norsys, Inc.) to create a model that represents potential influences on distribution response, numerical response, and overall population response of polar bears under multiple stressors, which include anthropogenic stressors, natural disturbances, and other key environmental correlates to polar bear population amount and distribution.

The BN population stressor model was created to represent the knowledge and judgment of one polar bear biologist (S. Amstrup) with guidance from an ecologist modeler (B. Marcot). The general underlying influence diagram for the BN model is shown in Plate 2, and the full model is in Plate 3. A BN model consists of a series of variables represented as "nodes" (boxes in Plate 3) that interact through links (arrows in Plate 3). Nodes that have no incoming arrows are "input nodes" (the yellow boxes in Plate 3, e.g., node T Parasites and Disease). Nodes with both incoming and outgoing arrows are summary nodes (or latent variables, e.g., node L2 Vital Rates). In our model, we also specified four of the summary nodes as listing factors used by the U.S. Fish and Wildlife Service (S. Morey, personal communication, 2007). Nodes with incoming arrows but no outgoing arrows are output nodes (node D1 Overall Population Outcome). The model (Plate 3) was run by specifying conditions of all input nodes for each combination of ecoregion (node M), time period (node Q), and climate data source or GCM run (node R). In running the model, specifying ecoregion automatically adjusts two of its summary nodes (D and U). The input data set (Tables D1a and D1b) also specifies time period, GCM run, and all other inputs. Nodes Q and R appear in the model unconnected as visual placeholders for displaying the basis of each model run.

Each node in this model consists of a short node name (e.g., node D1), a longer node title (e.g., Overall Population Outcome), a set of states (e.g., larger, same as now, smaller, rare, and extinct), and an underlying probability table. The probability tables consist of unconditional (or prior) probabilities in the input nodes, or conditional probabilities in all other nodes, the latter representing probabilities of each state as a function of (conditional upon) the states of all nodes that directly influence it.

Table A1 presents a complete list of all nodes in the model with their short code letters, titles, description, possible states, and the group (Node Set, in Netica® parlance) to which it belongs (input nodes, output node, summary node, or summary listing factor node).

**Table A1.** Complete List of All Nodes in the Bayesian Network Population Stressor Model

| Node Name | Node Title | Node Description | States |
|---|---|---|---|
| | | *Input Nodes* | |
| T | Parasites and Disease | As the climate warms, regions of the Arctic are hospitable to parasites and disease agents which formerly did not survive there. Polar bears have always been free of most disease and parasite agents. *Trichinella* is one notable exception, but even rabies, common in the Arctic has had no significance to polar bears. Changes in other species disease vulnerability suggest that similar changes could occur in polar bears so that they could move from a position where parasites and disease are not influential on a population level to where they are influential. | influential not |

**Table A1.** (continued)

| Node Name | Node Title | Node Description | States |
|---|---|---|---|
| T2 | Predation | Predation on polar bears by other species is very uncommon partly because bears spend almost all of their time on the ice. With more time on land, polar bears, especially young, will be subject to increased levels of predation from wolves, and perhaps grizzly bears.<br>This will vary by region as some regions where polar bears occur have few other predators.<br>Intraspecific predation is one behavior which is known to occur in bears. It has rarely been observed in polar bears and historically is not thought to have been influential. Recent observations of predation on other bears by large males, in regions where it has not been observed before, are consistent with the hypothesis that this sort of behavior may increase in frequency if polar bears are nutritionally stressed. At present, intraspecific predation is not thought to be influential at the population level anywhere in the polar bear range. It appears, however, that its frequency may be on the increase. At some point, it therefore could become influential. At very low population levels, even a minor increase in predation could be influential. | influential<br>not |
| E | Intentional Takes | This node represents direct mortalities including hunting, and collection for zoos, and management actions. It also includes research deaths even though they are not intentional.<br>These are mortality sources that are very much controllable by regulation. | increased<br>same_as_now<br>decreased |
| T1 | Contaminants | Increased precipitation and glacial melt have recently resulted in greater influx of contaminants into the Arctic region from the interior of Eurasia via the large northward flowing rivers. Similarly, differing atmospheric circulation patterns have altered potential pathways for contaminants from lower latitudes. This node reflects the possible increase or decrease of contamination in the Arctic as a result of modified pathways. These contaminants can act to make habitat less suitable and directly affect things like survival and reproduction.<br>The greatest likelihood seems to be that such contaminants will increase in Arctic regions (and indeed worldwide) as increasing numbers of chemicals are developed and as their persistence in the environment is belatedly determined. Some contaminants have been reduced and we have the ability to reduce others, but the record of reduction and the persistence of many of these chemicals in the environment suggests the greatest likelihood is for elevated levels in the short to medium term with some probability of stability or even declines far in the future. | elevated<br>same_as_now<br>reduced |
| R4 | Hydrocarbons / Oil Spill | This refers to the release of oil or oil related products into polar bear habitat. Such action would result in direct mortality of bears, direct mortality of prey, and could result in displacement of bears from areas they formerly occupied. Hence, it has ramifications for both habitat quality and population dynamics directly.<br>Hydrocarbon exploration and development are expanding and proposed to expand farther in the Arctic. Greater levels of such activity are most likely to increase the probability of oil spills.<br>Also, increased shipping will result in higher levels of hydrocarbon release into Arctic waters. | increased_occurrence<br>same_as_now<br>decreased_occurrence |
| J1 | Tourism | As sea ice extent declines spatially and temporally, access and opportunities for Arctic tourism also will increase. Increased tourism could lead to direct disturbances of polar bears as well as to increased levels of contamination. Here, we address only the physical presence of more tourism and the conveyances used by tourists (vessels, land vehicles, aircraft).<br>The greatest likelihood seems to be that tourism will increase. It could decline, however, if governments take actions to reduce interactions with increasingly stressed polar bears. However, as tourism currently accounts for essentially no limitation to polar bears this effect only comes into play when it is noted to increase.<br>I believe that tourism will increase in all areas of the Arctic until such time as fuel | increased<br>same_as_now<br>decreased |

**Table A1.** (continued)

| Node Name | Node Title | Node Description | States |
|---|---|---|---|
| | | becomes too expensive for people to venture to such remote areas or in the polar basin divergent unit, when it is essentially devoid of ice, it may not attract many tourists and such activity may surge and then decline in that region. The arctic areas with more interesting coastlines etc., however, will probably see nothing but increases in tourism. Contamination that may accompany such activities, and biological effects from introduced organisms that may compete with residents of the food web or cause disease are covered under the nodes for contamination and parasites and disease. | |
| B1 | Bear-human interactions | This includes nonlethal takes which may increase as a result of increased human-bear interactions because of food-stressed bears more frequently entering Arctic communities. Such takes can displace bears from their preferred locations and reduce habitat quality. This is separate from the similar interactions that may occur around oil and gas or other industrial sites which also can displace bears and lower habitat quality. These interactions also, however, can result in deaths as when problem bears are shot in defense of life and property. So, this node includes a component of both habitat quality and direct mortality. I believe that bear-human interactions will increase until such time as areas are devoid of bears or climate cools again and ice returns. | increased same_as_now decreased |
| R1 | Oil and Gas Activity | This refers to the spatial effects of oil and gas activity. It refers to activities and infrastructure which may physically displace bears from habitat that was formally available to them. It also can result in direct killings of bears which become a persistent safety problem around industrial facilities. Oil companies etc. have great resources to prevent these events from leading to mortalities, but such mortalities cannot be totally avoided and are likely to increase as habitat base shrinks. I think oil and gas activity will increase in the polar basin region through mid century and then decline because resources will have been tapped. We may see some increase in exploration and development in the archipelago, however, as it becomes increasingly accessible. | increase no_change decrease |
| J | Shipping | As sea ice extent declines spatially and temporally it is predicted that shipping in Arctic regions will increase. Increased shipping could lead to direct disturbances of polar bears as well as to increased levels of contamination. Here, we address only the physical presence of more vessel traffic. Contamination (bilge oil, etc.), and biological effects from introduced organisms that may compete with residents of the food web or cause disease are covered under the nodes for contamination and parasites and disease. We allow only two states here: increased and same as now, because we can think of no reason why shipping will decrease in the foreseeable future. Even if international shipping does not increase, local shipping will, because barges and vessels are more efficient ways to move fuel and freight into remote Arctic locations than aircraft. | increased same_as_now |
| F | Alternate Regions Available | Are there geographic ecoregions to which bears from the subject region may effectively be able to relocate. This ability is contingent on other ecoregions with suitable habitats being contiguous with regions where habitat quantity or quality have degraded to the point they will not support polar bears on a seasonal or annual basis. For example, if the sea ice is deteriorating throughout the polar basin including the Beaufort Sea and the last vestiges of ice are along the Alaskan Coast, there may be no where else to go if the ice deteriorates to an unsatisfactory state. If, however, the ice retreats to the northeast as its extent reduces, bears remaining on the ice may have access to suitable habitats in the archipelago or in NE Greenland. I believe that bears in the seasonal ice region and in the polar basin will be able to collapse into the archipelago. Ice patterns suggest that the remaining ice in the Arctic is | yes no |

**Table A1.** (continued)

| Node Name | Node Title | Node Description | States |
|---|---|---|---|
| | | likely to converge on the archipelago rather than form disjunct chunks of ice (although some GCMs do predict the latter, this is contrary to the historical record and the paleo-record). Yes = other suitable areas are contiguous No = other suitable regions are not contiguous | |
| R2 | Relative Ringed Seal Availability | This node expresses changes in prey availability that are likely to occur as sea ice cover declines and its character changes. This node specifically includes only the possibility that ringed seals, the mainstay of polar bears over most of their range, might change in abundance and availability. This is specific to the amount of remaining ice. That is, as sea ice declines in coverage (which is the only way it seems possible for it to go) will the remaining habitat be more productive? Availability here refers to the combined effects of abundance and accessibility, recognizing that seals may occupy areas that make them less available to polar bears even if the seals are still relatively abundant. Examples of this are the recent observations of failed bear attempts to dig through solid ice (a result of the thinner ice that deforms and rafts more easily) that predominates now, and the fact that seals may simply stay in open water all summer and not be available to bears even if the seal numbers are stable. My judgment is that only in the northern part of the ice convergent zone of the polar basin and in portions of the archipelago are conditions likely to improve for ringed seal availability. And, there, such improvements are likely to be transient perhaps through mid century. increase = greater abundance or availability of ringed seals compared to now decrease = less abundance or availability | increase same_as_now decrease |
| R3 | Alternate Prey Availability | This node expresses changes in prey availability that are likely to occur as sea ice cover declines and its character changes. This is largely expert opinion because there is little to go on to suggest prey base change possibilities in the future. With very different ice and other ecological differences that may accompany global warming things could occur which are totally unforeseen. Today's experience, however, suggests that little in the way of significant alternate prey is likely to emerge to allow bears to replace traditional prey that may be greatly reduced in the future. Where alternate prey could become important is in the seasonal ice regions and the archipelago. Now, harp (*Phoca groenlandicus*) and hooded (*Cystophora cristata*) seals have become important to polar bears as they have moved farther north than historically. As the ice retreats into the archipelago it is reasonable to expect that these animals may penetrate deeper into the archipelago and provide at least a transient improvement in alternate prey. It is unclear, however, that such changes could persist as bears prey on these seals which are forced onto smaller and smaller areas of ice. So, I project only transient improvements followed by decline. This node specifically addresses the possibility that alternate prey, either marine or terrestrial, might change in a way that would allow polar bears to take advantage of it. increase = greater availability of alternate prey decrease = less opportunity for access to prey items other than ringed seals | increase same_as_now decrease |
| S1 | Foraging Habitat Character | This node expresses a subjective assessment of the quality of sea ice for foraging by polar bears. Recent observations of the changes in sea ice character in the southern Beaufort Sea suggest that the later freeze-up, warmer winters, and earlier ice retreat in summer have resulted in thinner ice that more easily deforms and more frequently rafts over itself. These changes have reduced the quality of ice as a denning substrate, and may have reduced its quality as a foraging substrate since the extensive ice deformation can result in ice covered refugia for ringed seals which are less likely for polar bears to get into. Also, it can result in very rough sharp pressure ridges that are hugely | more_optimal same_as_now less_optimal |

**Table A1.** (continued)

| Node Name | Node Title | Node Description | States |
|---|---|---|---|
| | | expansive compared to earlier years. This rough ice may also provide refuge for seals, and it also is surely difficult for polar bear COYs to negotiate as they attempt to move out onto the ice after den emergence in spring.<br><br>More optimal ice is somewhat heavier and not as rough, with pressure ridges composed of larger ice blocks. However, it can go the other way. Very heavy stable ice in the Beaufort Sea in the past may have been limiting polar bears. This is also probably currently true in portions of the AE and in the northern part of the PBCE. So, in those areas, I expect that ice quality will at first improve with global warming and then decline.<br><br>Because my only sense of this ice quality is in the polar basin, I am leaving all priors uniform for the other ice regions. | |
| C | Foraging Habitat Absence Change | This node expresses the length, in months, of ice absence from the continental shelf regions currently preferred by polar bears. It corresponds to the value "proportional ice-free months" from D. Douglas' calculations based on GCMs. This is the number of months during which the continental shelf was ice free where ice free is defined as fewer than 50% of the pixels over the shelf having less than 50% ice cover.<br><br>We express this as a change from now. So the figures in this node represent the difference in months between the forecasted number of ice-free months for four future time periods and the number of ice-free months for the present which is defined as the GCM model outputs for the period 2000–2009.<br><br>The bears in some regions already experience protracted ice-free periods. In other regions they do not. The impact of the length of the ice-free period is dependent mainly upon the productivity of the environment, and has a different impact in the Beaufort Sea, for example, than it does in the currently seasonal ice environments which are, for the most part, very productive.<br><br>For example, a difference in the amount of time ice was absent of GT 3 months means a mean absence of 7 or 8 months in the PBDE, and 8, 9 or 10 months in the seasonal ice zone, but only 3+ months in portions of the AE or the PBCE where ice is now present year-round. | −1 to 0<br>0 to 1<br>1 to 3<br>>=3 |
| B | Foraging Habitat Quantity Change | This node expresses the proportional change in the area of polar bear habitat over time. Polar bear habitat is expressed as the number of square km months of optimal RSF habitat in the two polar basin geographic units, and as square km months of ice over continental shelf in the other regions. Because the other regions are almost entirely shallow water areas, the habitat in those areas boils down to essentially the ice extent months over each region.<br><br>We further express this as the percent change in quantity of these ice habitats, from the baseline now which is defined as the period 1996–2006.<br><br>Interpreting the percent difference must take into account that a given percent change in the archipelago or the PB convergent region is a very different thing than it might be in the other two units. The absolute change in the archipelago, for example, may be very small, but because it is measured from essentially 0, it may look like a great% . These measurements are derived from the satellite record for the observational period and from the GCM outputs of sea ice for future periods. | 0 to 20<br>−20 to 0<br>−40 to −20<br>< −40 |
| N | Shelf Distance Change (km) | This node expresses the distance that the ice retreats from traditional autumn/winter foraging areas which are over the continental shelves and other shallow water areas within the polar basin. It is calculated by extracting the largest contiguous chunk of ice whose pixels have >50% concentration and determining the mean of the measured distances between all cells in the subpopulation unit and the nearest point within that chunk of ice. It is expressed as the difference between this mean distance calculated for the period 1996–2006 and the same mean distance calculated for the other time periods of interest. These distances are derived from the satellite record for the observational period and from the GCM outputs of sea ice for future periods. | −200 to 0<br>0 to 200<br>200 to 800<br>>= 800 |

**Table A1.** (continued)

| Node Name | Node Title | Node Description | States |
|---|---|---|---|
| | | Expressing this value as a change from the current time allows the model to show that conditions improve in a hind cast back to the period of 1985–1995.<br>This measurement is available only from the polar basin ecoregions because all other management units occur in areas that are essentially all shelf. Hence, the measurement of distance to shelf means nothing. | |
| M | Ecoregion | Geographic region used for combining populations of polar bears. | Polar_Basin_ Divergence Polar_Basin_ Convergence Archipelago Seasonal_Ice |

*Output Nodes*[a]

| Node Name | Node Title | Node Description | States |
|---|---|---|---|
| D1 | Overall Population Outcome | Composite influence of numerical response and distribution response. | larger same_as_now smaller rare extinct |
| C4 | Numerical Response | This node represents the anticipated numerical response of polar bears based upon the sum total of the identified factors which are likely to have affected numbers of polar bears in any particular area. | increased_density same_as_now reduced_density rare absent |
| C3 | Distribution Response | This is the sum total of ecological and human factors that predict the future distribution of polar bears.<br>Reduced but Resident: habitat has changed in a way that would likely lead to a reduced spatial distribution (e.g., because of avoidance of a human development or because sea ice is still present in the area but in more limited quantity). Bears would still occur in the area, but their distribution would be more limited. Transient = habitat is seasonally limited or human activities have resulted in a situation where available ice is precluded from use on a seasonal basis. | same_as_now reduced_but_resident transient_visitors extirpated |

*Summary Nodes*

| Node Name | Node Title | Node Description | States |
|---|---|---|---|
| C2 | Pollution | This is the sum of pollution effects from hydrocarbon discharges directly into arctic waters and from other pollutants brought to the Arctic from other parts of the world. The FWS listing proposal included pollution as one of the "other factors" along with direct human bear interactions that may displace bears or otherwise make habitats less satisfactory. I viewed the main effect of pollution as a potential effect on population dynamics. Clearly, severe pollution as in an oil spill, for example, could make habitats unsatisfactory and result in direct displacement. The main effect, however, is likely to be how pollution affects immune systems, reproductive performance, and survival. Hence, I have included input from this node as well as from the human disturbance node into both the habitat and the abundance side of the network by including input from factor E into both population effects and habitat effects. | reduced same_as_now elevated greatly_elevated |
| C1 | Human Disturbance | This node expresses the combination of the changes in "other" direct human disturbances to polar bears. This does not include changes in sea ice habitat. Nor does it include the contamination possibilities from hydrocarbon exploration. Those are covered elsewhere. It does cover the direct bear-human interactions that can occur in association with industrial development. | reduced same_as_now elevated greatly_elevated |
| H | Crowding Tolerance | The degree to which polar bears may tolerate increased densities that may result from migration of bears from presently occupied regions that become unsuitable to other regions already occupied by polar bears. | none moderate high |

**Table A1.** (continued)

| Node Name | Node Title | Node Description | States |
|---|---|---|---|
| | | In essence, this is the tolerance of bears to live in more crowded conditions than those at which they presently live. And, it is a function of food availability. I believe that bears have a reasonable tolerance of crowding if food is abundant or if they are in good condition while waiting for sea ice to return etc. Examples of these situations include (1) portions of the high Arctic-like near Resolute, where bear densities on the sea ice in spring are apparently much higher than they are in most of the polar basin, and (2) the high densities at which polar bears occur on land in Hudson Bay in summer when they are loafing and waiting for the sea ice to return. I assumed that crowding tolerance has little or no effect on outcome likelihoods until habitat quantity was reduced substantially requiring bears from one area to either perish or find some place else to go on at least a seasonal basis. Thereafter, if relocations of members of some subpopulations meant invading the areas occupied by other bears, crowding tolerance entered an assessment of whether or not relocation was a practical solution. | |
| G | Relocation Possible | Is it likely that polar bears displaced from one region could either seasonally or permanently relocate to another region in order to persist. This is a function of foraging effects (e.g., prey availability) in the alternative area (here I am specifically focusing on prey availability in the alternative area rather than the area from which the bears may have been displaced) crowding tolerance, and contiguity of habitats. | yes no |
| A | Foraging Habitat Value | This node expresses the sum total of things which may work to alter the quality of habitats available to polar bears in the future. The idea here is that sea ice is retreating spatially and temporally, but is the ice that remains of comparable, better or worse quality as polar bear habitat. Our RSF values are projected into the future with the assumption that a piece of ice in 2090 that looks the same as piece of ice in 1985 has the same value to a polar bear. Perhaps because of responses we cannot foresee, it may be better seal habitat, or it may be habitat for an alternate prey. Conversely, it may be worse because of atmospheric and oceanic processes (e.g., the epontic community is less vibrant because of thinner ice which is not around for as long each year). Or it may be worse habitat because of oil and gas development, tourism, shipping etc. | better same_as_now worse |
| D | Change in Foraging Habitat Distribution | This node expresses the combination of the quantitative ways the retreat of sea ice may affect use of continental shelf habitats. Our analyses indicate, in addition to reductions of total ice (and RSF Optimum ice) extent (expressed under habitat quantity), we will see seasonal retreats of the sea ice away from coastal areas now preferred by polar bears, and these retreats are projected to progressively become longer. These changes will affect polar bears by reducing the total availability of ice substrate for bears. They also will make ice unavailable for extended periods in some regions where bears now occur year-round. This will result in the opportunity for seasonal occupancy but not year-round occupancy as they have had in the past. Note that in the PBCE because it includes the North Beaufort and Queen Elizabeth and East Greenland each of which has different starting points, the values in the CPT express kind of an average. Similarly, in the Seasonal region, there is a huge difference between Hudson Bay and Foxe Basin or Baffin Bay. So, again the CPT values are a sort of an average, trying to reflect these differences. | improved_availability same_as_now reduced_availability Gr_reduced_availability Unavailable |
| L2 | Vital Rates | This expresses the combined effect of changes in survival of adult females and of young and reproductive patterns. The probabilities assigned each of the states reflects the relative importance to polar bear population dynamics of each of these vital rates to the growth of the population. This node does not reflect human influences on population growth such as hunting, or mortalities resulting from bear-human interactions. Those things, along with effects of parasites, contaminants, etc. are brought in as modifiers at the level of the next node. | improve same_as_now decline |

**Table A1.** (continued)

| Node Name | Node Title | Node Description | States |
|---|---|---|---|
| U | Reproduction | The sum of trends in numbers of cubs produced and the effect of retreating sea ice on the ability of females to reach traditional denning areas. | increased<br>same_as_now<br>decreased |
| V1 | Cub production per event | This node describes the number of cubs produced per denning attempt. | fewer_than_now<br>same_as_now<br>more_than_now |
| L | Juvenile Survival | Annual natural survival rate of cubs and yearlings. Note that this is conditional on survival of the mother. This is the survival rate for juveniles that would occur in absence of hunting or other anthropogenic factors. Those anthropogenic factors that would influence survival are included in node F. | increase<br>no_change<br>decrease |
| L1 | Adult Female Survival | Annual natural survival rate of sexually mature females. This is the survival rate for adult females that would occur in absence of hunting or other anthropogenic factors. Those anthropogenic factors that would influence survival are included in node F. | increase<br>no_change<br>decrease |
| K | Adult Body Condition | Body mass index or other indicator of ability of bears to secure resources. Our analysis suggests body stature has been declining in the SBS and is inversely correlated with ice extent. Also recent analyses indicate that body condition may soon be an important predictor of survival of polar bears in SHB. | increase<br>same_as_now<br>decrease |

*Summary Nodes – USFWS Listing Factors*[b]

| | | | |
|---|---|---|---|
| F2 | Factor A. Habitat Threats | This node summarizes the combined information about changes in habitat quantity and quality. It approximately reflects factor A of the proposal to list polar bears as threatened. | improvement<br>no_effect<br>minor_restriction<br>major_restriction |
| A1 | Factor B. Overutilization | This node approximates the FWS listing factor B. It includes the combination of hunting (harvest), take for scientific purposes, and take for zoos. It also includes mortalities from bear-human interactions etc. brought in from factor E. These all are factors which serve to modify the population changes that would be brought about without the direct local interference of humans. | fewer<br>same_as_now<br>more |
| A4 | Factor C. Disease, predation | This node expresses probability of changing vulnerability of polar bears to diseases and parasites, and to potential increases of intraspecific predation/cannibalism. | same_as_now<br>worse |
| A6 | Factor E. Other factors (natural or man-made) | This node approximately corresponds to factor E of the listing proposal. It includes factors (other than the changes in sea ice quality and quantity) which may affect habitat suitability for polar bears. Also, its effects can be directly on population dynamics features. Hence, it applies directly to both the habitat and population sides of our network. Included here are effects of a variety of contaminants, including: petroleum hydrocarbons, persistent organic pollutants, and metals. Although we do not know much quantitatively about effects of these contaminants at the population level, we know qualitatively that effects on immune systems and steroid levels etc. will ultimately have such effects. We also know that oil spills will have immediate and dire effects.<br>It also includes effects of human activities and developments which may directly affect habitat quality, including: shipping and transportation activities, habitat change, noise, spills, ballast discharge, and ecotourism. This includes disturbance but not direct killing of bears by humans as a result of DLP cases (direct killing is included under node A1). I viewed human disturbances as the most predictable in their negative effects until pollution levels reached their greatly elevated stage at which time, their import to future populations was judged to be great. | improvement<br>no_effect<br>minor_restriction<br>major_restriction |

*Descriptive (Disconnected) Nodes*[c]

| | | | |
|---|---|---|---|
| Q | Time Period | The states for this node correspond to years −10 (historic), 0 (now), 25 (early century), 50 (mid century), 75 (late century), and 95 (end of century). | historic (1985–1995)<br>now (1996–2006)<br>early century |

**Table A1.** (continued)

| Node Name | Node Title | Node Description | States |
|---|---|---|---|
| | | | (2020–2029) mid century (2045–2055) late century (2070–2080) end of century (2090–2099) |
| R | GCM run | The states for this node correspond to the data source (either "satellite" for year −10 and 0 runs) and GCM modeling scenario (minimum, ensemble mean, or maximum) basis for a given condition. | GCM_minimum ensemble_mean GCM_maximum satellite |

[a]Output nodes here include the Numerical Response and Distribution Response nodes that provide summary output conditions.
[b]U.S. Fish and Wildlife Service (USFWS) lists five listing factors. Listing factor D pertains to inadequacy of existing regulatory mechanisms, and was not included in the BN population stressor model because it does not correspond to any specific environmental stressor.
[c]These two nodes are included in the model to help denote the basis for a given model run. They are not included as environmental stressors per se.

## APPENDIX B:  PROBABILITY TABLES FOR EACH NODE IN THE BAYESIAN NETWORK MODEL

Tables B1–B19 are probability tables for each node in the BN model. (These were generated in the Netica® software.) Not included here are all input nodes (yellow coded nodes in Plate 3) because each of their prior probability tables was set to uniform distributions.

**Table B1.** Node A: Foraging Habitat Value

| Node S1: Foraging Habitat Character | Node G: Relocation Possible | Value of Foraging Habitat Better | Same as Now | Worse |
|---|---|---|---|---|
| More optimal | yes | 0.7 | 0.3 | 0.0 |
| More optimal | no | 0.2 | 0.6 | 0.2 |
| Same as now | yes | 0.1 | 0.8 | 0.1 |
| Same as now | no | 0.0 | 0.8 | 0.2 |
| Less optimal | yes | 0.0 | 0.3 | 0.7 |
| Less optimal | no | 0.0 | 0.0 | 1.0 |

**Table B2.** Node A1: Factor B Overutilization

| Node E: Intentional Takes | Node A6: Factor E—Other Factors (Natural or Man-Made) | Level of Overutilization Fewer | Same as Now | More |
|---|---|---|---|---|
| Increased | improvement | 0.0 | 0.4 | 0.6 |
| Increased | no effect | 0.0 | 0.0 | 1.0 |
| Increased | minor restriction | 0.0 | 0.0 | 1.0 |
| Increased | major restriction | 0.0 | 0.0 | 1.0 |
| Same as now | improvement | 1.0 | 0.0 | 0.0 |
| Same as now | no effect | 0.0 | 1.0 | 0.0 |
| Same as now | minor restriction | 0.0 | 0.6 | 0.4 |
| Same as now | major restriction | 0.0 | 0.3 | 0.7 |
| Decreased | improvement | 1.0 | 0.0 | 0.0 |
| Decreased | no effect | 1.0 | 0.0 | 0.0 |
| Decreased | minor restriction | 0.0 | 0.8 | 0.2 |
| Decreased | major restriction | 0.0 | 0.6 | 0.4 |

**Table B3.** Node A4: Factor C Disease and Predation

| Node T : Parasites and Disease | Node T2: Predation | Level of Disease and Predation | |
|---|---|---|---|
| | | Same as Now | Worse |
| Influential | influential | 0.0 | 1.0 |
| Influential | not | 0.3 | 0.7 |
| Not | influential | 0.7 | 0.3 |
| Not | not | 1.0 | 0.0 |

**Table B4.** Node A6: Factor E Other Factors Natural or Man-Made

| Node C1: Human Disturbance | Node C2: Pollution | Level of Other Factors | | | |
|---|---|---|---|---|---|
| | | Improvement | No Effect | Minor Restriction | Major Restriction |
| Reduced | reduced | 1.0 | 0.0 | 0.0 | 0.0 |
| Reduced | same as now | 1.0 | 0.0 | 0.0 | 0.0 |
| Reduced | elevated | 0.3 | 0.4 | 0.3 | 0.0 |
| Reduced | greatly elevated | 0.0 | 0.3 | 0.3 | 0.4 |
| Same as now | reduced | 0.6 | 0.4 | 0.0 | 0.0 |
| Same as now | same as now | 0.0 | 1.0 | 0.0 | 0.0 |
| Same as now | elevated | 0.0 | 0.4 | 0.6 | 0.0 |
| Same as now | greatly elevated | 0.0 | 0.2 | 0.2 | 0.6 |
| Elevated | reduced | 0.0 | 0.2 | 0.5 | 0.3 |
| Elevated | same as now | 0.0 | 0.0 | 0.5 | 0.5 |
| Elevated | elevated | 0.0 | 0.0 | 0.4 | 0.6 |
| Elevated | greatly elevated | 0.0 | 0.0 | 0.3 | 0.7 |
| Greatly elevated | reduced | 0.0 | 0.0 | 0.3 | 0.7 |
| Greatly elevated | same as now | 0.0 | 0.0 | 0.2 | 0.8 |
| Greatly elevated | elevated | 0.0 | 0.0 | 0.1 | 0.9 |
| Greatly elevated | greatly elevated | 0.0 | 0.0 | 0.0 | 1.0 |

**Table B5.** Node C1: Human Disturbance

| Node B1: Bear-Human Interactions | Node J: Shipping | Node R1: Oil and Gas Activity | Node J1: Tourism | Level of Human Disturbance | | | |
|---|---|---|---|---|---|---|---|
| | | | | Reduced | Same as Now | Elevated | Greatly Elevated |
| Increased | increased | increase | increased | 0.0 | 0.0 | 0.0 | 1.0 |
| Increased | increased | increase | same as now | 0.0 | 0.0 | 0.0 | 1.0 |
| Increased | increased | increase | decreased | 0.0 | 0.0 | 0.1 | 0.9 |
| Increased | increased | no change | increased | 0.0 | 0.0 | 0.0 | 1.0 |
| Increased | increased | no change | same as now | 0.0 | 0.0 | 0.1 | 0.9 |
| Increased | increased | no change | decreased | 0.0 | 0.0 | 0.2 | 0.8 |
| Increased | increased | decrease | increased | 0.0 | 0.0 | 0.3 | 0.7 |
| Increased | increased | decrease | same as now | 0.0 | 0.0 | 0.6 | 0.4 |
| Increased | increased | decrease | decreased | 0.0 | 0.0 | 0.5 | 0.5 |

**Table B5.** (continued)

| Node B1:<br>Bear-Human<br>Interactions | Node J:<br>Shipping | Node R1:<br>Oil and Gas<br>Activity | Node J1:<br>Tourism | Level of Human Disturbance | | | |
|---|---|---|---|---|---|---|---|
| | | | | Reduced | Same<br>as Now | Elevated | Greatly<br>Elevated |
| Increased | same as now | increase | increased | 0.0 | 0.0 | 0.0 | 1.0 |
| Increased | same as now | increase | same as now | 0.0 | 0.0 | 0.2 | 0.8 |
| Increased | same as now | increase | decreased | 0.0 | 0.0 | 0.3 | 0.7 |
| Increased | same as now | no change | increased | 0.0 | 0.0 | 0.5 | 0.5 |
| Increased | same as now | no change | same as now | 0.0 | 0.0 | 0.7 | 0.3 |
| Increased | same as now | no change | decreased | 0.0 | 0.2 | 0.6 | 0.2 |
| Increased | same as now | decrease | increased | 0.0 | 0.0 | 0.7 | 0.3 |
| Increased | same as now | decrease | same as now | 0.0 | 0.2 | 0.7 | 0.1 |
| Increased | same as now | decrease | decreased | 0.0 | 0.4 | 0.6 | 0.0 |
| Same as now | increased | increase | increased | 0.0 | 0.0 | 0.2 | 0.8 |
| Same as now | increased | increase | same as now | 0.0 | 0.0 | 0.5 | 0.5 |
| Same as now | increased | increase | decreased | 0.0 | 0.2 | 0.6 | 0.2 |
| Same as now | increased | no change | increased | 0.0 | 0.2 | 0.8 | 0.0 |
| Same as now | increased | no change | same as now | 0.0 | 0.3 | 0.7 | 0.0 |
| Same as now | increased | no change | decreased | 0.0 | 0.5 | 0.5 | 0.0 |
| Same as now | increased | decrease | increased | 0.0 | 0.3 | 0.7 | 0.0 |
| Same as now | increased | decrease | same as now | 0.0 | 0.5 | 0.5 | 0.0 |
| Same as now | increased | decrease | decreased | 0.0 | 0.6 | 0.4 | 0.0 |
| Same as now | same as now | increase | increased | 0.0 | 0.2 | 0.8 | 0.0 |
| Same as now | same as now | increase | same as now | 0.0 | 0.4 | 0.6 | 0.0 |
| Same as now | same as now | increase | decreased | 0.0 | 0.5 | 0.5 | 0.0 |
| Same as now | same as now | no change | increased | 0.0 | 0.8 | 0.2 | 0.0 |
| Same as now | same as now | no change | same as now | 0.0 | 1.0 | 0.0 | 0.0 |
| Same as now | same as now | no change | decreased | 0.1 | 0.9 | 0.0 | 0.0 |
| Same as now | same as now | decrease | increased | 0.3 | 0.7 | 0.0 | 0.0 |
| Same as now | same as now | decrease | same as now | 0.5 | 0.5 | 0.0 | 0.0 |
| Same as now | same as now | decrease | decreased | 0.6 | 0.4 | 0.0 | 0.0 |
| Decreased | increased | increase | increased | 0.0 | 0.0 | 0.6 | 0.4 |
| Decreased | increased | increase | same as now | 0.0 | 0.2 | 0.6 | 0.2 |
| Decreased | increased | increase | decreased | 0.0 | 0.3 | 0.7 | 0.0 |
| Decreased | increased | no change | increased | 0.1 | 0.6 | 0.3 | 0.0 |
| Decreased | increased | no change | same as now | 0.2 | 0.6 | 0.2 | 0.0 |
| Decreased | increased | no change | decreased | 0.3 | 0.5 | 0.2 | 0.0 |
| Decreased | increased | decrease | increased | 0.2 | 0.6 | 0.2 | 0.0 |
| Decreased | increased | decrease | same as now | 0.3 | 0.7 | 0.0 | 0.0 |
| Decreased | increased | decrease | decreased | 0.4 | 0.6 | 0.0 | 0.0 |
| Decreased | same as now | increase | increased | 0.0 | 0.5 | 0.5 | 0.0 |
| Decreased | same as now | increase | same as now | 0.2 | 0.6 | 0.2 | 0.0 |
| Decreased | same as now | increase | decreased | 0.3 | 0.6 | 0.1 | 0.0 |
| Decreased | same as now | no change | increased | 0.5 | 0.5 | 0.0 | 0.0 |
| Decreased | same as now | no change | same as now | 0.7 | 0.3 | 0.0 | 0.0 |
| Decreased | same as now | no change | decreased | 0.8 | 0.2 | 0.0 | 0.0 |
| Decreased | same as now | decrease | increased | 0.9 | 0.1 | 0.0 | 0.0 |
| Decreased | same as now | decrease | same as now | 1.0 | 0.0 | 0.0 | 0.0 |
| Decreased | same as now | decrease | decreased | 1.0 | 0.0 | 0.0 | 0.0 |

**Table B6.** Node C2: Pollution

| Node R4: Hydrocarbons/ Oil Spill | Node T1: Contaminants | Level of Pollution | | | |
|---|---|---|---|---|---|
| | | Reduced | Same as Now | Elevated | Greatly Elevated |
| Increased occur | elevated | 0.0 | 0.0 | 0.0 | 1.0 |
| Increased occur | same as now | 0.0 | 0.0 | 0.6 | 0.4 |
| Increased occur | reduced | 0.0 | 0.4 | 0.4 | 0.2 |
| Same as now | elevated | 0.0 | 0.3 | 0.7 | 0.0 |
| Same as now | same as now | 0.0 | 1.0 | 0.0 | 0.0 |
| Same as now | reduced | 0.4 | 0.6 | 0.0 | 0.0 |
| Decreased occur | elevated | 0.3 | 0.5 | 0.2 | 0.0 |
| Decreased occur | same as now | 0.8 | 0.2 | 0.0 | 0.0 |
| Decreased occur | reduced | 1.0 | 0.0 | 0.0 | 0.0 |

**Table B7.** Node C3: Distribution Response

| Node F2: Factor A— Habitat Threats | Node A6: Factor E—Other Factors (Natural or Man-Made) | Node G: Relocation Possible | Distribution Response | | | |
|---|---|---|---|---|---|---|
| | | | Same as Now | Reduced but Resident | Transient Visitor | Extirpated |
| Improvement | improvement | yes | 1.0 | 0.0 | 0.0 | 0.0 |
| Improvement | improvement | no | 1.0 | 0.0 | 0.0 | 0.0 |
| Improvement | no effect | yes | 1.0 | 0.0 | 0.0 | 0.0 |
| Improvement | no effect | no | 1.0 | 0.0 | 0.0 | 0.0 |
| Improvement | minor restriction | yes | 0.9 | 0.1 | 0.0 | 0.0 |
| Improvement | minor restriction | no | 0.9 | 0.1 | 0.0 | 0.0 |
| Improvement | major restriction | yes | 0.8 | 0.1 | 0.1 | 0.0 |
| Improvement | major restriction | no | 0.8 | 0.2 | 0.0 | 0.0 |
| No effect | improvement | yes | 1.0 | 0.0 | 0.0 | 0.0 |
| No effect | improvement | no | 1.0 | 0.0 | 0.0 | 0.0 |
| No effect | no effect | yes | 1.0 | 0.0 | 0.0 | 0.0 |
| No effect | no effect | no | 1.0 | 0.0 | 0.0 | 0.0 |
| No effect | minor restriction | yes | 0.8 | 0.1 | 0.1 | 0.0 |
| No effect | minor restriction | no | 0.8 | 0.2 | 0.0 | 0.0 |
| No effect | major restriction | yes | 0.5 | 0.2 | 0.3 | 0.0 |
| No effect | major restriction | no | 0.5 | 0.5 | 0.0 | 0.0 |
| Minor restriction | improvement | yes | 0.5 | 0.25 | 0.25 | 0.0 |
| Minor restriction | improvement | no | 0.5 | 0.5 | 0.0 | 0.0 |
| Minor restriction | no effect | yes | 0.4 | 0.3 | 0.3 | 0.0 |
| Minor restriction | no effect | no | 0.4 | 0.6 | 0.0 | 0.0 |
| Minor restriction | minor restriction | yes | 0.3 | 0.3 | 0.4 | 0.0 |
| Minor restriction | minor restriction | no | 0.3 | 0.6 | 0.0 | 0.1 |
| Minor restriction | major restriction | yes | 0.2 | 0.2 | 0.6 | 0.0 |
| Minor restriction | major restriction | no | 0.2 | 0.5 | 0.0 | 0.3 |
| Major restriction | improvement | yes | 0.0 | 0.3 | 0.35 | 0.35 |
| Major restriction | improvement | no | 0.0 | 0.3 | 0.0 | 0.7 |
| Major restriction | no effect | yes | 0.0 | 0.2 | 0.4 | 0.4 |
| Major restriction | no effect | no | 0.0 | 0.2 | 0.0 | 0.8 |
| Major restriction | minor restriction | yes | 0.0 | 0.1 | 0.45 | 0.45 |
| Major restriction | minor restriction | no | 0.0 | 0.1 | 0.0 | 0.9 |
| Major restriction | major restriction | yes | 0.0 | 0.0 | 0.3 | 0.7 |
| Major restriction | major restriction | no | 0.0 | 0.0 | 0.0 | 1.0 |

**Table B8.** Node C4: Numerical Response

| Node L2: Vital Rates | Node A1: Factor B— Overutilization | Node A4: Factor C— Disease and Predation | Numerical Response | | | | |
|---|---|---|---|---|---|---|---|
| | | | Increased Density | Same as Now | Reduced Density | Rare | Absent |
| Improve | fewer | same as now | 1.0 | 0.0 | 0.0 | 0.0 | 0.0 |
| Improve | fewer | worse | 0.5 | 0.25 | 0.25 | 0.0 | 0.0 |
| Improve | same as now | same as now | 0.8 | 0.2 | 0.0 | 0.0 | 0.0 |
| Improve | same as now | worse | 0.5 | 0.25 | 0.25 | 0.0 | 0.0 |
| Improve | more | same as now | 0.3 | 0.35 | 0.35 | 0.0 | 0.0 |
| Improve | more | worse | 0.1 | 0.4 | 0.5 | 0.0 | 0.0 |
| Same as now | fewer | same as now | 0.2 | 0.8 | 0.0 | 0.0 | 0.0 |
| Same as now | fewer | worse | 0.0 | 0.8 | 0.2 | 0.0 | 0.0 |
| Same as now | same as now | same as now | 0.0 | 1.0 | 0.0 | 0.0 | 0.0 |
| Same as now | same as now | worse | 0.0 | 0.3 | 0.7 | 0.0 | 0.0 |
| Same as now | more | same as now | 0.0 | 0.2 | 0.8 | 0.0 | 0.0 |
| Same as now | more | worse | 0.0 | 0.0 | 1.0 | 0.0 | 0.0 |
| Decline | fewer | same as now | 0.0 | 0.5 | 0.5 | 0.0 | 0.0 |
| Decline | fewer | worse | 0.0 | 0.3 | 0.7 | 0.0 | 0.0 |
| Decline | same as now | same as now | 0.0 | 0.0 | 1.0 | 0.0 | 0.0 |
| Decline | same as now | worse | 0.0 | 0.0 | 0.75 | 0.25 | 0.0 |
| Decline | more | same as now | 0.0 | 0.0 | 0.4 | 0.4 | 0.2 |
| Decline | more | worse | 0.0 | 0.0 | 0.2 | 0.4 | 0.4 |

**Table B9.** Node D: Change in Foraging Habitat Distribution

| Node M: Ecoregion | Node C: Foraging Habitat Absence Change | Node N: Shelf Distance Change | Distribution of Foraging Habitat | | | | |
|---|---|---|---|---|---|---|---|
| | | | Improved Availability | Same as Now | Reduced Availability | Greatly Reduced Availability | Unavailable |
| Polar basin divergent | −1 to 0 | −200 to 0 | 1.0 | 0.0 | 0.0 | 0.0 | 0.0 |
| Polar basin divergent | −1 to 0 | 0 to 200 | 0.8 | 0.2 | 0.0 | 0.0 | 0.0 |
| Polar basin divergent | −1 to 0 | 200 to 800 | 0.2 | 0.6 | 0.2 | 0.0 | 0.0 |
| Polar basin divergent | −1 to 0 | >= 800 | 0.0 | 0.4 | 0.6 | 0.0 | 0.0 |
| Polar basin divergent | 0 to 1 | −200 to 0 | 0.5 | 0.5 | 0.0 | 0.0 | 0.0 |
| Polar basin divergent | 0 to 1 | 0 to 200 | 0.0 | 0.2 | 0.8 | 0.0 | 0.0 |
| Polar basin divergent | 0 to 1 | 200 to 800 | 0.0 | 0.0 | 0.5 | 0.5 | 0.0 |
| Polar basin divergent | 0 to 1 | >= 800 | 0.0 | 0.0 | 0.25 | 0.5 | 0.25 |
| Polar basin divergent | 1 to 3 | −200 to 0 | 0.2 | 0.4 | 0.4 | 0.0 | 0.0 |
| Polar basin divergent | 1 to 3 | 0 to 200 | 0.0 | 0.0 | 0.5 | 0.3 | 0.2 |
| Polar basin divergent | 1 to 3 | 200 to 800 | 0.0 | 0.0 | 0.2 | 0.4 | 0.4 |
| Polar basin divergent | 1 to 3 | >= 800 | 0.0 | 0.0 | 0.0 | 0.2 | 0.8 |
| Polar basin divergent | >= 3 | −200 to 0 | 0.0 | 0.3 | 0.5 | 0.2 | 0.0 |
| Polar basin divergent | >= 3 | 0 to 200 | 0.0 | 0.0 | 0.2 | 0.4 | 0.4 |
| Polar basin divergent | >= 3 | 200 to 800 | 0.0 | 0.0 | 0.0 | 0.1 | 0.9 |
| Polar basin divergent | >= 3 | >= 800 | 0.0 | 0.0 | 0.0 | 0.0 | 1.0 |
| Polar basin convergent | −1 to 0 | −200 to 0 | 1.0 | 0.0 | 0.0 | 0.0 | 0.0 |
| Polar basin convergent | −1 to 0 | 0 to 200 | 0.8 | 0.2 | 0.0 | 0.0 | 0.0 |
| Polar basin convergent | −1 to 0 | 200 to 800 | 0.6 | 0.4 | 0.0 | 0.0 | 0.0 |
| Polar basin convergent | −1 to 0 | >= 800 | 0.4 | 0.6 | 0.0 | 0.0 | 0.0 |
| Polar basin convergent | 0 to 1 | −200 to 0 | 1.0 | 0.0 | 0.0 | 0.0 | 0.0 |
| Polar basin convergent | 0 to 1 | 0 to 200 | 0.6 | 0.4 | 0.0 | 0.0 | 0.0 |
| Polar basin convergent | 0 to 1 | 200 to 800 | 0.2 | 0.4 | 0.4 | 0.0 | 0.0 |
| Polar basin convergent | 0 to 1 | >= 800 | 0.0 | 0.2 | 0.8 | 0.0 | 0.0 |
| Polar basin convergent | 1 to 3 | −200 to 0 | 0.6 | 0.4 | 0.0 | 0.0 | 0.0 |

**Table B9.** (continued)

| Node M: Ecoregion | Node C: Foraging Habitat Absence Change | Node N: Shelf Distance Change | Distribution of Foraging Habitat | | | | |
|---|---|---|---|---|---|---|---|
| | | | Improved Availability | Same as Now | Reduced Availability | Greatly Reduced Availability | Unavailable |
| Polar basin convergent | 1 to 3 | 0 to 200 | 0.1 | 0.5 | 0.4 | 0.0 | 0.0 |
| Polar basin convergent | 1 to 3 | 200 to 800 | 0.0 | 0.3 | 0.7 | 0.0 | 0.0 |
| Polar basin convergent | 1 to 3 | >= 800 | 0.0 | 0.0 | 1.0 | 0.0 | 0.0 |
| Polar basin convergent | >= 3 | −200 to 0 | 0.4 | 0.6 | 0.0 | 0.0 | 0.0 |
| Polar basin convergent | >= 3 | 0 to 200 | 0.1 | 0.3 | 0.5 | 0.1 | 0.0 |
| Polar basin convergent | >= 3 | 200 to 800 | 0.0 | 0.2 | 0.6 | 0.2 | 0.0 |
| Polar basin convergent | >= 3 | >= 800 | 0.0 | 0.0 | 0.7 | 0.3 | 0.0 |
| Archipelago | −1 to 0 | −200 to 0 | 0.0 | 1.0 | 0.0 | 0.0 | 0.0 |
| Archipelago | −1 to 0 | 0 to 200 | 0.0 | 1.0 | 0.0 | 0.0 | 0.0 |
| Archipelago | −1 to 0 | 200 to 800 | 0.0 | 1.0 | 0.0 | 0.0 | 0.0 |
| Archipelago | −1 to 0 | >= 800 | 0.0 | 1.0 | 0.0 | 0.0 | 0.0 |
| Archipelago | 0 to 1 | −200 to 0 | 0.6 | 0.4 | 0.0 | 0.0 | 0.0 |
| Archipelago | 0 to 1 | 0 to 200 | 0.6 | 0.4 | 0.0 | 0.0 | 0.0 |
| Archipelago | 0 to 1 | 200 to 800 | 0.6 | 0.4 | 0.0 | 0.0 | 0.0 |
| Archipelago | 0 to 1 | >= 800 | 0.6 | 0.4 | 0.0 | 0.0 | 0.0 |
| Archipelago | 1 to 3 | −200 to 0 | 0.4 | 0.6 | 0.0 | 0.0 | 0.0 |
| Archipelago | 1 to 3 | 0 to 200 | 0.4 | 0.6 | 0.0 | 0.0 | 0.0 |
| Archipelago | 1 to 3 | 200 to 800 | 0.4 | 0.6 | 0.0 | 0.0 | 0.0 |
| Archipelago | 1 to 3 | >= 800 | 0.4 | 0.6 | 0.0 | 0.0 | 0.0 |
| Archipelago | >= 3 | −200 to 0 | 0.0 | 0.6 | 0.4 | 0.0 | 0.0 |
| Archipelago | >= 3 | 0 to 200 | 0.0 | 0.6 | 0.4 | 0.0 | 0.0 |
| Archipelago | >= 3 | 200 to 800 | 0.0 | 0.6 | 0.4 | 0.0 | 0.0 |
| Archipelago | >= 3 | >= 800 | 0.0 | 0.6 | 0.4 | 0.0 | 0.0 |
| Seasonal ice | −1 to 0 | −200 to 0 | 0.5 | 0.5 | 0.0 | 0.0 | 0.0 |
| Seasonal ice | −1 to 0 | 0 to 200 | 0.5 | 0.5 | 0.0 | 0.0 | 0.0 |
| Seasonal ice | −1 to 0 | 200 to 800 | 0.5 | 0.5 | 0.0 | 0.0 | 0.0 |
| Seasonal ice | −1 to 0 | >= 800 | 0.5 | 0.5 | 0.0 | 0.0 | 0.0 |
| Seasonal ice | 0 to 1 | −200 to 0 | 0.0 | 0.2 | 0.6 | 0.2 | 0.0 |
| Seasonal ice | 0 to 1 | 0 to 200 | 0.0 | 0.2 | 0.6 | 0.2 | 0.0 |
| Seasonal ice | 0 to 1 | 200 to 800 | 0.0 | 0.2 | 0.6 | 0.2 | 0.0 |
| Seasonal ice | 0 to 1 | >= 800 | 0.0 | 0.2 | 0.6 | 0.2 | 0.0 |
| Seasonal ice | 1 to 3 | −200 to 0 | 0.0 | 0.0 | 0.2 | 0.4 | 0.4 |
| Seasonal ice | 1 to 3 | 0 to 200 | 0.0 | 0.0 | 0.2 | 0.4 | 0.4 |
| Seasonal ice | 1 to 3 | 200 to 800 | 0.0 | 0.0 | 0.2 | 0.4 | 0.4 |
| Seasonal ice | 1 to 3 | >= 800 | 0.0 | 0.0 | 0.2 | 0.4 | 0.4 |
| Seasonal ice | >= 3 | −200 to 0 | 0.0 | 0.0 | 0.0 | 0.1 | 0.9 |
| Seasonal ice | >= 3 | 0 to 200 | 0.0 | 0.0 | 0.0 | 0.1 | 0.9 |
| Seasonal ice | >= 3 | 200 to 800 | 0.0 | 0.0 | 0.0 | 0.1 | 0.9 |
| Seasonal ice | >= 3 | >= 800 | 0.0 | 0.0 | 0.0 | 0.1 | 0.9 |

**Table B10.** Node D1: Overall Population Outcome

| Node C4: Numerical Response | Node C3: Distribution Response | Overall Population Outcome | | | | |
|---|---|---|---|---|---|---|
| | | Larger | Same as Now | Smaller | Rare | Extinct |
| Increased density | same as now | 1.0 | 0.0 | 0.0 | 0.0 | 0.0 |
| Increased density | reduced but resident | 0.3 | 0.5 | 0.2 | 0.0 | 0.0 |
| Increased density | transient visitor | 0.1 | 0.3 | 0.3 | 0.3 | 0.0 |
| Increased density | extirpated | 0.0 | 0.0 | 0.0 | 0.0 | 1.0 |

**Table B10.** (continued)

| Node C4: Numerical Response | Node C3: Distribution Response | Overall Population Outcome | | | | |
|---|---|---|---|---|---|---|
| | | Larger | Same as Now | Smaller | Rare | Extinct |
| Same as now | same as now | 0.0 | 1.0 | 0.0 | 0.0 | 0.0 |
| Same as now | reduced but resident | 0.0 | 0.3 | 0.7 | 0.0 | 0.0 |
| Same as now | transient visitor | 0.0 | 0.0 | 0.6 | 0.4 | 0.0 |
| Same as now | extirpated | 0.0 | 0.0 | 0.0 | 0.0 | 1.0 |
| Reduced density | same as now | 0.0 | 0.0 | 1.0 | 0.0 | 0.0 |
| Reduced density | reduced but resident | 0.0 | 0.0 | 0.7 | 0.3 | 0.0 |
| Reduced density | transient visitor | 0.0 | 0.0 | 0.3 | 0.7 | 0.0 |
| Reduced density | extirpated | 0.0 | 0.0 | 0.0 | 0.0 | 1.0 |
| Rare | same as now | 0.0 | 0.0 | 0.0 | 1.0 | 0.0 |
| Rare | reduced but resident | 0.0 | 0.0 | 0.0 | 0.8 | 0.2 |
| Rare | transient visitor | 0.0 | 0.0 | 0.0 | 0.7 | 0.3 |
| Rare | extirpated | 0.0 | 0.0 | 0.0 | 0.0 | 1.0 |
| Absent | same as now | 0.0 | 0.0 | 0.0 | 0.0 | 1.0 |
| Absent | reduced but resident | 0.0 | 0.0 | 0.0 | 0.0 | 1.0 |
| Absent | transient visitor | 0.0 | 0.0 | 0.0 | 0.0 | 1.0 |
| Absent | extirpated | 0.0 | 0.0 | 0.0 | 0.0 | 1.0 |

**Table B11.** Node F2: Factor A Habitat Threats

| Node B: Foraging Habitat Quantity Change | Node D: Change in Foraging Habitat Distribution | Node A: Foraging Habitat Value | Level of Habitat Threat | | | |
|---|---|---|---|---|---|---|
| | | | Improvement | No Effect | Minor Restriction | Major Restriction |
| 0 to 20 | improved availability | better | 1.0 | 0.0 | 0.0 | 0.0 |
| 0 to 20 | improved availability | same as now | 1.0 | 0.0 | 0.0 | 0.0 |
| 0 to 20 | improved availability | worse | 0.8 | 0.2 | 0.0 | 0.0 |
| 0 to 20 | same as now | better | 1.0 | 0.0 | 0.0 | 0.0 |
| 0 to 20 | same as now | same as now | 0.8 | 0.2 | 0.0 | 0.0 |
| 0 to 20 | same as now | worse | 0.3 | 0.5 | 0.2 | 0.0 |
| 0 to 20 | reduced availability | better | 0.4 | 0.4 | 0.2 | 0.0 |
| 0 to 20 | reduced availability | same as now | 0.2 | 0.6 | 0.2 | 0.0 |
| 0 to 20 | reduced availability | worse | 0.0 | 0.2 | 0.6 | 0.2 |
| 0 to 20 | greatly reduced availability | better | 0.0 | 0.2 | 0.4 | 0.4 |
| 0 to 20 | greatly reduced availability | same as now | 0.0 | 0.0 | 0.4 | 0.6 |
| 0 to 20 | greatly reduced availability | worse | 0.0 | 0.0 | 0.2 | 0.8 |
| 0 to 20 | unavailable | better | 0.0 | 0.0 | 0.0 | 1.0 |
| 0 to 20 | unavailable | same as now | 0.0 | 0.0 | 0.0 | 1.0 |
| 0 to 20 | unavailable | worse | 0.0 | 0.0 | 0.0 | 1.0 |
| −20 to 0 | improved availab | better | 0.8 | 0.2 | 0.0 | 0.0 |
| −20 to 0 | improved availab | same as now | 0.2 | 0.6 | 0.2 | 0.0 |
| −20 to 0 | improved availab | worse | 0.2 | 0.4 | 0.4 | 0.0 |
| −20 to 0 | same as now | better | 0.2 | 0.6 | 0.2 | 0.0 |
| −20 to 0 | same as now | same as now | 0.0 | 0.2 | 0.6 | 0.2 |
| −20 to 0 | same as now | worse | 0.0 | 0.0 | 0.6 | 0.4 |
| −20 to 0 | reduced availability | better | 0.1 | 0.5 | 0.2 | 0.2 |
| −20 to 0 | reduced availability | same as now | 0.0 | 0.1 | 0.5 | 0.4 |
| −20 to 0 | reduced availability | worse | 0.0 | 0.0 | 0.4 | 0.6 |
| −20 to 0 | greatly reduced availability | better | 0.0 | 0.0 | 0.5 | 0.5 |
| −20 to 0 | greatly reduced availability | same as now | 0.0 | 0.0 | 0.2 | 0.8 |
| −20 to 0 | greatly reduced availability | worse | 0.0 | 0.0 | 0.0 | 1.0 |

**Table B11.** (continued)

| Node B: Foraging Habitat Quantity Change | Node D: Change in Foraging Habitat Distribution | Node A: Foraging Habitat Value | Level of Habitat Threat | | | |
|---|---|---|---|---|---|---|
| | | | Improvement | No Effect | Minor Restriction | Major Restriction |
| −20 to 0 | unavailable | better | 0.0 | 0.0 | 0.0 | 1.0 |
| −20 to 0 | unavailable | same as now | 0.0 | 0.0 | 0.0 | 1.0 |
| −20 to 0 | unavailable | worse | 0.0 | 0.0 | 0.0 | 1.0 |
| −40 to −20 | improved availability | better | 0.4 | 0.4 | 0.2 | 0.0 |
| −40 to −20 | improved availability | same as now | 0.1 | 0.5 | 0.4 | 0.0 |
| −40 to −20 | improved availability | worse | 0.0 | 0.3 | 0.5 | 0.2 |
| −40 to −20 | same as now | better | 0.1 | 0.4 | 0.4 | 0.1 |
| −40 to −20 | same as now | same as now | 0.0 | 0.1 | 0.4 | 0.5 |
| −40 to −20 | same as now | worse | 0.0 | 0.0 | 0.4 | 0.6 |
| −40 to −20 | reduced availability | better | 0.0 | 0.1 | 0.6 | 0.3 |
| −40 to −20 | reduced availability | same as now | 0.0 | 0.0 | 0.5 | 0.5 |
| −40 to −20 | reduced availability | worse | 0.0 | 0.0 | 0.2 | 0.8 |
| −40 to −20 | greatly reduced availability | better | 0.0 | 0.0 | 0.3 | 0.7 |
| −40 to −20 | greatly reduced availability | same as now | 0.0 | 0.0 | 0.0 | 1.0 |
| −40 to −20 | greatly reduced availability | worse | 0.0 | 0.0 | 0.0 | 1.0 |
| −40 to −20 | unavailable | better | 0.0 | 0.0 | 0.0 | 1.0 |
| −40 to −20 | unavailable | same as now | 0.0 | 0.0 | 0.0 | 1.0 |
| −40 to −20 | unavailable | worse | 0.0 | 0.0 | 0.0 | 1.0 |
| < −40 | improved availability | better | 0.2 | 0.6 | 0.2 | 0.0 |
| < −40 | improved availability | same as now | 0.0 | 0.2 | 0.6 | 0.2 |
| < −40 | improved availability | worse | 0.0 | 0.0 | 0.5 | 0.5 |
| < −40 | same as now | better | 0.0 | 0.1 | 0.6 | 0.3 |
| < −40 | same as now | same as now | 0.0 | 0.0 | 0.3 | 0.7 |
| < −40 | same as now | worse | 0.0 | 0.0 | 0.2 | 0.8 |
| < −40 | reduced availability | better | 0.0 | 0.1 | 0.2 | 0.7 |
| < −40 | reduced availability | same as now | 0.0 | 0.0 | 0.1 | 0.9 |
| < −40 | reduced availability | worse | 0.0 | 0.0 | 0.0 | 1.0 |
| < −40 | greatly reduced availability | better | 0.0 | 0.0 | 0.0 | 1.0 |
| < −40 | greatly reduced availability | same as now | 0.0 | 0.0 | 0.0 | 1.0 |
| < −40 | greatly reduced availability | worse | 0.0 | 0.0 | 0.0 | 1.0 |
| < −40 | unavailable | better | 0.0 | 0.0 | 0.0 | 1.0 |
| < −40 | unavailable | same as now | 0.0 | 0.0 | 0.0 | 1.0 |
| < −40 | unavailable | worse | 0.0 | 0.0 | 0.0 | 1.0 |

**Table B12.** Node G: Relocation Possible

| Node F: Alternative Regions Available | Node H: Crowding Tolerance | Possibility of Relocation | |
|---|---|---|---|
| | | Yes | No |
| Yes | none | 0.0 | 1.0 |
| Yes | moderate | 0.8 | 0.2 |
| Yes | high | 1.0 | 0.0 |
| No | none | 0.0 | 1.0 |
| No | moderate | 0.0 | 1.0 |
| No | high | 0.0 | 1.0 |

**Table B13.** Node H: Crowding Tolerance

| Node R2: Alternative Prey Availability | Node R3: Relative Ringed Seal Availability | Level of Crowding Tolerance | | |
|---|---|---|---|---|
| | | None | Moderate | High |
| Increase | increase | 0.0 | 0.2 | 0.8 |
| Increase | same as now | 0.0 | 0.4 | 0.6 |
| Increase | decrease | 0.1 | 0.5 | 0.4 |
| Same as now | increase | 0.0 | 0.4 | 0.6 |
| Same as now | same as now | 0.1 | 0.8 | 0.1 |
| Same as now | decrease | 0.3 | 0.6 | 0.1 |
| Decrease | increase | 0.1 | 0.5 | 0.4 |
| Decrease | same as now | 0.3 | 0.5 | 0.2 |
| Decrease | decrease | 0.5 | 0.5 | 0.0 |

**Table B14.** Node K: Adult Body Condition

| Node F2: Factor A—Habitat Threats | Quality of Adult Body Condition | | |
|---|---|---|---|
| | Increase | Same as Now | Decrease |
| Improvement | 1.0 | 0.0 | 0.0 |
| No effect | 0.0 | 1.0 | 0.0 |
| Minor restriction | 0.0 | 0.5 | 0.5 |
| Major restriction | 0.0 | 0.0 | 1.0 |

**Table B15.** Node L: Juvenile Survival

| Node K: Adult Body Condition | Node L1: Adult Female Survival | Juvenile Survival | | |
|---|---|---|---|---|
| | | Increase | No Change | Decrease |
| Increase | increase | 1.0 | 0.0 | 0.0 |
| Increase | no change | 0.7 | 0.3 | 0.0 |
| Increase | decrease | 0.0 | 0.4 | 0.6 |
| Same as now | increase | 0.8 | 0.2 | 0.0 |
| Same as now | no change | 0.0 | 1.0 | 0.0 |
| Same as now | decrease | 0.0 | 0.2 | 0.8 |
| Decrease | increase | 0.0 | 0.6 | 0.4 |
| Decrease | no change | 0.0 | 0.3 | 0.7 |
| Decrease | decrease | 0.0 | 0.0 | 1.0 |

**Table B16.** Node L1: Adult Female Survival

| Node K: Adult Body Condition | Node F2: Factor A—Habitat Threats | Adult Female Survival | | |
|---|---|---|---|---|
| | | Increase | No Change | Decrease |
| Increase | improvement | 1.0 | 0.0 | 0.0 |
| Increase | no effect | 0.8 | 0.2 | 0.0 |
| Increase | minor restriction | 0.1 | 0.6 | 0.3 |
| Increase | major restriction | 0.0 | 0.5 | 0.5 |
| Same as now | improvement | 0.5 | 0.5 | 0.0 |
| Same as now | no effect | 0.0 | 1.0 | 0.0 |
| Same as now | minor restriction | 0.0 | 0.6 | 0.4 |
| Same as now | major restriction | 0.0 | 0.3 | 0.7 |
| Decrease | improvement | 0.0 | 0.4 | 0.6 |
| Decrease | no effect | 0.0 | 0.2 | 0.8 |
| Decrease | minor restriction | 0.0 | 0.1 | 0.9 |
| Decrease | major restriction | 0.0 | 0.0 | 1.0 |

**Table B17.** Node L2: Vital Rates

| Node L1: Adult Female Survival | Node L: Juvenile Survival | Node U: Reproduction | Vital Rates | | |
|---|---|---|---|---|---|
| | | | Improve | Same as Now | Decline |
| Increase | increase | increased | 1.0 | 0.0 | 0.0 |
| Increase | increase | same as now | 1.0 | 0.0 | 0.0 |
| Increase | increase | decreased | 0.6 | 0.4 | 0.0 |
| Increase | no change | increased | 0.9 | 0.1 | 0.0 |
| Increase | no change | same as now | 0.8 | 0.2 | 0.0 |
| Increase | no change | decreased | 0.7 | 0.2 | 0.1 |
| Increase | decrease | increased | 0.3 | 0.5 | 0.2 |
| Increase | decrease | same as now | 0.2 | 0.5 | 0.3 |
| Increase | decrease | decreased | 0.0 | 0.4 | 0.6 |
| No change | increase | increased | 0.7 | 0.3 | 0.0 |
| No change | increase | same as now | 0.6 | 0.4 | 0.0 |
| No change | increase | decreased | 0.2 | 0.5 | 0.3 |
| No change | no change | increased | 0.2 | 0.8 | 0.0 |
| No change | no change | same as now | 0.0 | 1.0 | 0.0 |
| No change | no change | decreased | 0.0 | 0.8 | 0.2 |
| No change | decrease | increased | 0.0 | 0.6 | 0.4 |
| No change | decrease | same as now | 0.0 | 0.5 | 0.5 |
| No change | decrease | decreased | 0.0 | 0.3 | 0.7 |
| Decrease | increase | increased | 0.2 | 0.4 | 0.4 |
| Decrease | increase | same as now | 0.0 | 0.6 | 0.4 |
| Decrease | increase | decreased | 0.0 | 0.5 | 0.5 |
| Decrease | no change | increased | 0.1 | 0.5 | 0.4 |
| Decrease | no change | same as now | 0.0 | 0.4 | 0.6 |
| Decrease | no change | decreased | 0.0 | 0.3 | 0.7 |
| Decrease | decrease | increased | 0.0 | 0.2 | 0.8 |
| Decrease | decrease | same as now | 0.0 | 0.0 | 1.0 |
| Decrease | decrease | decreased | 0.0 | 0.0 | 1.0 |

**Table B18.** Node U: Reproduction

| Node M: Ecoregion | Node V1: Cub Production per Event | Node N: Shelf Distance Change (km) | Rate of Reproduction | | |
|---|---|---|---|---|---|
| | | | Increased | Same as Now | Decreased |
| Polar basin divergent | fewer than now | −200 to 0 | 0.0 | 0.3 | 0.7 |
| Polar basin divergent | fewer than now | 0 to 200 | 0.0 | 0.2 | 0.8 |
| Polar basin divergent | fewer than now | 200 to 800 | 0.0 | 0.0 | 1.0 |
| Polar basin divergent | fewer than now | >= 800 | 0.0 | 0.0 | 1.0 |
| Polar basin divergent | same as now | −200 to 0 | 0.7 | 0.3 | 0.0 |
| Polar basin divergent | same as now | 0 to 200 | 0.0 | 1.0 | 0.0 |
| Polar basin divergent | same as now | 200 to 800 | 0.0 | 0.3 | 0.7 |
| Polar basin divergent | same as now | >= 800 | 0.0 | 0.0 | 1.0 |
| Polar basin divergent | more than now | −200 to 0 | 1.0 | 0.0 | 0.0 |
| Polar basin divergent | more than now | 0 to 200 | 0.5 | 0.5 | 0.0 |
| Polar basin divergent | more than now | 200 to 800 | 0.0 | 0.5 | 0.5 |
| Polar basin divergent | more than now | >= 800 | 0.0 | 0.0 | 1.0 |
| Polar basin convergent | fewer than now | −200 to 0 | 0.0 | 0.5 | 0.5 |
| Polar basin convergent | fewer than now | 0 to 200 | 0.0 | 0.4 | 0.6 |
| Polar basin convergent | fewer than now | 200 to 800 | 0.0 | 0.3 | 0.7 |
| Polar basin convergent | fewer than now | >= 800 | 0.0 | 0.2 | 0.8 |
| Polar basin convergent | same as now | −200 to 0 | 1.0 | 0.0 | 0.0 |
| Polar basin convergent | same as now | 0 to 200 | 0.5 | 0.5 | 0.0 |
| Polar basin convergent | same as now | 200 to 800 | 0.2 | 0.6 | 0.2 |
| Polar basin convergent | same as now | >= 800 | 0.0 | 0.5 | 0.5 |
| Polar basin convergent | more than now | −200 to 0 | 1.0 | 0.0 | 0.0 |
| Polar basin convergent | more than now | 0 to 200 | 0.8 | 0.2 | 0.0 |
| Polar basin convergent | more than now | 200 to 800 | 0.4 | 0.4 | 0.2 |
| Polar basin convergent | more than now | >= 800 | 0.2 | 0.4 | 0.4 |
| Archipelago | fewer than now | −200 to 0 | 0.0 | 0.2 | 0.8 |
| Archipelago | fewer than now | 0 to 200 | 0.0 | 0.2 | 0.8 |
| Archipelago | fewer than now | 200 to 800 | 0.0 | 0.2 | 0.8 |
| Archipelago | fewer than now | >= 800 | 0.0 | 0.2 | 0.8 |
| Archipelago | same as now | −200 to 0 | 0.2 | 0.6 | 0.2 |
| Archipelago | same as now | 0 to 200 | 0.2 | 0.6 | 0.2 |
| Archipelago | same as now | 200 to 800 | 0.2 | 0.6 | 0.2 |
| Archipelago | same as now | >= 800 | 0.2 | 0.6 | 0.2 |
| Archipelago | more than now | −200 to 0 | 0.8 | 0.2 | 0.0 |
| Archipelago | more than now | 0 to 200 | 0.8 | 0.2 | 0.0 |
| Archipelago | more than now | 200 to 800 | 0.8 | 0.2 | 0.0 |
| Archipelago | more than now | >= 800 | 0.8 | 0.2 | 0.0 |
| Seasonal ice | fewer than now | −200 to 0 | 0.0 | 0.2 | 0.8 |
| Seasonal ice | fewer than now | 0 to 200 | 0.0 | 0.2 | 0.8 |
| Seasonal ice | fewer than now | 200 to 800 | 0.0 | 0.2 | 0.8 |
| Seasonal ice | fewer than now | >= 800 | 0.0 | 0.2 | 0.8 |
| Seasonal ice | same as now | −200 to 0 | 0.2 | 0.6 | 0.2 |
| Seasonal ice | same as now | 0 to 200 | 0.2 | 0.6 | 0.2 |
| Seasonal ice | same as now | 200 to 800 | 0.2 | 0.6 | 0.2 |
| Seasonal ice | same as now | >= 800 | 0.2 | 0.6 | 0.2 |
| Seasonal ice | more than now | −200 to 0 | 0.8 | 0.2 | 0.0 |
| Seasonal ice | more than now | 0 to 200 | 0.8 | 0.2 | 0.0 |
| Seasonal ice | more than now | 200 to 800 | 0.8 | 0.2 | 0.0 |
| Seasonal ice | more than now | >= 800 | 0.8 | 0.2 | 0.0 |

**Table B19.** Node V1: Cub Production per Event

| Node F2: Factor A—Habitat Threats | Cub Production per Event | | |
|---|---|---|---|
| | Fewer Than Now | Same as Now | More Than Now |
| Improvement | 0.0 | 0.3 | 0.7 |
| No effect | 0.0 | 1.0 | 0.0 |
| Minor restriction | 0.6 | 0.4 | 0.0 |
| Major restriction | 1.0 | 0.0 | 0.0 |

## APPENDIX C: RESULTS OF SENSITIVITY ANALYSES OF THE BAYESIAN NETWORK POPULATION STRESSOR MODEL

Tables C1 and C2 present the results of conducting a series of sensitivity analyses of the Bayesian network population stressor model discussed in the text (also see Plate 3). Sensitivity analysis reveals the degree to which selected input or summary variables influence the calculated values of a specified output variable. Tables C1 and C2 present results of 10 sensitivity tests on various summary and output nodes in the model (see text for explanation of calculations). Note that mutual information is also called entropy reduction. All tests were conducted using the Bayesian network modeling software package Netica® (Norsys, Inc.).

**Table C1.** Sensitivity Group 1: Sensitivity of Overall Population Outcome

| Node | Mutual Information | Node Title |
|---|---|---|
| *Test 1: Sensitivity of Node D1—Overall Population Outcome to All Input Nodes* | | |
| B | 0.12974 | Foraging Habitat Quantity Change |
| C | 0.04876 | foraging habitat absence change |
| M | 0.04166 | ecoregion |
| F | 0.02590 | alternate regions available |
| E | 0.01607 | intentional takes |
| N | 0.01393 | shelf distance change (km) |
| S1 | 0.01037 | foraging habitat character |
| B1 | 0.00821 | bear-human interactions |
| T | 0.00506 | parasites and disease |
| R4 | 0.00271 | hydrocarbons/oil spill |
| R1 | 0.00254 | oil and gas activity |
| J | 0.00198 | shipping |
| T2 | 0.00092 | predation |
| T1 | 0.00073 | contaminants |
| R3 | 0.00069 | alternate prey availability |
| R2 | 0.00065 | relative ringed seal availability |
| J1 | 0.00040 | tourism |
| *Test 2. Sensitivity of Node D1—Overall Population Outcome to Listing Factor Nodes* | | |
| F2 | 0.66422 | factor a: habitat threats |
| A1 | 0.05253 | factor b: direct mortalities |
| A6 | 0.03150 | factor e: other factors (natural or man-made) |
| A4 | 0.01039 | factor c: disease, predation |
| *Test 3: Sensitivity of Node D1—Overall Population Outcome to Intermediate Nodes*[a] | | |
| L2 | 0.56624 | vital rates |
| L | 0.54067 | juvenile survival |
| L1 | 0.54057 | adult female survival |
| K | 0.53353 | adult body condition |
| V1 | 0.44705 | cub production per event |

**Table C1.** (continued)

| Node | Mutual Information | Node Title |
|------|-------------------|------------|
| U | 0.24141 | reproduction |
| D | 0.19993 | change in foraging habitat distribution |
| G | 0.04235 | relocation possible |
| A | 0.02866 | foraging habitat value |
| C1 | 0.01856 | human disturbance |
| H | 0.00537 | crowding tolerance |
| C2 | 0.00432 | pollution |

*Test 4: Sensitivity of Node D1—Overall Population Outcome to Selected Intermediate Nodes*[b]

| | | |
|------|-------------------|------------|
| F2 | 0.66422 | factor a: habitat threats |
| L2 | 0.56624 | vital rates |
| A1 | 0.05253 | factor b: direct mortalities |
| G | 0.04235 | relocation possible |
| A6 | 0.03150 | factor e: other factors (natural or man-made) |
| A4 | 0.01039 | factor c: disease and predation |

[a]This does not include the listing factor nodes included in test 2 above.
[b]This includes all (6) nodes that are two links distant from the output node.

**Table C2.** Sensitivity Group 2: Sensitivity of Submodels

| Node | Mutual Information | Node Title |
|------|-------------------|------------|
| | *Test 5: Sensitivity of Node A4—Factor C Disease and Predation* | |
| T | 0.39016 | parasites and disease |
| T2 | 0.06593 | predation |
| | *Test 6: Sensitivity of Node C2—Pollution* | |
| R4 | 0.69005 | hydrocarbons/oil spill |
| T1 | 0.13542 | contaminants |
| | *Test 7: Sensitivity of Node C1—Human Disturbance* | |
| B1 | 0.45796 | bear-human interactions |
| R1 | 0.12450 | oil and gas activity |
| J | 0.08941 | shipping |
| J1 | 0.01729 | tourism |
| | *Test 8: Sensitivity of Node A—Foraging Habitat Value* | |
| S1 | 0.51589 | foraging habitat character |
| F | 0.04028 | alternate regions available |
| R3 | 0.00105 | alternate prey availability |
| R2 | 0.00100 | relative ringed seal availability |
| | *Test 9: Sensitivity of Node D—Change in Foraging Habitat Distribution* | |
| M | 0.33239 | ecoregion |
| C | 0.32674 | foraging habitat absence change |
| N | 0.06131 | shelf distance change (km) |

**Table C2.** (continued)

| Node | Mutual Information | Node Title |
|------|--------------------|------------|
| *Test 10: Sensitivity of Node L2—Vital Rates* | | |
| L1 | 1.04302 | adult female survival |
| L | 1.04048 | juvenile survival |
| F2 | 0.93484 | factor a: habitat threats |
| K | 0.92047 | adult body condition |
| V1 | 0.64819 | cub production per event |
| U | 0.34420 | reproduction |
| M | 0.04217 | ecoregion |
| N | 0.01843 | shelf distance change (km) |

## APPENDIX D:  INPUT VALUES USED IN THE BAYESIAN NETWORK POLAR BEAR POPULATION STRESSOR MODEL FOR EACH OF FOUR POLAR BEAR ECOREGIONS

Tables D1a and D1b present input data values used in the Bayesian network polar bear population stressor model for each of four polar bear ecoregions. Separate input values were provided for each time period projected and for the ensemble mean of general circulation model outputs as well as for individual GCMs that projected the maximum and minimum sea ice remaining in each time period.

**Table D1a.** Input Data Values for Nodes B, C, N, S1, R3, R2, and F Used in the Bayesian Network Polar Bear Population Stressor Model for Each of Four Polar Bear Ecoregions

| Time Period[a] | Sea Ice Data Source | B: Foraging Habitat Quantity Change (%) | C: Foraging Habitat Absence Change | N: Shelf Distance Change[c] (km) | S1: Foraging Habitat Character | R3: Alternative Prey Availability | R2: Relative Ringed Seal Availability | F: Alternative Regions Available |
|---|---|---|---|---|---|---|---|---|
| *Seasonal Ice Ecoregion* | | | | | | | | |
| Year −10 | Satellite data | 17.14 | −0.7 | NA | more_optimal | decrease | increase | yes |
| Year 0 | Satellite data | 0.00 | 0.0 | NA | same_as_now | same_as_now | same_as_now | yes |
| Year 25 | GCM minimum | −4.17 | 0.1 | NA | same_as_now | same_as_now | decrease | yes |
| Year 50 | GCM minimum | −10.36 | 1.0 | NA | same_as_now | decrease | decrease | yes |
| Year 75 | GCM minimum | −31.89 | 2.5 | NA | less_optimal | decrease | decrease | yes |
| Year 95 | GCM minimum | −32.11 | 2.7 | NA | less_optimal | decrease | decrease | yes |
| Year 25 | Ensemble mean | −4.65 | 0.3 | NA | same_as_now | same_as_now | decrease | yes |
| Year 50 | Ensemble mean | −14.62 | 1.0 | NA | same_as_now | decrease | decrease | yes |
| Year 75 | Ensemble mean | −25.75 | 1.6 | NA | less_optimal | decrease | decrease | yes |
| Year 95 | Ensemble mean | −27.83 | 1.8 | NA | less_optimal | decrease | decrease | yes |
| Year 25 | GCM maximum | −0.05 | 0.1 | NA | same_as_now | same_as_now | decrease | yes |
| Year 50 | GCM maximum | −6.71 | 0.7 | NA | same_as_now | decrease | decrease | yes |
| Year 75 | GCM maximum | −21.16 | 1.3 | NA | same_as_now | decrease | decrease | yes |
| Year 95 | GCM maximum | −21.69 | 1.7 | NA | same_as_now | decrease | decrease | yes |
| *Archipelago Ecoregion* | | | | | | | | |
| Year −10 | Satellite data | 3.21 | −0.5 | NA | less_optimal | same_as_now | decrease | no |
| Year 0 | Satellite data | 0.00 | 0.0 | NA | same_as_now | same_as_now | same_as_now | no |
| Year 25 | GCM minimum | −6.16 | 0.6 | NA | more_optimal | same_as_now | increase | no |
| Year 50 | GCM minimum | −13.79 | 1.1 | NA | more_optimal | increase | increase | no |
| Year 75 | GCM minimum | −20.71 | 2.0 | NA | same_as_now | decrease | decrease | no |

**Table D1a.** (continued)

| Time Period[a] | Sea Ice Data Source | B: Foraging Habitat Quantity Change (%) | C: Foraging Habitat Absence Change | N: Shelf Distance Change[c] (km) | S1: Foraging Habitat Character | R3: Alternative Prey Availability | R2: Relative Ringed Seal Availability | F: Alternative Regions Available |
|---|---|---|---|---|---|---|---|---|
| | | | | | Node and Variable Name[b] | | | |
| Year 95 | GCM minimum | −24.30 | 2.3 | NA | same_as_now | decrease | decrease | no |
| Year 25 | Ensemble mean | −2.35 | 0.2 | NA | more_optimal | same_as_now | increase | no |
| Year 50 | Ensemble mean | −11.93 | 1.5 | NA | more_optimal | increase | increase | no |
| Year 75 | Ensemble mean | −20.06 | 2.4 | NA | same_as_now | increase | decrease | no |
| Year 95 | Ensemble mean | −22.16 | 2.5 | NA | same_as_now | decrease | decrease | no |
| Year 25 | GCM maximum | −0.08 | 0.0 | NA | more_optimal | same_as_now | increase | no |
| Year 50 | GCM maximum | −3.43 | 0.0 | NA | more_optimal | increase | increase | no |
| Year 75 | GCM maximum | −18.02 | 2.7 | NA | more_optimal | increase | increase | no |
| Year 95 | GCM maximum | −20.85 | 2.3 | NA | same_as_now | decrease | decrease | no |
| *Polar Basin Divergent Ecoregion* | | | | | | | | |
| Year −10 | Satellite data | 5.33 | −0.3 | −83 | more_optimal | same_as_now | increase | yes |
| Year 0 | Satellite data | 0.00 | 0.0 | 0 | same_as_now | same_as_now | same_as_now | yes |
| Year 25 | GCM minimum | −9.76 | 0.9 | 183 | less_optimal | same_as_now | decrease | yes |
| Year 50 | GCM minimum | −32.16 | 2.1 | 1359 | less_optimal | same_as_now | decrease | yes |
| Year 75 | GCM minimum | −41.28 | 2.9 | 2006 | less_optimal | same_as_now | decrease | yes |
| Year 95 | GCM minimum | −46.30 | 3.2 | 2177 | less_optimal | same_as_now | decrease | yes |
| Year 25 | Ensemble mean | −5.25 | 0.7 | 114 | same_as_now | same_as_now | decrease | yes |
| Year 50 | Ensemble mean | −19.31 | 1.8 | 631 | less_optimal | same_as_now | decrease | yes |
| Year 75 | Ensemble mean | −31.68 | 2.6 | 1034 | less_optimal | same_as_now | decrease | yes |
| Year 95 | Ensemble mean | −35.77 | 3.0 | 1275 | less_optimal | same_as_now | decrease | yes |
| Year 25 | GCM maximum | −5.12 | 0.7 | 42 | same_as_now | same_as_now | same_as_now | yes |
| Year 50 | GCM maximum | −15.68 | 2.2 | 234 | less_optimal | same_as_now | decrease | yes |
| Year 75 | GCM maximum | −24.23 | 2.4 | 233 | less_optimal | same_as_now | decrease | yes |
| Year 95 | GCM maximum | −21.33 | 2.7 | 315 | less_optimal | same_as_now | decrease | yes |
| *Polar Basin Convergent Ecoregion* | | | | | | | | |
| Year −10 | Satellite data | 4.34 | −0.5 | −41 | same_as_now | same_as_now | same_as_now | no |
| Year 0 | Satellite data | 0.00 | 0.0 | 0 | same_as_now | same_as_now | same_as_now | no |
| Year 25 | GCM minimum | 2.65 | 0.3 | 26 | more_optimal | same_as_now | increase | no |
| Year 50 | GCM minimum | −4.60 | 0.9 | 831 | same_as_now | increase | same_as_now | no |
| Year 75 | GCM minimum | −23.19 | 1.9 | 1542 | less_optimal | decrease | decrease | no |
| Year 95 | GCM minimum | −30.33 | 2.5 | 1478 | less_optimal | decrease | decrease | no |
| Year 25 | Ensemble mean | −2.76 | 0.7 | 83 | more_optimal | increase | increase | no |
| Year 50 | Ensemble mean | −13.85 | 2.0 | 464 | same_as_now | increase | increase | no |
| Year 75 | Ensemble mean | −22.65 | 3.0 | 847 | less_optimal | decrease | same_as_now | no |
| Year 95 | Ensemble mean | −25.02 | 3.3 | 795 | less_optimal | decrease | decrease | no |
| Year 25 | GCM maximum | −6.68 | 0.9 | 109 | more_optimal | increase | increase | no |
| Year 50 | GCM maximum | −26.76 | 2.9 | 334 | same_as_now | increase | increase | no |
| Year 75 | GCM maximum | −34.08 | 3.5 | 434 | less_optimal | increase | increase | no |
| Year 95 | GCM maximum | −34.88 | 3.7 | 510 | less_optimal | decrease | same_as_now | no |

[a]Time period is expressed as the central year in each decade for which projections were made.

[b]Units of measure at each node are B, percentile change from present in the annual sum of habitat quantity; C, difference between present and future number of ice-free months; N, difference between present and future distance between the edge of the continental shelf and the edge of the pack ice; and discrete states for all other nodes. See Figure 3 for allowable states at each node.

[c]NA stands for not applicable; shelf distance change only applies to the polar basin ecoregions.

**Table D1b.** Input Data Values for Nodes J1, B1, R1, J, R4, T1, E, T, and T2 Used in the Bayesian Network Polar Bear Population Stressor Model for Each of Four Polar Bear Ecoregions[a]

| Time Period[a] | Sea Ice Data Source | J1: Tourism | B1: Bear-Human Interactions | R1: Oil and Gas Activity | J: Shipping |
|---|---|---|---|---|---|
| | | | BN Node and Variable Name[b] | | |
| | | | *Seasonal Ice Ecoregion* | | |
| Year −10 | Satellite data | decreased | decreased | no_change | same_as_now |
| Year 0 | Satellite data | same_as_now | same_as_now | no_change | same_as_now |
| Year 25 | GCM minimum | increased | increased | no_change | increased |
| Year 50 | GCM minimum | increased | increased | no_change | increased |
| Year 75 | GCM minimum | increased | increased | no_change | increased |
| Year 95 | GCM minimum | increased | increased | no_change | increased |
| Year 25 | Ensemble mean | increased | increased | no_change | increased |
| Year 50 | Ensemble mean | increased | increased | no_change | increased |
| Year 75 | Ensemble mean | increased | increased | no_change | increased |
| Year 95 | Ensemble mean | increased | increased | no_change | increased |
| Year 25 | GCM maximum | increased | increased | no_change | increased |
| Year 50 | GCM maximum | increased | increased | no_change | increased |
| Year 75 | GCM maximum | increased | increased | no_change | increased |
| Year 95 | GCM maximum | increased | increased | no_change | increased |
| | | | *Archipelago Ecoregion* | | |
| Year −10 | Satellite data | decreased | increased | no_change | same_as_now |
| Year 0 | Satellite data | same_as_now | same_as_now | no_change | same_as_now |
| Year 25 | GCM minimum | increased | same_as_now | no_change | same_as_now |
| Year 50 | GCM minimum | increased | increased | no_change | same_as_now |
| Year 75 | GCM minimum | increased | increased | increase | increased |
| Year 95 | GCM minimum | increased | increased | increase | increased |
| Year 25 | Ensemble mean | increased | same_as_now | no_change | same_as_now |
| Year 50 | Ensemble mean | increased | increased | no_change | same_as_now |
| Year 75 | Ensemble mean | increased | increased | increase | same_as_now |
| Year 95 | Ensemble mean | increased | increased | increase | increased |
| Year 25 | GCM maximum | increased | same_as_now | no_change | same_as_now |
| Year 50 | GCM maximum | increased | increased | no_change | same_as_now |
| Year 75 | GCM maximum | increased | increased | increase | same_as_now |
| Year 95 | GCM maximum | increased | increased | increase | same_as_now |
| | | | *Polar Basin Divergent Ecoregion* | | |
| Year −10 | Satellite data | decreased | decreased | decrease | same_as_now |
| Year 0 | Satellite data | same_as_now | same_as_now | no_change | same_as_now |
| Year 25 | GCM minimum | increased | increased | increase | increased |
| Year 50 | GCM minimum | increased | increased | increase | increased |
| Year 75 | GCM minimum | decreased | increased | increase | increased |
| Year 95 | GCM minimum | decreased | increased | decrease | increased |
| Year 25 | Ensemble mean | increased | increased | increase | increased |
| Year 50 | Ensemble mean | increased | increased | increase | increased |
| Year 75 | Ensemble mean | same_as_now | increased | increase | increased |
| Year 95 | Ensemble mean | decreased | increased | decrease | increased |
| Year 25 | GCM maximum | increased | increased | increase | increased |
| Year 50 | GCM maximum | increased | increased | increase | increased |
| Year 75 | GCM maximum | same_as_now | increased | increase | increased |
| Year 95 | GCM maximum | same_as_now | increased | decrease | increased |

**Table D1b.** (continued)

| | | BN Node and Variable Name[b] | | | |
|---|---|---|---|---|---|
| Time Period[a] | Sea Ice Data Source | J1: Tourism | B1: Bear-Human Interactions | R1: Oil and Gas Activity | J: Shipping |
| | | *Polar Basin Convergent Ecoregion* | | | |
| Year –10 | Satellite data | decreased | decreased | decrease | same_as_now |
| Year 0 | Satellite data | same_as_now | same_as_now | no_change | same_as_now |
| Year 25 | GCM minimum | increased | same_as_now | no_change | same_as_now |
| Year 50 | GCM minimum | increased | increased | increase | increased |
| Year 75 | GCM minimum | increased | increased | increase | increased |
| Year 95 | GCM minimum | increased | increased | increase | increased |
| Year 25 | Ensemble mean | increased | same_as_now | same_as_now | same_as_now |
| Year 50 | Ensemble mean | increased | increased | increase | increased |
| Year 75 | Ensemble mean | increased | increased | increase | increased |
| Year 95 | Ensemble mean | increased | increased | increase | increased |
| Year 25 | GCM maximum | increased | same_as_now | same_as_now | same_as_now |
| Year 50 | GCM maximum | increased | increased | increase | increased |
| Year 75 | GCM maximum | increased | increased | increase | increased |
| Year 95 | GCM maximum | increased | increased | increase | increased |

| | | BN Node and Variable Name[b] | | | |
|---|---|---|---|---|---|
| Time Period[a] | R4: Hydrocarbons/ Oil Spill | T1: Contaminants | E: Intentional Takes | T: Parasites and Disease | T2: Predation |
| | | *Seasonal Ice Ecoregion* | | | |
| Year –10 | same_as_now | reduced | decreased | not | not |
| Year 0 | same_as_now | same_as_now | same_as_now | not | not |
| Year 25 | same_as_now | elevated | decreased | influential | not |
| Year 50 | same_as_now | elevated | decreased | influential | influential |
| Year 75 | increased_occurrence | elevated | decreased | influential | influential |
| Year 95 | increased_occurrence | elevated | decreased | influential | influential |
| Year 25 | same_as_now | elevated | decreased | influential | not |
| Year 50 | same_as_now | elevated | decreased | influential | influential |
| Year 75 | increased_occurrence | elevated | decreased | influential | influential |
| Year 95 | increased_occurrence | elevated | decreased | influential | influential |
| Year 25 | same_as_now | elevated | decreased | influential | not |
| Year 50 | same_as_now | elevated | decreased | influential | influential |
| Year 75 | increased_occurrence | elevated | decreased | influential | influential |
| Year 95 | increased_occurrence | elevated | decreased | influential | influential |
| | | *Archipelago Ecoregion* | | | |
| Year –10 | same_as_now | reduced | same_as_now | not | not |
| Year 0 | same_as_now | same_as_now | same_as_now | not | not |
| Year 25 | same_as_now | elevated | same_as_now | not | not |
| Year 50 | same_as_now | elevated | increased | influential | not |
| Year 75 | increased_occurrence | elevated | same_as_now | influential | influential |
| Year 95 | increased_occurrence | elevated | decreased | influential | influential |
| Year 25 | same_as_now | elevated | same_as_now | not | not |
| Year 50 | same_as_now | elevated | increased | influential | not |
| Year 75 | increased_occurrence | elevated | same_as_now | influential | influential |
| Year 95 | increased_occurrence | elevated | decreased | influential | influential |
| Year 25 | same_as_now | elevated | same_as_now | not | not |
| Year 50 | same_as_now | elevated | increased | influential | not |

**Table D1b.** (continued)

| Time Period[a] | R4: Hydrocarbons/ Oil Spill | T1: Contaminants | E: Intentional Takes | T: Parasites and Disease | T2: Predation |
|---|---|---|---|---|---|
| | BN Node and Variable Name[b] | | | | |
| Year 75 | increased_occurrence | elevated | increased | influential | not |
| Year 95 | increased_occurrence | elevated | decreased | influential | influential |
| | | | | | |
| | *Polar Basin Divergent Ecoregion* | | | | |
| Year −10 | same_as_now | reduced | decreased | not | not |
| Year 0 | same_as_now | same_as_now | same_as_now | not | not |
| Year 25 | increased_occurrence | elevated | decreased | influential | influential |
| Year 50 | increased_occurrence | elevated | decreased | influential | influential |
| Year 75 | increased_occurrence | elevated | decreased | influential | influential |
| Year 95 | increased_occurrence | elevated | decreased | influential | influential |
| Year 25 | increased_occurrence | elevated | decreased | influential | influential |
| Year 50 | increased_occurrence | elevated | decreased | influential | influential |
| Year 75 | increased_occurrence | elevated | decreased | influential | influential |
| Year 95 | increased_occurrence | elevated | decreased | influential | influential |
| Year 25 | increased_occurrence | elevated | decreased | influential | influential |
| Year 50 | increased_occurrence | elevated | decreased | influential | influential |
| Year 75 | increased_occurrence | elevated | decreased | influential | influential |
| Year 95 | increased_occurrence | elevated | decreased | influential | influential |
| | | | | | |
| | *Polar Basin Convergent Ecoregion* | | | | |
| Year −10 | same_as_now | reduced | same_as_now | not | not |
| Year 0 | same_as_now | same_as_now | same_as_now | not | not |
| Year 25 | same_as_now | same_as_now | same_as_now | not | not |
| Year 50 | increased_occurrence | elevated | decreased | influential | influential |
| Year 75 | increased_occurrence | elevated | decreased | influential | influential |
| Year 95 | increased_occurrence | elevated | decreased | influential | influential |
| Year 25 | same_as_now | same_as_now | same_as_now | not | not |
| Year 50 | increased_occurrence | elevated | decreased | influential | influential |
| Year 75 | increased_occurrence | elevated | decreased | influential | influential |
| Year 95 | increased_occurrence | elevated | decreased | influential | influential |
| Year 25 | same_as_now | same_as_now | same_as_now | not | not |
| Year 50 | increased_occurrence | elevated | decreased | influential | influential |
| Year 75 | increased_occurrence | elevated | decreased | influential | influential |
| Year 95 | increased_occurrence | elevated | decreased | influential | influential |

[a]Time period is expressed as the central year in each decade for which projections were made.
[b]Units of measure are discrete states at each node. See Figure 3 for allowable states at each node.

*Acknowledgments.* Principal funding for this project was provided by the U.S. Geological Survey. We thank G. S. York and K. S. Simac for logistical support on this project and for keeping the office going while we were preoccupied. We thank G. M. Durner, E. V. Regehr, M. Runge, and S. Morey for valuable discussions and K. Oakley for effective and insightful project management. We are grateful to W. L. Thompson, T. Starfield, N. Lunn, and B. Taylor for helpful reviews of earlier versions of our model and this paper. We acknowledge the modeling groups for making their sea ice simulations and projections available for analysis, the Program for Climate Model Diagnosis and Intercomparison (PCMDI) for collecting and archiving the CMIP3 model output, and the WCRP's Working Group on Coupled Modeling (WGCM) for organizing the model data analysis activity. The WCRP CMIP3 multimodel data set is supported by the Office of Science, U.S. Department of Energy. Any use of trade, product, or company names in this publication is for descriptive purposes only and does not imply endorsement by the U.S. Government.

# REFERENCES

Aalders, I. (2008), Modeling land-use decision behavior with Bayesian belief networks. *Ecol. Soc.*, *13*(1), article 16.

Aars, J., N. J. Lunn, and A. E. Derocher (2006), Polar bears: Proceedings of the fourteenth working meeting of the IUCN/SSC Polar Bear Specialist Group, *Occas. Pap. IUCN Species Surv. Comm. 32*, 189 pp., Int. Union for Conserv. Nature and Nat. Resour., Gland, Switzerland.

Ainley, D. G., C. T. Tynan, and I. Stirling (2003), Sea ice: A critical habitat for polar marine mammals and birds, in *Sea Ice: An Introduction to Its Physics, Chemistry, Biology and Geology*, edited by D. N. Thomas and G. S. Dieckmann, pp. 240–266, Blackwell Sci., Malden, Mass.

Amstrup, S. C. (1986), Polar bear, in *Audubon Wildlife Report*, edited by R. L. DiSilvestro, pp. 790–804, Natl. Audubon Soc., New York.

Amstrup, S. C. (2000), Polar bear, in *The Natural History of an Oil Field: Development and Biota*, edited by J. C. Truett and S. R. Johnson, pp. 133–157, Academic, New York.

Amstrup, S. C. (2003), Polar bear, in *Wild Mammals of North America. Biology, Management, and Conservation*, 2nd ed., edited by G. A. Feldhammer, B. C. Thompson, and J. A. Chapman, pp. 587–610, Johns Hopkins Univ. Press, Baltimore, Md.

Amstrup, S. C., and D. P. DeMaster (1988), Polar bear—*Ursus maritimus*, in *Selected Marine Mammals of Alaska: Species Accounts With Research and Management Recommendations*, edited by J. W. Lentfer, pp. 39–56, Mar. Mammal Comm., Washington, D. C.

Amstrup, S. C., and C. Gardner (1994), Polar bear maternity denning in the Beaufort Sea, *J. Wildl. Manage.*, *58*(1), 1–10.

Amstrup, S. C., G. M. Durner, I. Stirling, N. J. Lunn, and F. Messier (2000), Movements and distribution of polar bears in the Beaufort Sea, *Can. J. Zool.*, *78*, 948–966.

Amstrup, S. C., T. L. McDonald, and G. M. Durner (2004), Using satellite radiotelemetry data to delineate and manage wildlife populations, *Wildl. Soc. Bull.*, *32*(3), 661–679.

Amstrup, S. C., G. M. Durner, I. Stirling, and T. L. McDonald (2005), Allocating harvests among polar bear stocks in the Beaufort Sea, *Arctic*, *58*(3), 247–259.

Amstrup, S. C., I. Stirling, T. S. Smith, C. Perham, and G. W. Thiemann (2006), Recent observations of intraspecific predation and cannibalism among polar bears in the southern Beaufort Sea, *Polar Biol.*, *29*(11), 997–1002, doi:10.1007/s00300-006-0142-5.

Amstrup, S. C., B. G. Marcot, and D. C. Douglas (2007), Forecasting the range-wide status of polar bears at selected times in the 21st century, administrative report, 123 pp., U.S. Geol. Surv., Alaska Sci. Cent., Anchorage, Alaska. (Available at http://www.usgs.gov/newsroom/special/polar_bears/)

Arthur, S. M., B. F. J. Manly, L. L. McDonald, and G. W. Garner (1996), Assessing habitat selection when availability changes, *Ecology*, *77*(1), 215–227.

Barrett, G. W. (1981), Stress ecology: An integrative approach, in *Stress Effects on Natural Ecosystems*, edited by G. W. Barrett and R. Rosenberg, pp. 3–12, John Wiley, New York.

Belchansky, G. I., D. C. Douglas, and N. G. Platonov (2005), Spatial and temporal variations in the age structure of Arctic sea ice, *Geophys. Res. Lett.*, *32*, L18504, doi:10.1029/2005GL023976.

Belchansky, G. I., D. C. Douglas, and N. G. Platonov (2008), Fluctuating Arctic sea ice thickness changes estimated by an in situ learned and empirically forced neural network model, *J. Clim.*, *21*(4), 716–729, doi:10.1175/2007JCLI1787.1

Best, R. C. (1984), Digestibility of ringed seals by the polar bear, *Can. J. Zool.*, *63*, 1033–1036.

Bluhm, B. A., and R. Gradinger (2008), Regional variability in food availability for Arctic marine mammals, *Ecol. Appl.*, *18*, suppl., S77–S96.

Bollen, K. A. (1989), *Structural Equations With Latent Variables*, 528 pp., John Wiley, New York.

Brook, R. K., and E. S. Richardson (2002), Observations of polar bear predatory behaviour toward caribou, *Arctic*, *55*(2), 193–196.

Cain, J. (2001), Planning improvements in natural resources management: Guidelines for using Bayesian networks to support the planning and management of development programmes in the water sector and beyond, report, 124 pp., Crowmarsh Gifford, Cent. for Ecol. and Hydrol., Wallingford, U. K.

Calvert, W., and I. Stirling (1990), Interactions between polar bears and overwintering walruses in the central Canadian High Arctic, *Int. Conf. Bear Res. Manage.*, *8*, 351–356.

Cavalieri, D., C. Parkinson, P. Gloersen, and H. J. Zwally (1996), Sea ice concentrations from Nimbus-7 SMMR and DMSP SSM/I passive microwave data, http://nsidc.org/data/nsidc-0051.html, Natl. Snow and Ice Data Cent., Boulder, Colo. (updated 2006)

Comiso, J. C. (2002), A rapidly declining perennial sea ice cover in the Arctic, *Geophys. Res. Lett.*, *29*(20), 1956, doi:10.1029/2002GL015650.

Comiso, J. C. (2006), Abrupt decline in the Arctic winter sea ice cover, *Geophys. Res. Lett.*, *33*, L18504, doi:10.1029/2006GL027341.

Das, B. (2000), Representing uncertainty using Bayesian networks, report, 58 pp., Def. Sci. and Technol. Organ., Dep. of Def., Salisbury, S. Aust., Australia.

Derocher, A. E., D. Andriashek, and I. Stirling (1993), Terrestrial foraging by polar bears during the ice-free period in western Hudson Bay, *Arctic*, *46*(3), 251–254.

Derocher, A. E., Ø. Wiig, and G. Bangjord (2000), Predation of Svalbard reindeer by polar bears, *Polar Biol.*, *23*(10), 675–678.

Derocher, A. E., Ø. Wiig, and M. Andersen (2002), Diet composition of polar bears in Svalbard and the western Barents Sea, *Polar Biol.*, *25*(6), 448–452.

Derocher, A. E., N. J. Lunn, and I. Stirling (2004), Polar bears in a warming climate, *Integr. Comp. Biol.*, *44*, 163–176.

DeWeaver, E. (2007), Uncertainty in climate model projections of Arctic sea ice decline, administrative report, 47 pp., U.S. Geol. Surv., Anchorage, Alaska.

Durner, G. M., and S. C. Amstrup (1995), Movements of a polar bear from northern Alaska to northern Greenland, *Arctic*, *48*(4), 338–341.

Durner, G. M., D. C. Douglas, R. M. Nielson, and S. C. Amstrup (2006), Model for autumn pelagic distribution of adult female polar bears in the Chukchi Sea, 1987–1994, final report to U.S. Fish and Wildlife Service, 67 pp., U.S. Geol. Surv., Alaska Sci. Cent., Anchorage, Alaska.

Durner, G. M., et al. (2008), Predicting 21st century polar bear habitat distribution from global climate models, *Ecol. Monogr.*, in press.

Dyck, M. G., and S. Romber (2007), Observations of a wild polar bear (*Ursus maritimus*) successfully fishing Artic charr (*Salvelinus alpinus*) and Fourhorn sculpin (*Myoxocephalus quadricornis*), *Polar Biol.*, *30*(12), 1625–1628, doi:10.1007/s00300-007-0338-3.

Ferguson, S. H., M. K. Taylor, E. W. Born, A. Rosing-Asvid, and F. Messier (1999), Determinants of home range size for polar bears (*Ursus maritimus*), *Ecol. Lett.*, *2*, 311–318.

Ferguson, S. H., M. K. Taylor, and F. Messier (2000), Influence of sea ice dynamics on habitat selection by polar bears, *Ecology*, *81*(3), 761–772.

Ferguson, S. H., I. Stirling, and P. McLoughlin (2005), Climate change and ringed seal (*Phoca hispida*) recruitment in western Hudson Bay, *Mar. Mammal Sci.*, *21*(1), 121–135.

Fischbach, A. S., S. C. Amstrup, and D. C. Douglas (2007), Landward and eastward shift of Alaskan polar bears denning associated with recent sea ice changes, *Polar Biol.*, *30*(11), 1395–1405, doi:1007/s00300-007-0300-4.

Furnell, D. J., and D. Oolooyuk (1980), Polar bear predation on ringed seals in ice-free water, *Can. Field Nat.*, *94*, 88–89.

Garner, G. W., S. T. Knick, and D. C. Douglas (1990), Seasonal movements of adult female polar bears in the Bering and Chukchi Seas, *Int. Conf. Bear Res. Manage*, *8*, 219–226.

Garner, G. W., S. C. Amstrup, I. Stirling, and S. E. Belikov (1994), Habitat considerations for polar bears in the North Pacific Rim, *Trans. North Am. Wildl. Nat. Resour. Conf.*, *59*, 111–120.

Ginzburg, L. R., L. B. Slobodkin, K. Johnson, and A. G. Bindman (1982), Quasiextinction probabilities as a measure of impact on population growth, *Risk Anal.*, *2*, 171–181.

Guisan, A., A. Lehmann, S. Ferrier, M. Austin, J. M. C. Overton, R. Aspinall, and T. Hastie (2006), Making better biogeographical predictions of species' distributions, *J. Appl. Ecol.*, *43*(3), 386–392.

Hammill, M. O., and T. G. Smith (1991), The role of predation in the ecology of the ringed seal in Barrow Strait, Northwest Territories, Canada, *Mar. Mammal Sci.*, *7*(2), 123–135.

Holland, M. M., C. M. Bitz, and B. Tremblay (2006), Future abrupt reductions in the summer Arctic sea ice, *Geophys. Res. Lett.*, *33*, L23503, doi:10.1029/2006GL028024.

Hufthammer, A. K. (2001), The Weichselian (c. 115,000–10,000 B.P.) vertebrate fauna of Norway, *Boll. Soc. Paleontol. Ital.*, *40*(2), 201–208.

Hunter, C. M., H. Caswell, M. C. Runge, E. V. Regehr, S. C. Amstrup, and I. Stirling (2007), Polar bears in the southern Beaufort Sea II: Demography and population growth in relation to sea ice conditions, administrative report, 46 pp., U.S. Geol. Surv., Anchorage, Alaska. (Available at http://www.usgs.gov/newsroom/special/polar_bears/)

Ingolfsson, O., and O. Wiig (2007), Fossil find on Svalbard highlights the natural history of the polar bear (*Ursus maritimus*), *Eos Trans. AGU*, *88*(52), Fall Meet. Suppl., Abstract GC11B-01.

Iverson, S. J., I. Stirling, and S. L. C. Land (2006), Spatial, temporal, and individual variation in the diets of polar bears across the Canadian Arctic: Indicators of changes in prey populations and environment, in Top Predators in Marine Ecosystems, edited by I. Boyd, S. W. Wanless, and C. J. Camphuysen, pp. 98–117, Cambridge Univ. Press, Cambridge, U. K.

Jensen, F. V. (2001), *Bayesian Networks and Decision Graphs*, 284 pp., Springer, New York.

Johnson, C. J., and M. P. Gillingham (2004), Mapping uncertainty: Sensitivity of wildlife habitat ratings to expert opinion, *J. Appl. Ecol.*, *41*(6), 1032–1041.

Kingsley, M. C. S. (1998), The numbers of ringed seals (*Phoca hispida*) in Baffin Bay and associated waters, in *Ringed Seals in the North Atlantic*, edited by M. P. Heide-Jorgensen and C. Lydersen, pp. 181–196, *NAAMCO Sci. Publ.*, vol. 1, N. Atl. Mar. Mammal Comm., Tromsø, Norway.

Lindsay, R. W., and J. Zhang (2005), The thinning of Arctic sea ice, 1988–2003: Have we passed a tipping point?, *J. Clim.*, *18*(22), 4879–4894.

Lunn, N. J., and I. Stirling (1985), The significance of supplemental food to polar bears during the ice-free period of Hudson Bay, *Can. J. Zool.*, *63*, 2291–2297.

Lunn, N. J., I. Stirling, and D. Andriashek (1995), Movements and distribution of polar bears in the northeastern Beaufort Sea and western M'Clure Strait, final report to the Inuvialuit Wildlife Management Advisory Committee, Can. Wildlife Serv., Edmonton, Alberta, Canada.

Lunn, N. J., S. Schliebe, and E. W. Born (Eds.) (2002), Polar bears, in *Proceedings of the Thirteenth Working Meeting of the IUCN/SSC Polar Bear Specialist Group*, Occas. Pap. IUCN Species Survival Comm. 26, 153 pp., Int. Union for Conserv. of Nature and Nat. Resour., Gland, Switzerland.

Manly, B. F. J., L. L. McDonald, D. L. Thomas, T. L. McDonald, and W. P. Erickson (2002), *Resource Selection by Animals: Sta-*

tistical Design and Analysis for Field Studies, Kluwer Acad., Norwell, Mass.

Marcot, B. G. (1990), Testing your knowledge base, in Expert Systems: A Software Methodology for Modern Applications, edited by P. G. Raeth, pp. 438–443, IEEE Comput. Soc. Press, Los Alamitos, Calif.

Marcot, B. G. (2006), Characterizing species at risk I: Modeling rare species under the Northwest Forest Plan, Ecol. Soc., 11(2), article 10.

Marcot, B. G., M. G. Raphael, and K. H. Berry (1983), Monitoring wildlife habitat and validation of wildlife-habitat relationships models, Trans. North Am. Wildl. Nat. Resour. Conf., 48, 315–329.

Marcot, B. G., R. S. Holthausen, M. G. Raphael, M. M. Rowland, and M. J. Wisdom (2001), Using Bayesian belief networks to evaluate fish and wildlife population viability under land management alternatives from an environmental impact statement, For. Ecol. Manage., 153(1–3), 29–42.

Marcot, B. G., J. D. Steventon, G. D. Sutherland, and R. K. Mc-Cann (2006), Guidelines for developing and updating Bayesian belief networks applied to ecological modeling and conservation, Can. J. For. Res., 36, 3063–3074.

Maslanik, J. A., S. Drobot, C. Fowler, W. Emery, and R. Barry (2007), On the Arctic climate paradox and the continuing role of atmospheric circulation in affecting sea ice conditions, Geophys. Res. Lett., 34, L03711, doi:10.1029/2006GL028269.

McCann, R., B. G. Marcot, and R. Ellis (2006), Bayesian belief networks: Applications in natural resource management, Can. J. For. Res., 36, 3053–3062.

McConkey, K. R., and D. R. Drake (2006), Flying foxes cease to function as seed dispersers long before they become rare, Ecology, 87(2), 271–276.

Meehl, G. A., C. Covey, T. Delworth, M. Latif, B. McAvaney, J. F. B. Mitchell, R. J. Stouffer, and K. E. Taylor (2007a), The WCRP CMIP3 multimodel dataset: A new era in climate change research, Bull. Am. Meteorol. Soc., 88(9), 1383–1394.

Meehl, G. A., et al. (2007b), Global climate projections, in Climate Change 2007: The Physical Science Basis: Contribution of Working Group I to the Fourth Assessment Report of the Intergovernmental Panel on Climate Change, edited by S. Solomon et al., pp. 747–845, Cambridge Univ. Press, Cambridge, U. K.

Meier, W. N., J. Stroeve, and F. Fetterer (2007), Whither Arctic sea ice? A clear signal of decline regionally, seasonally and extending beyond the satellite record, Ann. Glaciol., 46, 428–434.

Messier, F., M. K. Taylor, and M. A. Ramsay (1992), Seasonal activity patterns of female polar bears (Ursus maritimus) in the Canadian Arctic as revealed by satellite telemetry, J. Zool., 226, 219–229.

Miller, S. D., et al. (1997), Brown and black bear density estimation in Alaska using radiotelemetry and replicated mark-resight techniques, Wildl. Monogr., 133, 1–55.

Nakićenović, N., et al. (2000), Special Report on Emissions Scenarios: A Special Report of Working Group III of the Intergovernmental Panel on Climate Change, 599 pp., Cambridge Univ. Press, Cambridge, U. K.

Neopolitan, R. E. (2003), Learning Bayesian Networks, 674 pp., Prentice Hall, New York.

Obbard, M. E., M. R. L. Cattet, T. Moody, L. R. Walton, D. Potter, J. Inglis, and C. Chenier (2006), Temporal trends in the body condition of southern Hudson Bay polar bears, Res. Inf. Note 3, 8 pp., Ontario Minist. of Nat. Resour., Peterborough, Ont., Canada.

Ogi, M., and J. M. Wallace (2007), Summer minimum Arctic sea ice extent and associated summer atmospheric circulation, Geophys. Res. Lett., 34, L12705, doi:10.1029/2007GL029897.

Otway, N. M., C. J. A. Bradshaw, and R. G. Harcourt (2004), Estimating the rate of quasiextinction of the Australian grey nurse shark (Carcharias taurus) population using deterministic age- and stage-classified models, Biol. Conserv., 119, 341–350.

Ovsyanikov, N. (1996), Polar Bears: Living With the White Bear, 144 pp., Voyager, Stillwater, Minn.

Parovshchikov, V. Y. (1964), A study on the population of polar bear, Ursus (Thalarctos) maritimus Phipps, of Franz Joseph Land, Acta Soc. Zool. Bohemoslov., 28, 167–177.

Pomeroy, L. R. (1997), Primary production in the Arctic Ocean estimated from dissolved oxygen, J. Mar. Syst., 10, 1–8.

Pourret, O., P. Naïm, and B. G. Marcot (Eds.) (2008), Bayesian Belief Networks: A Practical Guide to Applications, 432 pp., John Wiley, Chichester, U. K.

Ramsay, M. A., and K. A. Hobson (1991), Polar bears make little use of terrestrial food webs: Evidence from stable-carbon isotope analysis, Oecologia, 86, 598–600.

Ramsay, M. A., and I. Stirling (1984), Interactions of wolves and polar bears in northern Manitoba, J. Mammal., 65, 693–694.

Raphael, M. G., M. J. Wisdom, M. M. Rowland, R. S. Holthausen, B. C. Wales, B. G. Marcot, and T. D. Rich (2001), Status and trends of habitats of terrestrial vertebrates in relation to land management in the interior Columbia River Basin, For. Ecol. Manage., 153(1–3), 63–87.

Rayner, N. A., D. E. Parker, E. B. Horton, C. K. Folland, L. V. Alexander, D. P. Rowell, E. C. Kent, and A. Kaplan (2003), Global analyses of sea surface temperature, sea ice, and night marine air temperature since the late nineteenth century, J. Geophys. Res., 108(D14), 4407, doi:10.1029/2002JD002670.

Regehr, E. V., C. M. Hunter, H. Caswell, S. C. Amstrup, and I. Stirling (2007a), Polar bears in the Southern Beaufort Sea I: Survival and breeding in relation to sea ice conditions, 2001-2006, administrative report, 45 pp., U.S. Geol. Surv., Anchorage, Alaska. (Available at http://www.usgs.gov/newsroom/special/polar_bears/)

Regehr, E. V., N. J. Lunn, S. C. Amstrup, and I. Stirling (2007b), Effects of earlier sea ice breakup on survival and population size of polar bears in western Hudson Bay, J. Wildlife Manage., 71(8), 2673–2683.

Rigor, I. G., and J. M. Wallace (2004), Variations in the age of Arctic sea-ice and summer sea-ice extent, Geophys. Res. Lett., 31, L09401, doi:10.1029/2004GL019492.

Rigor, I. G., J. M. Wallace, and R. L. Colony (2002), Response of sea ice to the Arctic Oscillation, J. Clim., 15(18), 2648–2663.

Robbins, C. T., C. C. Schwartz, and L. A. Felicetti (2004), Nutritional ecology of ursids: A review of newer methods and management implications, Ursus, 15(2), 161–171.

Rode, K. D., C. T. Robbins, and L. A. Shipley (2001), Constraints on herbivory by grizzly bears, *Oecologia*, *128*, 62–71.

Rode, K. D., S. C. Amstrup, and E. V. Regehr (2007), Polar bears in the southern Beaufort Sea III: Stature, mass, and cub recruitment in relationship to time and sea ice extent between 1982 and 2007, administrative report, 28 pp., U.S. Geol. Surv., Anchorage, Alaska. (Available at http://www.usgs.gov/newsroom/special/polar_bears/)

Russell, R. H. (1975), The food habits of polar bears of James Bay and southwest Hudson Bay in summer and autumn, *Arctic*, *28*(2), 117–129.

Sakshaug, E. (2004), Primary and secondary production in the Arctic seas, in *Organic Carbon Cycle in the Arctic Ocean*, edited by R. Stein and R. W. Macdonald, pp. 57–81, Springer, New York.

Schweinsburg, R. E., and L. J. Lee (1982), Movement of four satellite-monitored polar bears in Lancaster Sound, Northwest Territories, *Arctic*, *35*(4), 504–511.

Sekercioglu, C. H., G. C. Daily, and P. R. Ehrlich (2004), Ecosystem consequences of bird declines, *Proc. Natl. Acad. Sci. U. S. A.*, *101*(52), 18,042–18,047.

Serreze, M. C., M. M. Holland, and J. Stroeve (2007), Perspectives on the Arctic's shrinking sea-ice cover, *Science*, *315*(5818), 1533–1536.

Smith, A. E., and M. R. J. Hill (1996), Polar bear, *Ursus maritimus*, depredation of Canada goose, *Branta canadensis*, nests, *Can. Field Nat.*, *110*, 339–340.

Smith, T. G. (1980), Polar bear predation of ringed and bearded seals in the land-fast sea ice habitat, *Can. J. Zool.*, *58*, 2201–2209.

Smith, T. G. (1985), Polar bears, *Ursus maritimus*, as predators of belugas, *Delphinapterus leucas*, *Can. Field Nat.*, *99*, 71–75.

Smith, T. G., and B. Sjare (1990) Predation of belugas and narwals by polar bears in nearshore areas of the Canadian High Arctic, *Arctic*, *43*(2), 99–102.

Smith, T. G., and I. Stirling (1975), The breeding habitat of the ringed seal (*Phoca hispida*), The birth lair and associated structures, *Can. J. Zool.*, *53*, 1297–1305.

Stefansson, V. (1921), *The Friendly Arctic*, 361 pp., Macmillan, New York.

Stempniewicz, L. (1993), The polar bear *Ursus maritimus* feeding in a seabird colony in Frans Josef Land, *Polar Res.*, *12*(1), 33–36.

Stempniewicz, L. (2006), Polar bear predatory behaviour toward molting barnacle geese and nesting Glaucous Gulls on Spitsbergen, *Arctic*, *59*(3), 247–251.

Stirling, I. (1974), Midsummer observations on the behavior of wild polar bears (*Ursus maritimus*), *Can. J. Zool.*, *52*, 1191–1198.

Stirling, I. (1980), The biological importance of polynyas in the Canadian Arctic, *Arctic*, *33*(2), 303–315.

Stirling, I. (1990), Polar bears and oil: Ecologic perspectives, in *Sea Mammals and Oil: Confronting the Risks*, edited by J. R. Geraci and D. J. St. Aubin, pp. 223–234, Academic, San Diego, Calif.

Stirling, I., and A. E. Derocher (1993), Possible impacts of climatic warming on polar bears, *Arctic*, *46*(3), 240–245.

Stirling, I., and N. J. Lunn (1997), Environmental fluctuations in Arctic marine ecosystems as reflected by variability in reproduction of polar bears and ringed seals, in *Ecology of Arctic En-*vironments, *Br. Ecol. Soc. Spec. Publ.*, vol. 13, edited by S. J. Woodin and M. Marquiss, pp. 167–181, Blackwell Sci., Oxford, U. K.

Stirling, I., and N. A. Øritsland (1995), Relationships between estimates of ringed seal (*Phoca hispida*) and polar bear (*Ursus maritimus*) populations in the Canadian Arctic, *Can. J. Fish. Aquat. Sci.*, *52*, 2594–2612.

Stirling, I., and C. L. Parkinson (2006), Possible effects of climate warming on selected populations of polar bears (*Ursus maritimus*) in the Canadian Arctic, *Arctic*, *59*(3), 261–275.

Stirling, I., H. Cleator, and T. G. Smith (1981), Marine mammals, in *Polynyas in the Canadian Arctic*, edited by I. Stirling and H. Cleator, pp. 45–48, *Occas. Pap. 45*, Can. Wildlife Serv., Ottawa, Ont., Canada.

Stirling, I., N. J. Lunn, and J. Iacozza (1999), Long-term trends in the population ecology of polar bears in western Hudson Bay in relation to climatic change, *Arctic*, *52*(3), 294–306.

Stirling, I., T. L. McDonald, E. S. Richardson, and E. V. Regehr (2007), Polar bear population status in the Northern Beaufort Sea, administrative report, 33 pp., U.S. Geol. Surv., Anchorage, Alaska. (Available at http://www.usgs.gov/newsroom/special/polar_bears/)

Stirling, I., E. Richardson, G. W. Thiemann, and A. E. Derocher (2008), Unusual predation attempts of polar bears on ringed seals in the southern Beaufort Sea: Possible significance of changing spring ice conditions, *Arctic*, *61*(1), 14–22.

Stroeve, J., M. M. Holland, W. Meier, T. Scambos, and M. Serreze (2007), Arctic sea ice decline: Faster than forecast, *Geophys. Res. Lett.*, *34*, L09501, doi:10.1029/2007GL029703.

Talbot, S. L., and G. F. Shields (1996), Phylogeography of brown bears (*Ursus arctos*) of Alaska and paraphyly within the Ursidae, *Mol. Phylogenet. Evol.*, *5*, 477–494.

Townsend, C. W. (Ed.) (1911), *Captain Cartwright and His Labrador Journal*, 385 pp., Dana Estes, Boston, Mass.

U.S. Fish and Wildlife Service (2007), Endangered and threatened wildlife and plants; 12-month petition finding and proposed rule to list the polar bear (*Ursus maritimus*) as threatened throughout its range, *Fed. Regist.*, *72*(5), 1064–1099.

U.S. Fish and Wildlife Service (2008), Endangered and threatened wildlife and plants; determination of threatened status for the polar bear (*Ursus maritimus*) throughout its range, *Fed. Regist.*, *73*(95), 28,212–28,303.

Uusitalo, L. (2007), Advantages and challenges of Bayesian networks in environmental modeling, *Ecol. Modell.*, *203*, 312–318.

Watts, P. D., and S. E. Hansen (1987), Cyclic starvation as a reproductive strategy in the polar bear, *Symp. Zool. Soc. London*, *57*, 305–318.

Welch, C. A., J. Keay, K. C. Kendall, and C. T. Robbins (1997), Constraints on frugivory by bears, *Ecology*, *78*(4), 1105–1119.

S. C. Amstrup, Alaska Science Center, U.S. Geological Survey, Anchorage, AK 99508, USA. (samstrup@usgs.gov)

D. C. Douglas, Alaska Science Center, U.S. Geological Survey, Juneau, AK 99801, USA.

B. G. Marcot, Pacific Northwest Research Station, USDA Forest Service, Portland, OR 97204, USA.

# Index